D1236275

ARSENIC IN THE ENVIRONMENT

Volume

27

in the Wiley Series in

Advances in Environmental Science and Technology

JEROME O. NRIAGU, Series Editor

ARSENIC IN THE ENVIRONMENT

Part II: Human Health and Ecosystem Effects

Edited by

Jerome O. Nriagu

Department of Environmental and Industrial Health,
School of Public Health, The University of Michigan,
Ann Arbor, Michigan

A WILEY-INTERSCIENCE PUBLICATION

JOHN WILEY & SONS, INC.

New York • Chichester • Brisbane • Toronto • Singapore

This text is printed on acid-free paper.

Library of Congress Cataloging in Publication Data:

Arsenic in the environment / edited by Jerome O. Nriagu.
p. cm.—(Advances in environmental science and technology : v. 26–27)
"A Wiley-Interscience publication."
Includes bibliographical references and indexes.
Contents: pt. 1. Cycling and characterization—pt. 2. Human health and ecosystem effects.
ISBN 0-471-57929-7 (pt. 1 : alk. paper).—ISBN 0-471-30436-0 (pt. 2 : alk. paper)
1. Arsenic—Environmental aspects. I. Nriagu, Jerome O. II. Series.
TD180.A38 vol. 26–27
[TD196.A77]
622 s—dc20
[574.2′4] 93-23141
CIP

Printed in the United States of America

10 9 8 7 6 5 4 3 2

CONTRIBUTORS

A. Albores, Sección de Toxicología Ambiental, CINVESTAV-IPN, P.O. Box 14-740, México, D.F., 07000 Mexico

Bert Allard, Department of Water and Environmental Studies, Linköping University, S-581 83 Linköping, Sweden

M.E. Cebrián, Sección de Toxicología Ambiental, CINVESTAV-IPN, P.O. Box 14-740, México, D.F., 07000 Mexico

Chien-Jen Chen, Institute of Public Health, College of Public Health, National Taiwan University 10018; Institute of Biomedical Sciences, Academia Sinica 11592, Taipei, Taiwan

L.M. Del Razo, Sección de Toxicología Ambiental, CINVESTAV-IPN, P.O. Box 14-740, México, D.F., 07000 Mexico

David A. Dunnette, Department of Environmental Health, Portland State University, Portland, Oregon 97207

Ronald Eisler, U.S. National Biological Survey, Patuxent Wildlife Research Center, Laurel, Maryland 20708

Bruce A. Fowler, University of Maryland Baltimore County, The University Program in Toxicology, Baltimore, Maryland 21227

G. García-Vargas, Sección de Toxicología Ambiental, CINVESTAV-IPN, P.O. Box 14-740, México, D.F., 07000 Mexico

Michael S. Gorby, 126B Medical Drive, Palestine, Texas 75801

Anders Grimvall, Department of Water and Environmental Studies, Linköping University, S-581 83 Linköping, Sweden

Huang Jianzhong, Xinjiang Institute for Endemic Disease Control and Research, 141 Jianquan Street, Urumqi 830002, People's Republic of China

Wang Lianfang, Xinjiang Institute for Endemic Disease Control and Research, 141 Jianquan Street, Urumqi 830002, People's Republic of China

Li-Ju Lin, Institute of Public Health, College of Public Health, National Taiwan University 10018, Taipei, Taiwan

Ziqiang Meng, Department of Environmental Science, Shanxi University, Taiyuan 030006, People's Republic of China

WILLIAM E. MORTON, Department of Environmental Medicine, Oregon Health Sciences University, 3181 S.W. Sam Jackson Park Road, L352, Portland, Oregon 97201

SYED M. NAQVI, Department of Biology & Health Research Center, Southern University, Southern Branch Post Office, Baton Rouge, Louisiana 70813

SHOJI OKADA, Department of Radiobiochemistry, University of Shizuoka School of Pharmaceutical Sciences, 52-1 Yada, Shizuoka-shi 422, Japan

PATRICIA OSTROSKY-WEGMAN, Instituto de Investigaciones Biomedicas UNAM, P.O. Box 70228, México, D.F., 04510, Mexico

KAZUO SHIOMI, Department of Food Sciences and Technology, Tokyo University of Fisheries, Konan-4, Minato-ku, Tokyo 108, Japan

HARPAL SINGH, Department of Biology, Savannah State College, Savannah, Georgia 31404

CHETANA VAISHNAVI, Department of Experimental Medicine, Postgraduate Institute of Medical Education & Research, Chandigarh-16001, India

C.M. WAI, Department of Anesthesia, Research Laboratories, Harvard Medical School, 75 Francis Street, Boston, Massachusetts 02115

HAO XU, Groundwater Contamination Project, National Water Research Institute, 867 Lakeshore Road, P.O. Box 5050, Burlington, Ontario L7R 4A6, Canada

KENZO YAMANAKA, Department of Biochemical Toxicology, Nihon University College of Pharmacy, 7-7-1 Narashinodai, Funabashi-shi, Chiba 274, Japan

HIROSHI YAMAUCHI, University of Maryland Baltimore County, The University Program in Toxicology, TEC Building, 5202 Westland Boulevard, Baltimore, Maryland 21227

INTRODUCTION TO THE SERIES

The deterioration of environmental quality, which began when mankind first congregated into villages, has existed as a serious problem since the industrial revolution. In the second half of the twentieth century, under the ever-increasing impacts of exponentially growing population and of industrializing society, environmental contamination of the air, water, soil, and food has become a threat to the continued existence of many plant and animal communities of various ecosystems and may ultimately threaten the very survival of the human race. Understandably, many scientific, industrial, and governmental communities have recently committed large resources of money and human power to the problems of environmental pollution and pollution abatement by effective control measures.

Advances in Environmental Sciences and Technology deals with creative reviews and critical assessments of all studies pertaining to the quality of the environment and to the technology of its conservation. The volumes published in the series are expected to service several objectives: (1) stimulate interdisciplinary cooperation and understanding among the environmental scientists; (2) provide the scientists with a periodic overview of environmental developments that are of general concern or that are of relevance to their own work or interests; (3) provide the graduate student with a critical assessment of past accomplishment, which may help stimulate him or her toward the career opportunities in this vital area; and (4) provide the research manager and the legislative or administrative official with an assured awareness of newly developing research work on the critical pollutants and with the background information important to their responsibility.

As the skills and techniques of many scientific disciplines are brought to bear on the fundamental and applied aspects of the environmental issues, there is a heightened need to draw together the numerous threads and to present a coherent picture of the various research endeavors. This need and the recent tremendous growth in the field of environmental studies have clearly made some editorial adjustments necessary. Apart from the changes in style and format, each future volume in the series will focus on one particular theme or timely topic, starting with Volume 12. The author(s) of each pertinent section will be expected to critically review the literature and the most important recent developments in the particular field; to critically evaluate new concepts, methods, and data; and to focus attention on important unresolved or controversial questions and on probable future trends. Monographs embodying the results of unusually extensive and well-rounded investigations will also be published in the series. The net result of the new editorial policy should be more integrative and comprehensive volumes on key environmental issues and pollutants. Indeed, the development of realistic standards of environmental quality for many pollutants often entails such a holistic treatment.

JEROME O. NRIAGU, Series Editor

PREFACE

The long history of arsenic in science, medicine, and technology has been overshadowed by its notoriety as a preferred poison in homicides. In modern parlance, arsenic is often viewed as being synonymous with "toxic." The evil reputation is probably undeserved considering how well arsenic has served mankind. Salvarsan (an organoarsenic compound) was a highly renowned chemotherapeutic agent against syphilis and other venereal diseases, and for many years, Fowler's solution (1% solution of arsenic trioxide) remained one of the most dispensed medicaments in the Western culture. Arsenic still remains an important player in wood preservation, the cotton industry, electronics, especially in photocopying and high-speed computer machines, and in livestock and poultry farming.

The notoriety and lingering concern about the potential effects of arsenic on various fauna and flora has inevitably engendered a lot of research on the many facets of this element in the environment. This volume represents an attempt to bring together the key research results from the biological, chemical, geological, and clinical studies under one cover. Chapters in Part I focus on the sources, distribution, biotransformation, speciation, and fate of arsenic, especially in soils and the aquatic environment. Part II covers the human health and ecosystem effects of arsenic. The chapters have been written by leading experts in their fields and the transdisciplinary volume provides an integrated account of current knowledge on one of the "Big Four" metals of environmental concern (the other three are lead, mercury, and cadmium). The main emphasis in each chapter has been on the presentation of general principles rather than on the systematic compilation of published data. The volume should thus be of interest to graduate students and practicing scientists in the fields of environmental science and engineering, toxicology, public health, and environmental control. It is, however, my hope that it will be read by anyone who is concerned about the impact of metallic contaminants on our health and our life support system.

The success of this volume belongs to our distinguished authors. I thank Harriet Damon Shields & Associates and the staff of John Wiley & Sons, Inc. for invaluable editorial assistance.

JEROME O. NRIAGU

Ann Arbor, Michigan
January 1994

CONTENTS

CONTENTS—PART I

ADVANCES IN ENVIRONMENTAL SCIENCE AND TECHNOLOGY
Jerome O. Nriagu, Series Editor

1

ARSENIC IN HUMAN MEDICINE

Michael S. Gorby

126B Medical Drive, Palestine, Texas 75801

1. INTRODUCTION

Clinicians today are relatively unaware of the hazards of arsenicals. Advances in the treatment of syphilis and parasitism—which primarily used arsenicals—and in the development of industrial controls and environmental public health have dramatically decreased clinicians' exposure to cases of arsenic toxicity. Unfortu-

Arsenic in the Environment, Part II: Human Health and Ecosystem Effects,
Edited by Jerome O. Nriagu.
ISBN 0-471-30436-0

nately, the problem is still present. Arsenic remains the most common source of acute heavy metal and metalloid poisoning and is second only to lead in chronic ingestion (Goldfrank et al., 1986).

To understand arsenic in biology, one must first understand its chemistry. Arsenic is classified as a transition element or metalloid, reflecting the fact that it commonly forms complexes with metals, but arsenic also reacts readily to form covalent bonds with carbon, hydrogen, and oxygen. In fact, far more organic compounds of arsenic have been made than of any other trace element (Frost, 1967). Arsenic may exist in three different oxidation or valence states, namely, as the metalloid (0 oxidation state), as arsenite (trivalent or +3 oxidation state), and as arsenate (pentavalent or +5 oxidation state). Different arsenic-containing compounds vary substantially in their toxicity to mammals. Arsine gas (or arsenous hydride, AsH_3) is clearly the most toxic, followed in order of generally decreasing toxicity by inorganic trivalent compounds, organic trivalent compounds, inorganic pentavalent compounds, organic pentavalent compounds, and, finally, elemental arsenic. Some arsenic compounds are apparently not toxic at any dose. Toxicity also depends on other factors such as physical state—gas, solution, or powder—particle size, the rate of absorption into cells, the rate of elimination, the presence of impurities, the nature of chemical substituents in the toxic compound, and, of course, the preexisting state of the patient.

Because of the complex chemistry of arsenic, it is difficult to discuss arsenic poisoning as a single clinical disorder. The problems confronting the physician caring for a patient with exposure to arsenic vary greatly depending on the dose, duration, and type of exposure. Critically evaluating published work on this topic is hampered when it is not made clear what specific chemical compounds are being discussed. The use of the blanket term arsenic in case reporting is inappropriate. This review will briefly delineate the intriguing and important role this element has played throughout medical history, and will then describe the cellular toxicity, clinical manifestations, and treatment of arsenic poisoning. The specific clinical-pathological examples presented have previously been reported (Gorby, 1988).

2. SOURCES OF EXPOSURE

Exposure to arsenic may come from natural sources, from industrial sources, or from administered sources. Self-administration of arsenic, unintentionally by children or deliberately in suicide attempts by adults, represents the most common cause of poisoning (Fuortes, 1988). The source of such self-administration is typically an arsenic-containing insecticide, herbicide, or rodenticide. Some attempts are being made to restrict the easy availability of these products (Park and Currier, 1991), yet the clinical occurrence of this problem is likely to remain with us for some time, though hopefully to a lesser extent in the future. From a clinical perspective, massive exposures are now usually seen in suicidal or homicidal settings; accidental exposures, usually not serious yet largely preventable, are usually seen in children, and chronic or intermittent exposures often are the most diagnostically challenging.

Occupational and environmental health problems can result from the frequent commercial presence of arsenicals. To play on the title of a well-known play, this may be a problem of "arsenic and old waste." In 1973, the National Institute of Occupational Safety and Health estimated that 1.5 million people in the United States are potentially exposed to arsenic during the course of their work. Smelter workers appear to have an increased mortality—primarily from lung cancer—proportional to their arsenic exposure (Pinto et al., 1977). Large-scale accidental arsenic poisonings have occurred in the past. At the turn of this century, more than 6000 British beer drinkers were apparently poisoned with arsenic in the well-known Staffordshire beer epidemic. This was a manufacturing mishap in the Manchester area of England traced primarily to arsenic in iron pyrites used to make sulfuric acid, which, in turn, was used to make the glucose for brewing. Although some have incriminated selenium instead, the epidemic provided the impetus for the first government-mandated industrial regulations for arsenic (Frost, 1967). In 1955, more than 12,000 Japanese infants were poisoned, resulting in 130 deaths, when the sodium phosphate used as a stabilizer in infant formula preparations was found to be contaminated with arsenic (Valee et al., 1960). More recently, 11 cases of arsenic poisoning in a western Minnesota community were attributed to the consumption of well water contaminated from an adjacent arsenical insecticide storage dump (Feinglass, 1973). There is also a report of seasonal arsenic poisoning of a family in rural Wisconsin from extensive burning of marine plywood—treated with chromium–copper–arsenate—in a poorly ventilated cabin (Peters et al., 1984). In a more malicious vein, an acute massive epidemic of arsenic poisoning occurred in Argentina in 1987 when 718 subjects became symptomatic after eating meat laced with an acaricide by vandals in a butcher shop (Roses et al., 1991).

Exposure to arsine gas is also an environmental health hazard of concern in numerous occupational circumstances. Arsine is a colorless, nonirritating gas that causes a rapid and unique destruction of red blood cells and may result in kidney failure, which is uniformly fatal without proper therapy. Most cases of arsine poisoning have occurred with the use of acids and crude metals of which one or both have contained arsenic as an impurity (Fowler and Weissberg, 1974). Fewer than 250 cases of arsine gas poisoning have been reported in the past 65 years, half of which were fatal.

The use of arsenic commercially has steadily declined over the years, largely owing to its replacement by less toxic and equally or more effective substances. It was widely used in industry and agriculture in the past. As one of the elements making up the earth's crust, arsenic is ubiquitous in nature. It has also been used as an additive in poultry and other livestock feed because it is believed to serve as a growth promoter. In comparison to most other foods, unusually high concentrations of arsenic are found in many types of seafood. It is reported that fish may contain arsenic levels of 2 to 8 parts per million (ppm), oysters 3 to 10 ppm, and mussels as much as 120 ppm. Consequently, daily dietary intake of arsenic is greatly influenced by the proportion of seafood in the diet. No clinically important ill effects have been reported from such exposure.

3. HISTORY OF ARSENIC POISONING

Arsenic has been used as a medicine and as a poison since humans first became interested in chemistry. Instances of poisoning with arsenic compounds have occurred throughout recorded history, although the perpetrators of these acts have varied in their intent from gross malevolence to enthusiastic beneficence. The untoward effects of "medicinal" arsenic, primarily inorganic arsenite, have only recently been appreciated because their ill effects are of a chronic nature and large epidemiologic databases are needed to define deleterious outcomes. The toxic properties of all arsenic preparations are dose-dependent. Although this is an integral precept in toxicology today, it was a revolutionary concept in the past. Regarding the administration of arsenic, the dictum of Paracelsus (1493?–1541) is appropriate to remember: "All substances are poisons; there is none which is not a poison. The right dose differentiates a poison and a remedy."

3.1. Arsenic Administration with Malevolent Intent

Arsenic probably has the worst public relations record of any element, and its name has become almost synonymous with the word poison. The popularity of arsenic as a poison is due more to its availability, low cost, and the fact that it is tasteless and odorless than to its effectiveness. Common arsenicals simply are not that efficient as a poison, causing a slow and painful illness rather than instantaneous and certain death. In his play *Arsenic and Old Lace*, John Kesselring was careful to include cyanide and strychnine in the concoction with which the little old ladies swiftly and unerringly dispatched their gentleman callers. Arsenic simply would not have done it alone, although it made a far more intriguing title.

Aristotle and Socrates knew a few vegetable poisons such as hemlock and henbane. They were most familiar, however, with the inorganic poison arsenic trioxide, which, in the course of the following centuries, became the poison of poisons. One of the earliest documented cases of arsenic poisoning was Nero's poisoning of Britannicus to secure his Roman throne in the year 55 A.D. The poisoner became an integral part of social and political life in the early Middle Ages—arsenic was known to be the favorite of some. The records of the city councils of Florence and Venice contain ample testimony of the political use of poisons. Contracts were recorded, naming victims and prices, and when the deed was accomplished, the notation "*factum*" would be entered into the city archives (Doull and Bruce, 1986). Poisoners like the Borgia pope, Alexander VI, and his son, Cesare Borgia, became legendary, as did such sinister figures as Teofania di Adamo and Marie Madeleine, who in the 17th century killed left and right with arsenic solutions. White arsenic acquired so terrible a reputation that ultimately it was called *poudre de succession*, or inheritance powder. Arsenic was the favorite

Table 1 Poisoning Attempts in France per 1000

Arsenic	331	Opiates	12
Phosphorus	301	Mercurials	9
Copper	183	Antimonials	6
Mineral Acids	54	Cyanides	5
Cantharides	35	Iron	5
Strychnine	14	Other	45

Source: Blyth (1885).

poison of 19th century France (Table 1), as recorded by early forensic toxicologists (Blyth, 1885). A French woman, Catherine Deshayes, had a commercial service that grew to infamous proportions, earning her the title *La Vosine.* Her business dissolved with her execution after a special judicial commission established by Louis XIV convicted her of many poisonings, including more than 2000 infant victims (Doull and Bruce, 1986).

Poisoning seems in the past to have been accepted as one of the normal hazards of daily life. Devices and methods of poisoning proliferated at an alarming rate. Arsenic's popularity as a poison, however, declined dramatically in the latter half of the 19th century, at least partly due to the development of a highly reliable and sensitive assay for arsenic described by Marsh in 1836 (Thorwald, 1964). The importance of this test can be appreciated from its role as decisive evidence in the trial of a celebrated poisoner, Marie Lafarge, for the murder of her husband in 1842. Variations of this test (e.g., Reinsch's test and Gutzeit's test) that involve acidifying body tissue or fluids and detecting liberated arsine gas calorimetrically are still in use today as qualitative diagnostic tests.

3.2. Arsenic Administration with Benevolent Intent

The medicinal effects of arsenicals were written about by Hippocrates (460–357 B.C.), Aristotle (384–322 B.C.), and Pliny the Elder (23–79 A.D.). Hippocrates administered orpiment (As_2S_3) and realgar (As_2S_2) as escharotics and as remedies for ulcers. Paracelsus, one of the strangest and most paradoxical characters in medical history, is known to have used elemental arsenic and mercury extensively. Small doses of inorganic arsenicals were long believed to have a "tonic" or "alterative" property, which led to their enthusiastic use for more than two centuries. Initially these compounds cause a cutaneous capillary flush, giving a "milk and roses" complexion valued by many women. Local mountaineers in Austria, the so-called arsenic eaters, consumed large quantities of arsenical ores in the belief that it improved endurance at high altitudes, increased weight, strength, and appetite, and cleared the complexion (Schroeder and Balassa, 1966). Taken internally in either liquid or solid form, injected hypodermically, inhaled as a vapor, administered intravenously, and, on rare occasions, even given in enemas, arsenic proved to be one of the mainstays of 19th century *materia medica*

(Haller, 1975). Many arsenic-containing "medicinals"—Fowler's solution, Asiatic pills, Donovan's solution, DeValagin's elixir, to name but a few—came into use for a variety of ailments. These were considered at various times to be specific therapy for anorexia and other nutritional disturbances and for neuralgia, rheumatism, asthma, chorea, tuberculosis, diabetes, intermittent fever, skin disorders, and hematologic abnormalities. Perhaps the most widely prescribed of these agents was Fowler's solution, a 1% potassium arsenite concoction introduced by Thomas Fowler in the London Pharmacopoeia in 1809. It was withdrawn from the U.S. market in the 1950s after several case series linking its prolonged use to skin changes, cutaneous malignant lesions, and neuropathies appeared. Arsenic retains its mystique as a panacea in some cultures to this day.

By the early 1900s, physicians began using the less toxic organic preparations of arsenic, such as sodium cacodylate and sodium arsanilate, for the treatment of pellagra, as well as for malaria and sleeping sickness. In 1909, Ehrlich's experiments with arsenic led to the widespread use of arsphenamine (salvarsan), often called "606," which, until its replacement by penicillin, was the principal drug in the treatment of syphilis for nearly 40 years. Its success stimulated intense activity on the part of the organic chemists, who extended the list of synthesized arsenic compounds from 606 to an estimated 32,000. Arsenicals in medicine have now been largely replaced by antibiotics but are still used as antiparasitic agents in veterinary medicine and occasionally in patients with trypanosomiasis and amebiasis.

4. TOXICITY AND BIOCHEMISTRY

Much of the basis of our current understanding of the mechanisms of arsenic toxicity in living systems comes from work on arsenic toxicity in animals in the late 1800s, from the development of organic arsenical drugs during the early 1900s, and from work stimulated during the 1940s by the need to find effective antidotes to arsenical warfare agents. The difference in toxicity between trivalent and pentavalent arsenic compounds can best be understood by considering the biochemical mechanism of action of these two distinct families of compounds. Present knowledge of the biochemical mechanisms of arsenic toxicity in mammals is far from complete, although some aspects have been studied in considerable detail. Single-cell studies have established that specific forms of arsenic can be cytotoxic, and it is generally thought that the overt toxicity of arsenic is due to the inhibition of critical sulfhydryl-containing enzymes by trivalent arsenic (Squibb and Fowler, 1983). Many enzymes are susceptible to deactivation by arsenic. In most cases, the enzyme activity can be restored by adding an excess of a monothiol such as glutathione, suggesting that the inhibition is due to the reversible reaction of the arsenic with a single sulfhydryl group in the enzyme molecule. Glutathione depletion has been shown to increase organ toxicity and decrease methylation and renal excretion after ingestion of sodium arsenite by a living hamster model (Hirata et al., 1990). An important exception to this generalization, however, proved to be the pyruvate oxidase system, which could not be protected against trivalent arsenicals by even a 200% excess of monothiol. Such an apparent anomaly was clarified when it was shown that trivalent arsenicals can complex with two sulfhydryl groups in the same protein molecule, thereby forming a stable ring structure that is not easily ruptured by monothiols. This finding stimulated the testing of various dithiol compounds for their ability to block the action of arsenicals on pyruvate oxidase and led to the discovery of a 2,3-dimercaptopropanol, also known as British anti-lewisite (BAL). This agent eventually became a widely used antidote for arsenic poisoning, allowing the formation of a stable, soluble five-membered ring that is excreted in the urine (Fig. 1).

The simultaneous interaction of arsenic with two thiol groups led Peters (1955) to postulate the existence of a dithiol-containing cofactor in the pyruvate oxidase system. This idea was later verified with the identification of lipoic acid. The pyruvate oxidase complex is necessary for oxidative decarboxylation of pyruvate to acetyl coenzyme A and carbon dioxide before it enters the tricarboxylic acid

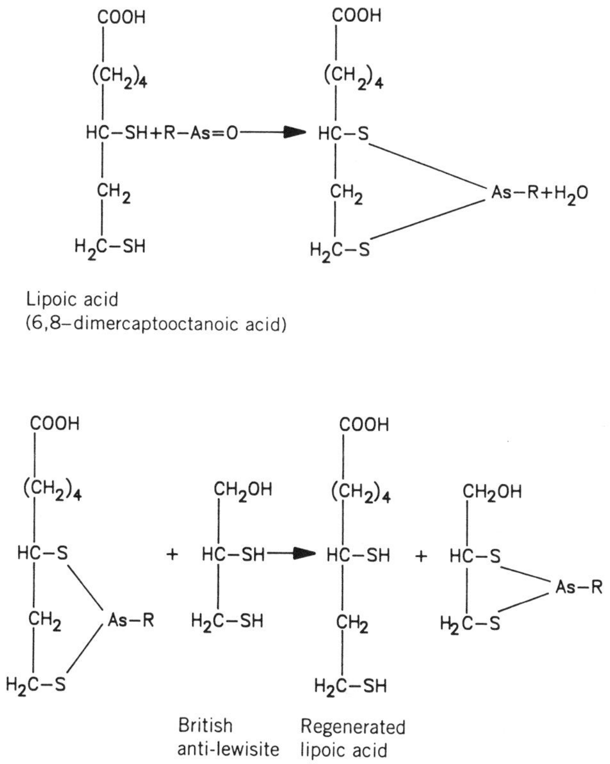

Lipoic acid
(6,8–dimercaptooctanoic acid)

British anti-lewisite

Regenerated lipoic acid

Figure 1. Reaction of lipoic acid with a trivalent monosubstituted arsenical and regeneration of lipoic acid by addition of BAL.

cycle. This enzyme system comprises several enzymes and cofactors. In the presence of trivalent arsenic, a dihydrolipoyl-arsenite chelate is formed, preventing the reoxidation of the dihydrolipoyl group necessary for continued enzymatic activity (Fig. 2), and this pivotal enzyme step is blocked (Gossel and Bricker, 1984).

These well-defined cellular effects of trivalent arsenic may be modified in more complex systems by in vivo oxidation–reduction reactions, by differences in uptake and loss from particular organ systems, and by differences in the natural susceptibility of various tissues. Little is known about the biotransformation of arsenicals in humans. Some pentavalent compounds are partly reduced in vivo to the more toxic trivalent forms, but the redox equilibrium in vivo favors the pentavalent state (Squibb and Fowler, 1983). Although pentavalent arsenic does not appear to lead to enzyme inhibition, it cannot be dismissed as nontoxic, because of its potential to uncouple oxidative phosphorylation. The mechanism is thought to be related to the competitive substitution of arsenate for phosphate, with which it is isoelectric and isosteric, and to the subsequent formation of an unstable arsenate ester bond that is rapidly hydrolyzed. Thus, the so-called high-energy bonds of adenosine triphosphate are not conserved in the presence of arsenate. This process is termed arsenolysis (Valee et al., 1960). Arsenic may therefore be doubly toxic to cellular respiration by inhibiting energy-linked functions of the mitochondria in two very different ways. Trivalent arsenic inhibits the reduction of nicotinamide adenine dinucleotide by deactivating critical enzymes in the tricarboxylic acid cycle, and pentavalent arsenic uncouples oxidative phosphorylation by arsenolysis.

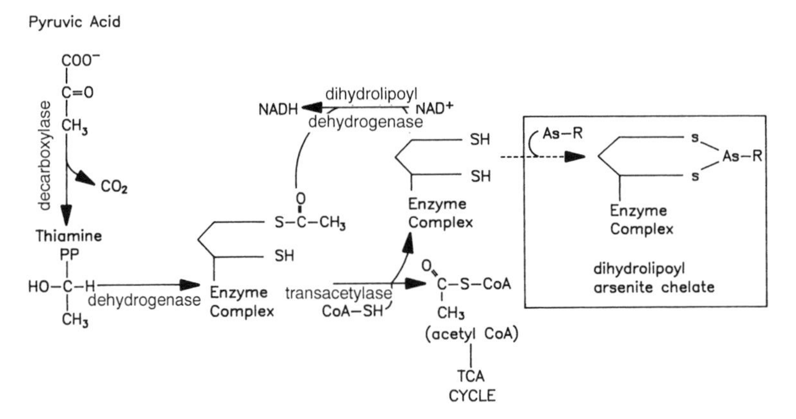

Figure 2. The effect of trivalent arsenic on sulfhydryl enzyme systems. Shown here is the inhibition of pyruvate oxidase by the formation of an arsenite chelate that prevents reoxidation of the dihydrolipoyl group necessary for continued enzymatic activity. Succinate oxidation is disrupted in an identical manner, as are many other enzymes. Abbreviations: CoA, coenzyme A; NAD^+, nicotinamide adenine dinucleotide; NADH, the reduced form of NAD^+; TCA, tricarboxylic acid.

Soluble forms of arsenic are nearly completely absorbed from the gastrointestinal and respiratory tract and probably also from the skin. Excretion is primarily through the urine in the form of methylated arsenic, although the organic arsenic in seafood is largely excreted unchanged (Vahten, 1983). Absorbed arsenic is initially bound to the protein portion of hemoglobin, yet it leaves the intravascular space within 24 hr and is concentrated in the liver, kidneys, spleen, lungs, and gastrointestinal tract. After 2 to 4 weeks, most arsenic remaining in the body is found in hair, nails, and skin due to the high sulfhydryl content of keratin, and it is slowly excreted in this manner. Renal dysfunction is a major impediment to the normal excretion of all arsenic compounds.

The rapid and fulminant intravascular hemolysis characteristic of arsine poisoning results from rapid fixation of arsine in a nonvolatile form by hemoglobin within the red cells. In the presence of oxygen, this leads to extensive lysis of the cell, yet the biochemical mechanisms for this effect are not completely understood.

In addition to these direct toxic effects, there is a growing realization that arsenic is a member of a class of carcinogens, the gene inducers or indirect carcinogens. Genes induced are involved in proliferation, recombination, amplification, and the activation of viruses. All of mankind has unwittingly been exposed to low levels of arsenic and has therefore participated in an experiment of nature. There appears to be a threshold dose for arsenical disease of around 400 μg daily of arsenic based on its action as a nongenotoxic gene inducer and on the epidemiologic evidence that has accumulated so far (Stohrer, 1991).

5. CLINICAL MANIFESTATIONS

The clinical manifestations of arsenic poisoning depend on the type of arsenical involved and on the time–dose relationship of exposure. A fatal dose of arsenic trioxide is probably in the 200 to 300 mg range, yet a dose of 20 mg has been life-threatening, and recovery from 10 g has occurred (Schoolmeester and White, 1980). Clinical manifestations may conveniently, although somewhat artificially, be described as either acute or chronic (Table 2).

5.1. Acute Poisoning

Symptoms of acute intoxication usually occur within 30 minutes of ingestion but may be delayed if arsenic is taken with food. Initially, a patient may have a metallic taste or notice a slight garlicky odor to the breath associated with a dry mouth and difficulty swallowing. Severe nausea and vomiting, colicky abdominal pain, and profuse diarrhea with rice-water stools abruptly ensue. In acute arsenic poisoning of massive proportions, almost always as an attempt at suicide, the fundamental lesion of endothelial cellular toxicity can be considered to account for the predominant clinical features. Capillary damage leads to generalized vasodilation, transudation of plasma, and shock. Arsenic's effect on the mucosal

Table 2 Clinicopathological Findings in Acute and Chronic Arsenic Poisoning

	Acute	Chronic
Dermatologic:	Capillary flush, contact dermatitis, folliculitis Hair: delayed loss Nails: Aldrich–Mees' lines (4–6 weeks postingestion)	Melanosis, Bowen's disease, facial edema, palmoplantar hyperkeratosis, cutaneous malignancies, hyper pigmentation, desquamation
Neurologic:	Hyperpyrexia, convulsions, tremor, coma, disorientation	Encephalopathy, headache, peripheral polyneuropathy, axonal degeneration (Fig. 5)
Gastrointestinal:	Abdominal pain, dysphagia, vomiting, bloody or rice-water diarrhea, garlicky odor to breath and stools, mucosal erosions (Fig. 4) fatty liver (Fig. 3)	Nausea, vomiting, diarrhea, anorexia, weight loss, hepatomegaly, jaundice, pancreatitis, cirrhosis
Renal:	Tubular and glomerular damage, oliguria, uremia	Nephritic findings, proteinuria
Hematologic:	Anemia, thrombocytopenia	Bone marrow hypoplasia, anemia, leukopenia, thrombocytopenia, impaired folate metabolism, basophilic stippling and karyorrhexis (Fig. 6)
Cardiovascular:	ST-T wave abnormalities, QT_c prolongation, ventricular fibrillation, atypical ventricular tachycardia	Arrhythmias, pericarditis, acrocyanosis, Raynaud's, gangrene (Blackfoot disease)
Respiratory:	Pulmonary edema, ARDS, bronchial pneumonia, tracheobronchitis	Cough, pulmonary fibrosis, lung cancer

vascular supply, not a direct corrosive action, leads to transudation of fluid in the bowel lumen, mucosal vesical formation, and sloughing of tissue fragments. The patient may complain of muscle cramps and intense thirst. In severe poisoning, the skin becomes cold and clammy, and some degree of circulatory collapse usually occurs along with kidney damage and decreased urine output. Drowsiness and confusion are often seen along with the development of a psychosis associated with paranoid delusions, hallucinations, and delirium. Finally, seizures, coma, and death, usually due to shock, may ensue.

Following the gastrointestinal phase, multisystem organ damage may occur. If death does not occur in the first 24 hr from irreversible circulatory insufficiency, it may result from hepatic or renal failure over the next several days. Cardiac manifestations include acute cardiomyopathy, subendocardial hemorrhages, and electrocardiographic changes. The most common changes on an electrocardiogram are prolonged QT intervals and nonspecific ST-segment changes (Glazener et al., 1968). A case of an atypical ventricular fibrillation resembling *torsades de pointes* has been reported (Goldsmith and From, 1980). The pathological lesions described in patients with rapidly fatal arsenic intoxication are fatty degeneration of the liver (Fig. 3), hyperemia and hemorrhages of the gastrointestinal tract (Fig. 4), renal tubular necrosis, and demyelination of peripheral nerves (Fig. 5).

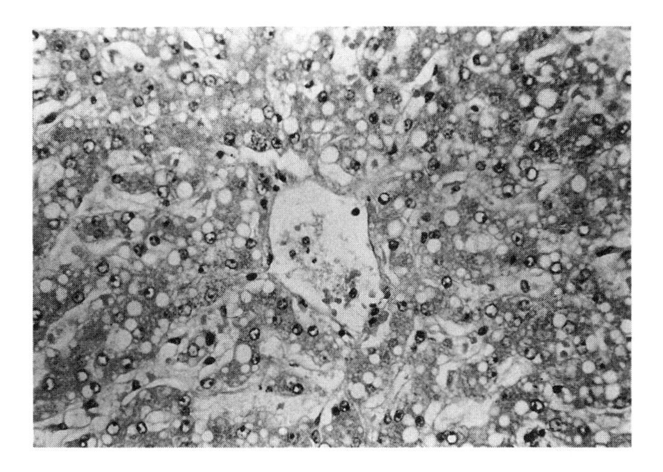

Figure 3. This photomicrograph shows acute fatty change in the liver of a patient who died 12 days after a massive (more than 7 g) overdose of arsenic trioxide (original magnification: × 40).

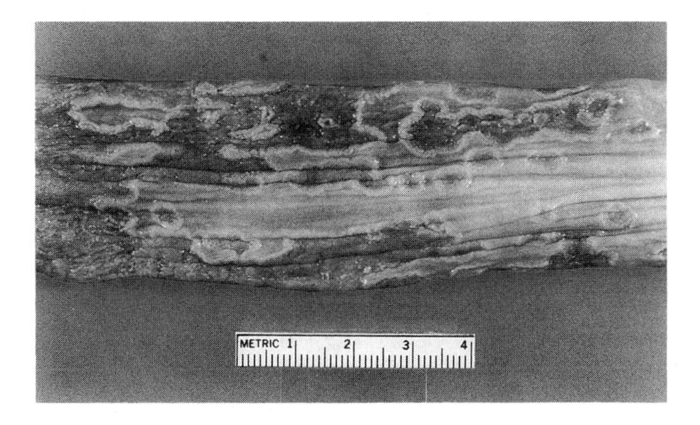

Figure 4. Gross specimen of the distal esophagus in the same patient as in Figure 3. Necrotic oral lesions and marked hemorrhagic gastritis were also noted.

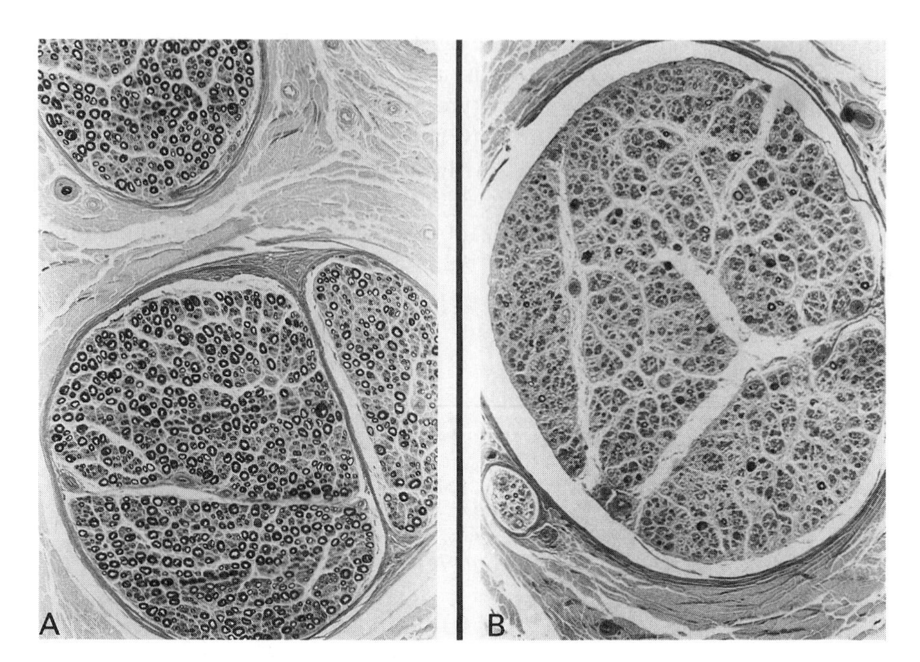

Figure 5. These photomicrographs show osmium-stained sections of sural nerves (original magnification: × 20). A, normal autopsy specimen; B, surgical specimen from a 64-year-old man with chronic arsenic poisoning. Loss of the myelin sheath (normally stained by osmium) and perineural fibrosis can be seen.

5.2. Chronic Poisoning

Chronic arsenic poisoning is much more insidious in nature, often involving multiple hospital admissions before the correct diagnosis is made. The source of arsenic exposure is discovered in fewer than 50% of cases. The most prominent chronic manifestations involve the skin, blood, and neurologic systems.

The cutaneous changes are characteristic yet nonspecific. An initial persistent erythematous flush slowly, over time, leads to melanosis, hyperkeratosis, and desquamation. The skin pigmentation is patchy and has been given the poetic description of "raindrops on a dusty road." The hyperkeratosis is frequently punctate and occurs on the distal extremities. A diffuse desquamation of the palms and soles is also seen. Long-term cutaneous complications include the development of multicentric basal cell and squamous cell carcinomas (Pershagen, 1981). Bowen's disease, a rare precancerous skin lesion, is associated with both arsenic and human papilloma virus (HPV). Both arsenic and HPV cause cancer of the epithelial tissues (McDonell et al., 1989), and one may speculate that arsenic causes cancer in human beings through the activation of an oncogenic virus like HPV. This would explain why arsenic promotes cancer of the epithelial tissues in human beings but not in rodents, which normally do not carry

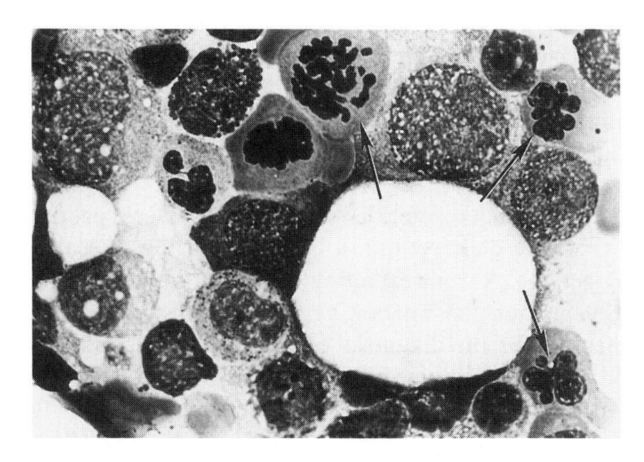

Figure 6. Photomicrograph of a bone marrow specimen from the patient in Figure 5 (original magnification: × 100). Most notable are bizarre nuclei in the erythrocytic precursors—karyorrhexis—indicated by arrows.

papilloma virus (Stohrer, 1991). Brittle nails, patchy alopecia, and facial edema can occur in arsenical skin disease. Transverse white bands across the nails (Aldrich–Mees' lines) are frequently seen 3 to 6 weeks after an acute exposure and may even be used to date the event (average nail growth rate is 0.10 to 0.12 mm per day).

Anemia and leukopenia are almost universal with chronic arsenic exposure; thrombocytopenia frequently occurs. The anemia is usually normochromic and normocytic and caused at least partially by hemolysis (Kyle and Pease, 1965). Interference with folate metabolism and DNA synthesis may result in megaloblastic changes (Westhoff et al., 1975). Karyorrhexis, an accelerated pyknosis of the normoblast nucleus, is characteristic of arsenic poisoning. This is manifested as bizarre nuclear forms seen on bone marrow examination (Fig. 6). Basophilic stippling is also seen. Aplastic anemia progressing to acute myelogenous leukemia has been reported (Kjeldsberg and Ward, 1972).

A peripheral neuropathy is the hallmark of chronic arsenic poisoning but may be seen within two hours of ingestion. Usually it is a symmetric polyneuropathy of both sensory and motor nerve fibers, often resembling the Landry–Guillain–Barré syndrome in its presentation (Chhuttani et al., 1967). On microscopic examination, there is resorption of myelin and destruction of axonal cylinders progressing to nerve atrophy and perineural fibrosis.

6. DIAGNOSIS AND TREATMENT OF ARSENIC POISONING

Because arsenic is rapidly cleared from the bloodstream and the major route of arsenic elimination is through the kidneys as methylated arsenic metabolites, the preferred sample for diagnostic analysis is a 24-hr urine collection. Arsenic exposure may also be assessed by analyzing the content in hair and nails, because arsenic tends to accumulate in these tissues over time. However, environmental arsenic may adhere to these exposed tissues and lead to false elevations due to external contamination. In some instances, the concentration of arsenic along the length of the hair is measured to obtain information about exposure over a period of time.

Persons on ordinary diets excrete less than 20 μg of arsenic per day through the urine, but those whose diets contain a large proportion of seafood excrete as much as 200 μg per day. Arsenic excretion of greater than 100 μg per day should be considered suspicious and warranting further investigation. A mobilization test is recommended for the diagnosis of mild or chronic arsenic exposure. A 24-hr urinary collection is obtained for baseline arsenic excretion, followed by a second 24-hr urinary collection while the patient receives four doses (250 mg every 6 hr) of oral penicillamine (Campbell and Alvarez, 1989). The test is positive if either urine collection reveals greater than 100 μg arsenic per 24 hr; a 5-day course of penicillamine is recommended.

Numerous more sophisticated methods are available for quantitative analysis of arsenic in biologic specimens (Malachowski, 1990). The most common and widely applicable is atomic absorption spectrometry. Neutron activation analysis is another sensitive analytic method, although of limited availability.

In patients with acute arsenic poisoning, supportive therapy and chelation treatment are indicated. Essential supportive measures, as with any poisoning, include scrupulous intestinal decontamination—charcoal lavage and the administration of cathartics—and adequate intravenous fluid therapy. Exchange transfusions may be of help in clearing the blood of elevated arsenic concentrations early in the course. In the conventional view, hemodialysis is thought to be useful only in cases of renal failure to maintain fluid and electrolyte homeostasis until renal function recovers (Vaziri et al., 1980). However, in cases of massive arsenic poisoning one should consider early hemodialysis as an elimination measure and the choice of BAL as a chelator when dialysis is required (Mathieu et al., 1992).

Dimercaprol (2, 3-dimercaptopropanol) is the traditional chelating agent used, but penicillamine has been used with some success (Peterson and Rumack, 1977). Parenteral dimercaprol is administered intramuscularly at an initial dose of 3 to 5 mg per kilogram of body weight every 4 hr. The dose should be tapered but administration continued until the urinary arsenic excretion is less than 50 μg per 24 hr. This therapy is frequently effective in preventing or neutralizing systemic toxicity. In most cases, the degree of recovery from neuropathy, aplastic anemia, encephalopathy, and jaundice is limited and directly related to the initial severity of the systemic involvement and the rapidity with which chelation therapy is initiated. Penicillamine, although only a monothiol agent, has been used successfully; its great advantage is that it may be orally administered. Both agents have a high frequency of side effects, although this is less of a problem in the presence of large amounts of body arsenic. A recently reintroduced drug that appears to be a promising agent for treating arsenic poisoning is 2, 3-dimercaptosuccinic acid. This is a dithiol agent that can be orally administered and has few reported side effects (Graziano, 1986). Experience, however, is limited with the use of this drug.

Chronic arsenic poisoning can also be treated with chelating agents, but the possible benefits must be weighed against the known side effects. There is little convincing evidence of the benefit of chelation therapy in most cases of chronic poisoning.

7. SUMMARY

Arsenic is a metalloid that has played an important and fascinating role in human medicine throughout recorded history. Exposure to toxic amounts of arsenic continues to occur. Certain asepcts of the metabolism and biochemistry of the arsenical compounds are now well understood, and the application of this knowledge has led to a rational basis of therapy for cases of acute poisoning. The clinical manifestations of arsenic poisoning are myriad, and the correct diagnosis depends largely on awareness of the problem.

ACKNOWLEDGMENTS

Photomicrographs of nerve sections with osmium stains were prepared by Dr. Rachel Kleinmen. Manuscript preparation was aided by Kathy Lamb.

REFERENCES

Blyth, A. W. (1885). *Poisons: Their Effects and Detection.* William Wood, New York, pp. 35–71.

Campbell, J. P., and Alvarez, J. A. (1989). Acute arsenic poisoning. *Am. Fam. Pract.* **40**, 93–97.

Chhuttani, P. N., Chawla, L. S., and Sharma, T. D. (1967). Arsenical neuropathy. *Neurology* **17**, 269–274.

Doull, J., and Bruce, M. C. (1986). Origin and scope of toxicity. In C. D. Klaassen, M. O. Amdur, and J. Doull (Eds.), *Toxicology: The Basic Science of Poisons.* Macmillian, New York, 3rd ed., pp. 3–10.

Feinglass, E. J. (1973). Arsenic intoxication from well water in the United States. *N. Engl. J. Med.* **288**, 828–830.

Fowler, B. A., and Weissberg, J. B. (1974). Arsine poisoning. *N. Engl. J. Med.* **291**, 1171–1174.

Frost, D. V. (1967). Arsenicals in biology—Retrospect and prospect. *Fed. Proc., Fed. Am. Soc. Exp. Biol.* **26**, 194–208.

Fuortes, L. (1988). Arsenic poisoning. *Postgrad. Med.* **83**(1), 233–244.

Glazener, F. S., Ellis, J. G., and Johnson, P. K. (1968). Electrocardiographic findings with arsenic poisoning. *Calif. Med.* **109**, 158–162.

Goldfrank, L. R., Howland, M. A., and Kirstein, R. H. (1986). Arsenic. In L. R. Goldfrank (Ed.), *Goldfrank's Toxicologic Emergencies: A Handbook in Problem Solving.* Appleton-Century-Crofts, New York, 2nd ed., pp. 609–618.

Goldsmith, S., and From, A. H. (1980). Arsenic-induced atypical ventricular tachycardia. *N. Engl. J. Med.* **303**, 1096–1098.

Gorby, M. S. (1988). Arsenic poisoning. *West. J. Med.* **149**, 308–315.

Gossel, A. A., and Bricker, J. D. (1984) Metals. *Principles of Clinical Toxicology*. Raven Press, New York, Chapter 10, pp. 154–187.

Graziano, J. H. (1986). Role of 2, 3-dimercaptosuccinic acid in the treatment of heavy metal poisoning, *Med. Toxicol.* **1**, 155–162.

Haller, J. S. (1975). Therapeutic mule: The use of arsenic in the 19th-century materia medica. *Pharm. Hist.* **17**, 87–100.

Hirata, M., Tanaka, A., Hisanaga, A., and Ishinishi, N. (1990). Effects of glutathione depletion on the acute nephrotoxic potential of arsenite and on arsenic metabolism in hamsters. *Toxicol. Appl. Pharmacol.* **106**, 469–481.

Kjeldsberg, C. R., and Ward, H. P. (1972). Leukemia in arsenic poisoning. *Ann. Intern. Med.* **77**, 935–937.

Kyle, R. A., and Pease, G. L. (1965). Hematologic aspects of arsenic intoxication. *N. Engl. J. Med.* **273**, 18–23.

Malachowski, M. E. (1990). An update on arsenic. *Clin. Lab. Med.* **10**(3), 459–472.

Mathieu, D., Mathieu-Nolf, M., Germain-Alonso, M., Neviere, R., Furon, D., and Wattel, F. (1992). Massive arsenic poisoning—effect of hemodialysis and dimercaprol on arsenic kinetics. *Intensive Care Med.* **18**, 47–50.

McDonnell, J. M., Mayr, A. J., and Martin, W. J. (1989). DNA of human papilloma virus type 16 in dysplastic and malignant lesions of the conjunctiva and cornea. *N. Engl. J. Med.* **320**, 1442–1446.

Park, M. J., and Currier, M. (1991). Arsenic exposures in Mississippi: A review of cases. *South. Med. J.* **84**(4), 461–464.

Pershagen, G. (1981). The carcinogenicity of arsenic. *Environ. Health Perspect.* **40**, 93–100.

Peters, H. A., Croft, W. A., Woolson, E. A., Darcey, B. A., and Olson, M. A. (1984). Seasonal arsenic exposure from burning chromium-copper-arsenate-treated wood. *J. Am. Med. Assoc.* **251**, 2393–2396.

Peters, R. A. (1955). Biochemistry of some toxic agents: Present state of knowledge of biochemical lesions induced by trivalent arsenical poisoning. *Bull. Johns Hopkins Hosp.* **97**, 1–20.

Peterson, R. G., and Rumack, B. H. (1977). D-Penicillamine therapy of acute arsenic poisoning. *J. Pediatr.* **91**, 661–666.

Pinto, S. S., Enterline, P. E., Henderson, V., and Varner, M. O. (1977). Mortality experience in relation to a measured arsenic trioxide exposure. *Environ. Health Perspect.* **19**, 127–130.

Roses, O. E., Garcia Fernandez, J. C., Villaamil, E. C., Camussa, N., Minetti, S. A., et al. (1991). Mass poisoning by sodium arsenite. *Clin. Toxicol.* **29**(2), 209–213.

Schoolmeester, W. L., and White, D. R. (1980). Arsenic poisoning. *South. Med. J.* **73**, 198–208.

Schroeder, H. A., and Balassa, J. J. (1966). Abnormal trace metals in man: Arsenic. *J. Chronic Dis.* **19**, 85–106.

Squibb, K. S., and Fowler, B. A. (1983). The toxicity of arsenic and its compounds. In B. A. Fowler (Ed.), *Biological and Environmental Effects of Arsenic*. Elsevier, Amsterdam, pp. 233–263.

Stöhrer, G. (1991). Arsenic: Opportunity for risk assessment. *Arch. Toxicol.* **65**, 525–531.

Thorwald, J. (1964). The winding road of forensic toxicology. *The Century of the Detective*. Harcourt, Brace & World, New York, pp. 267–292.

Vahten, M. (1983). Metabolism of arsenic. In B. A. Fowler (Ed.), *Biological and Environmental Effects of Arsenic*. Elsevier, Amsterdam, pp. 171–198.

Valee, B. L., Ulmer, D. D., and Wacker, W. E. C. (1960). Arsenic toxicology and biochemistry. *Arch. Ind. Health* **21**, 132–151.

Vaziri, N. D., Upham, T., and Barton, C. H. (1980). Hemodialysis clearance of arsenic. *Clin. Toxicol.* **17**, 451–456.

Westhoff, D. D., Samaha, R. J., and Barnes, A. (1975). Arsenic intoxication as a cause of megaloblastic anemia. *Blood* **45**, 241–246.

2

HEALTH EFFECTS OF ENVIRONMENTAL ARSENIC

William E. Morton

Oregon Health Sciences University, Portland, Oregon 97201

David A. Dunnette

Portland State University, Portland, Oregon 97207

Arsenic in the Environment, Part II: Human Health and Ecosystem Effects,
Edited by Jerome O. Nriagu.
ISBN 0-471-30436-0

---

1. INTRODUCTION

Arsenic is a common constituent of the earth's crust and a common contaminant in metallic ores and industrial materials. Aside from contamination, arsenical compounds have had a variety of uses in manufacture, wood preservation, agriculture, electronics, and medicine. Humans may encounter arsenic in water from wells drilled into arsenic-rich ground strata or in water contaminated by industrial or agrochemical wastes (Hughes et al., 1988). They may come in contact with arsenic in contaminated dusts, fumes, or mists (Pershagen et al., 1981). They may eat food contaminated with arsenical pesticides or grown with arsenic-contaminated water or in arsenic-rich soil (Nriagu and Azcue, 1990). The per capita daily total arsenic intake is usually less than 200 μg, with the daily inorganic arsenic intake not normally exceeding 60 μg, whereas a fatal dose of ingested As_2O_3 is about 1 to 2.5 mg As/kg body weight (Pershagen et al., 1981).

Arsenic binds readily to sulfhydryl groups of globulins and other large molecules, accumulating in mitochondria where energy-linked functions are inhibited (Goyer, 1991). Although trivalent arsenic can be oxidized to the less toxic pentavalent form in natural systems, the usual bodily biotransformation leads in the other direction, to the formation of trivalent arsenic, which is the form that can be methylated to monomethylarsonic acid (MMA) and dimethylarsenic acid (DMA) for renal excretion (Goyer, 1991). The organic arsenic ingested with seafoods is relatively harmless and is excreted as cacodylic acid. The various forms of arsenic, in descending order from most to least toxic, are arsines > arsenites (inorganic, trivalent) > arsenoxides (organic, trivalent) > arsenates (inorganic, pentavalent) > arsonium compounds > metallic arsenic (Hindmarsh and McCurdy, 1986).

2. TYPES OF EXPOSURE

The two usual routes of absorption of arsenic are by inhalation and/or ingestion, although there may be some degree of skin absorption of trivalent arsenic oxide since it is more lipid-soluble than the pentavalent form (Winship, 1984). If the initial contact is by ingestion, then symptoms caused by gastrointestinal irritation will dominate the early reaction. Ingested arsenic has a shorter half-life than inhaled arsenic due to more rapid biotransformation in the liver (Vahter, 1988). If inhalation is the route of initial contact, then respiratory irritation will be a major determinant of early symptoms. However, once the arsenic is absorbed, the vascular circulation will ensure contact with other organs, with a wide variety of potential symptoms reflecting the diversity of possible organ damage.

3. DOSE AND TYPE OF RESPONSE

The magnitude of the exposure dose will determine the overall response, but a number of temporal and host susceptibility factors are important modifiers. Relatively large exposures will cause acute symptoms involving many bodily systems, possibly even collapse and death. Small doses may have no immediately obvious effects, so that the appearance of symptoms would be the result of arsenic accumulation from repeated exposures. For exposures large enough to cause acute symptoms, the risks of illness and/or more severe outcome can be expected to be directly correlated with the size of the dose. But when chronic symptoms have appeared gradually, the dose size is usually unknown, and it would seem unrealistic to make a dose-response measurement a requirement for acceptance of an hypothesized exposure–manifestation correlation. Years of exposure is a common basis for estimation of relative exposure dose and has been used with some success both for arsenic water exposures and for arsenic dust exposures.

The particular nature of the chronic symptoms seen in an individual from continued or repeated small doses of arsenic depends at least in part on which organ systems are the most susceptible to arsenic effects in that person. For example, persons with certain preexisting neurologic, hematologic, hepatic, cardiovascular, or other problems would have nonspecific respective local tissue susceptibility to arsenic effects. Thus, variations in individual and tissue susceptibilities could account for a significant part of the tremendous variation in clinical combinations of symptoms reported.

In chronic arsenic poisoning, the sole symptoms may be nonspecific general effects such as chronic weakness, easy fatigue, absent motivation, anorexia, weight loss, and hair loss (Hindmarsh and McCurdy, 1986). The slow worsening of symptoms related to several organ systems may suggest to the physician that the patient could have any one of a variety of progressive chronic diseases, possibly accounting for arsenic's popularity as a surreptitious homicidal agent (Windhorst et al., 1977).

4. BIOMARKERS

Blood-borne arsenic tends to be associated with globulins from which it is cleared into tissues within 24 hr (Ellenhorn and Barceloux, 1988). Only in cases of continued arsenic exposure is the blood arsenic level likely to be a useful indicator of its presence in the body. Normal blood arsenic values generally range from 1 to 40 μg/L, while current arsenic exposures will elevate these levels at least 10-fold (Vahter, 1988).

Urinary arsenic level is the most valuable and frequently used biomarker for monitoring arsenic exposure (Vahter, 1988). In a 24-hr urine specimen, up to 20 μg of inorganic arsenic or 50 μg of total arsenic are within normal limits. After exposure ceases, the urinary arsenic may not return to normal limits for as long as 7 to 10 days.

Analysis of hair or nail clippings for arsenic may be useful in someone whose exposures ceased several weeks to several months earlier. It would be a more valid index in someone whose exposure had been to contaminated water rather than to contaminated dust, because exposure to the latter would enable adsorption onto hair as well as incorporation within. Arsenic concentrations in normal hair are said to be less than 1 mg As/g hair (Pershagen et al., 1981; Vahter, 1988), but in our region they are less than that.

The medical diagnosis of arsenic poisoning begins with the history of known exposure, followed by the onset of characteristic symptoms in appropriate relation to that exposure, and verified by laboratory evidence of the presence of excess arsenic.

5. TARGET ORGAN SYSTEMS

5.1. Gastrointestinal Effects

Since much of the published experience with arsenic poisoning has described the effects of oral ingestion by accident or with suicidal or homicidal intent, the direct effects on the gastrointestinal tract have been prominent and duly noted (Ellenhorn and Barceloux, 1988; Fuortes, 1988; Gorby, 1988; Hindmarsh and McCurdy, 1986; Pershagen et al., 1981; Schoolmeester and White, 1980; Vahter, 1988; Windhorst et al., 1977; Winship, 1984). Although arsenic may produce direct irritant effects on gastrointestinal tissues with which it comes in contact, the greatest degrees of damage are produced by local submucosal capillary damage from absorbed arsenic (Clarkson, 1991). The vascular damage is thought to be the cause of submucosal vesicles, whose rupture can create grossly visible erosions and major fluid and protein loss. Nausea and vomiting can be severe, as can colicky abdominal pain and marked diarrhea. If sufficiently severe, the acute gastroenteritis can lead to circulatory collapse with renal damage and shutdown.

Subacute arsenic poisoning from lesser doses of arsenic may manifest as dry mouth and throat, heartburn, nausea, abdominal pains and cramps, and moderate diarrhea. Chronic low-dose arsenic ingestion may be without symptomatic gastrointestinal irritation or may produce a mild esophagitis, gastritis, or colitis with respective upper and lower abdominal discomfort. Anorexia, malabsorption, and weight loss may be present.

5.2. Respiratory Effects

Situations in which the predominant arsenic contact is by dust or fume inhalation are more apt to be encountered in mining and milling of ores, in industrial processing, and, less frequently now, in agriculture (Clarkson, 1991; Gerhardsson et al., 1988; Pershagen et al., 1981; Pinto and McGill, 1953). There is often rhinitis, pharyngitis, laryngitis, and tracheobronchitis, which cause such symptoms as stuffy nose, sore throat, hoarseness, and chronic cough (Tsuda et al., 1992). The longer such irritation is present, the more secondary infections and scarring may occur. Arsenic can induce or aggravate reactive airway disease (asthma) in susceptible individuals. Unprotected workers may manifest perforated nasal septum after 1–3 weeks of exposure (Hamilton, 1925). A fatal case of arsenic trioxide inhalation manifested widespread tracheobronchial mucosal and submucosal hemorrhages, with mucosal sloughing, alveolar hemorrhages, and pulmonary edema (Gerhardsson et al., 1988).

5.3. Dermatologic Effects

Not only does the skin have characteristic acute and chronic manifestations of arsenicism, it is also a major organ of arsenic accumulation. If there is direct fume or dust exposure to the skin, its amount and duration will determine whether the immediate cutaneous response ranges from none to florid. It has been observed that some individuals with insufficient arsenic contact to induce a rash may experience eventual increased severity of minor bacterial skin infections, perhaps by enzyme damage, which inhibits the localizing power of inflammation (Stone and Willis, 1968). If sufficient in amount, direct skin contact with arsenic powder or dust can produce an irritant erythema of exposed surfaces, particularly the face and eyelids (Clarkson, 1991; Gerhardsson et al., 1988).

Chronic exposures to arsenic by either ingestion or inhalatioin will produce a variety of skin insignia of arsenic toxicity. The initial erythematous flush from arsenic may phase into an actinic keratosis, a hyperkeratosis of palms and soles, papillomatosis, recurrent episodes of pruritic urticaria, or even generalized pruritis without a visible rash (Cannon, 1936). Hyperpigmentation may occur, particularly in body areas where the skin tends to be a little darker (Shannon and Strayer, 1989). The darkening may be uniform or mottled. Some people develop diffuse or patchy hair loss, which may or may not be reversible when the arsenic effect dissipates (Tsuda et al., 1992). Arsenic can inhibit the growth of fingernails and toenails, producing transverse white indentations called Mees' or Aldrich–Mees' lines (Adams, 1990). A purplish-red flush of face and shoulders has been referred to as cyanosis and is thought to be due to capillary injury rather than systemic hypoxia (Clarkson, 1991).

In areas where the skin is or has been involved, neoplasia may develop as either basal cell carcinoma or squamous cell carcinoma, which are histologically indistinguishable from analogous nonarsenic tumors. Compared to other skin cancers, arsenic-induced skin cancers tend to occur at younger ages, to manifest simultaneous multiple occurrences, and to be located on the upper extremities and trunk in areas usually unexposed to the sun (Shannon and Strayer, 1989).

5.4. Hematologic Effects

Arsenic has a general depressant effect on the hematopoietic system, particularly in the presence of chronic exposures (Pershagen et al., 1981; Schoolmeester and

White, 1980; Winship, 1984). Normochromic normocytic anemia is relatively common. Granulocytopenia or thrombocytopenia may occur (Kyle and Pease, 1965). Pancytopenia and even aplastic anemia have been observed in arsenic poisoning. Arsenic-induced megaloblastic anemia has been reported (Feussner et al., 1979; Westhoff et al., 1975). Arsenical myelodysplasia has been described (Rezuke, 1991), and acute myeloid leukemia has followed the appearance of arsenic-induced aplastic anemia (Kjeldsberg and Ward, 1972).

Arsenic interferes with heme synthesis, so that prolonged exposure of mice and rats to arsenate will produce increased excretion of uroporphyrin and coproporphyrin (Fowler et al., 1987). However, humans chronically exposed to elevated arsenic levels in their drinking water have not manifested correlations between urinary arsenic and porphyrin levels (Garcia-Vargas et al., 1991).

Arsine gas (AsH_3) binds rapidly to sulfhydryl groups on enzymes and other proteins within red blood cells, producing, after a variable latent period, abrupt widespread hemolysis with anemia, hematuria, collapse, and risk of renal failure (Fowler and Weissberg, 1974; Landrigan et al., 1982; Teitelbaum and Kier, 1969). Although this is usually an acute syndrome, chronic arsine poisoning with cumulative damage has been described.

5.5. Hepatic Effects

Since the liver tends to accumulate arsenic with repeated exposures, hepatic involvement has been reported most commonly as a complication of chronic exposures over periods of months or years (Buchanan, 1962; Clarkson, 1991; Ellenhorn and Barceloux, 1988; Hindmarsh and McCurdy, 1986; Schoolmeester and White, 1980; Squibb and Fowler, 1983; Winship, 1984). Chronic arsenic-induced hepatic changes include cirrhosis, portal hypertension without cirrhosis, fatty degeneration, and primary hepatic neoplasia. Patients may first come to medical attention with bleeding esophageal varices, ascites, jaundice, or simply an enlarged tender liver. The occurrence of cirrhosis as a manifestation of chronic arsenicism may depend on the simultaneous use of excessive ethanol. Arsenic has been observed to produce mitochondrial damage and impaired mitochondrial functions, and accordingly might be expected to affect porphyrin metabolism (Fowler et al., 1987). Liver changes in acute arsenicism have included congestion, fatty infiltration, cholangitis, cholecystitis, and acute yellow atrophy (Schoolmeester and White, 1980; Squibb and Fowler, 1983; Winship, 1984).

5.6. Renal Effects

Like the liver, the kidneys will accumulate arsenic in the presence of repeated exposures. The kidneys are the major route of arsenic excretion, as well as a major site of conversion of pentavalent arsenic into the more toxic and less soluble trivalent arsenic. Sites of arsenic damage in the kidney include capillaries, tubules, and glomeruli (Schoolmeester and White, 1980; Sauibb and Fowler, 1983; Winship, 1984). As in other organs, capillary damage seems to be a basic event leading to other cellular manifestations. Glomerular arterioles dilate, permitting hematuria. Damaged proximal tubular cells lead to proteinuria and casts in the urine. Mitochondrial damage is also prominent in tubular cells. Oliguria is a common manifestation, but if acute arsenic poisoning is sufficiently severe to produce shock and dehydration, there is real risk of renal failure, although dialysis has been effective in overcoming this complication (Giberson et al., 1976). Acute cortical necrosis is an uncommon severe renal manifestation, which also benefits from dialysis (Gerhardt et al., 1978).

Arsine-induced hemolysis is likely to cause acute tubular necrosis with partial or complete renal failure, requiring hemodialysis for removal of the hemoglobin-bound arsenic (Fowler and Weissberg, 1974; Teitelbaum and Kier, 1969). Recovery may leave interstitial fibrosis and thickened glomerular basement membranes.

5.7. Cardiovascular Effects

Both the heart and peripheral arterial tree commonly manifest effects of arsenic toxicity (Hindmarsh and McCurdy, 1986; Pershagen et al., 1981; Schoolmeester and White, 1980; Winship, 1984). Arsenic interstitial myocarditis was discovered as a consequence of arsenical treatment of syphilis (Nelson, 1934), although electrocardiographic abnormalities indicating myocardial effects also have followed the consumption of arsenic-contaminated drinking water. Arsenical myocarditis may appear 6 to 20 days after the onset of the poisoning. The myocardial fibers remain relatively intact, though separated by eosinophilic and/or lymphocytic infiltrate, a pattern that has suggested an allergic rather than a direct toxic action on the myocardium (Brown and McNamara, 1940). Such myocarditis may be acutely fatal or may heal to fibrosis with chronic congestive heart failure (Edge, 1946). Cardiac involvement can be identified by characteristic electrocardiographic T-wave inversion (Weinberg, 1960). The inflamed myocardium can initiate ventricular tachycardia (Goldsmith and From, 1980), so that arsenic-poisoned patients with abnormal T waves should probably be monitored for early recognition of this complication.

General vascular effects of arsenic are dilatation and increased permeability of capillaries, which may be severe enough to cause hypovolemia, hypoproteinemia, and shock. According to observations on arsenic-poisoned young persons in Antofagasta, Chile, medium- and small-sized arteries show intimal thickenings that begin as subendothelial swelling and vacuolization and are associated with fibrosis and ischemic changes in the tissues supplied (Rosenberg, 1974; Zaldivar, 1980). Persons living in an area of southwestern Taiwan where wells are contaminated by arsenic may develop ischemic claudication, which, over periods from several months to several years, can progress to a black, mummified dry gangrene, usually involving the foot (Chi and Blackwell, 1968; Tseng, 1989). Taiwanese with high arsenic intake and blackfoot disease were found to have significantly higher mortality rates for peripheral vascular disease and ischemic heart disease, which corresponded with marked arteriosclerotic lesions in peripheral and coronary arteries (Chen et al., 1988). A study of inhabitants of 27 Taiwan villages showed that arsenic levels in the village wells from 1964 to 1966 had a significant direct effect on age-standardized mortality rates from 1973 to 1986 for peripheral vascular diseases and ischemic heart disease in both sexes (Wu et al., 1989). Mortality from ischemic heart disease was significantly increased among arsenic-exposed workers at the Anaconda copper smelter (Welch et al., 1982), but not at the Tacoma copper smelter (Pinto et al., 1978). Cardiovascular disease mortality risk was also increased among arsenic-exposed Swedish glassblowers (Wingren and Axelson, 1985) and copper smelter workers (Axelson et al., 1978). The morbidity of Raynaud's disease is increased among persons with higher arsenic intakes (Kraetzer, 1930; Lagerkvist et al., 1986; Tsuda et al., 1992).

5.8. Neurologic Effects

Like the cardiovascular system, both the peripheral and central components of the nervous system can be damaged by arsenic (Pershagen et al., 1981; Schoolmeester and White, 1980; Windhorst et al., 1977; Winship, 1984). Individuals with repeated arsenic exposures frequently contract sensorimotor polyneuropathy, which usually, but not always, displays symmetrical involvement and which may resemble Landry-Guillain-Barré syndrome in its presentation. Neuropathy may appear in 1 to 5 weeks after an acute exposure and is produced mainly by axonal degeneration, although myelin disruption is also present (Politis et al., 1980). Spinal anterior horn cells may be damaged and diminished in number. Severely involved patients may recover partially over a 2-year period after exposure stops, leaving some degree of permanent impairment and disability. Sensory symptoms, such as paresthesias and hypesthesias, usually appear first, followed in some cases by weakness or even paralysis starting in the distal lower extremities and occasionally involving the distal upper extremities. All types of sensation are affected, with vibration and position-sense thought to be more severely affected than pain and light touch. At their peak, motor symptoms may range from mild extensor weakness to paralysis and even muscle atrophy. Local muscle tenderness is often present. Cranial nerves are sometimes listed as being spared from arsenic damage but, in a 1955 epidemic among Japanese infants given powdered milk containing pentavalent arsenic, 18% of a follow-up group had severe hearing loss, and there was a case of optic nerve atrophy (described in Pershagen et al., 1981). Czechoslovakian children living near a plant burning arsenic-contaminated coal also had significant excess hearing loss (Bencko et al., 1977), but this could not be found in children living near a U.S. copper smelter (Milham, 1977).

Arsenic can also damage the central nervous system, but CNS impairment has been less frequently observed, so that some sources omit this possibility. Among individuals diagnosed with arsenic poisoning, the proportion who manifest encephalopathy may range from 5 to 9% (Jenkins, 1966; Reynolds, 1901) to as high as 40% among a group of former workers and neighbors of an abandoned Japanese arsenic mine and refinery (Hotta, 1989). Symptoms of chronic encephalopathy include persistent headache, diminished recent memory, distractibility, abnormal irritability, restless sleep, loss of libido, increased urinary urgency, and increased effects of small amounts of ethanol (Morton and Caron, 1989). Secondary depression, anxiety, panic attacks, and somatizations are common, in addition to the organic cognitive impairment documented by neuropsychological testing. Organic cognitive impairment induced by arsenic is reversible to a greater degree than that induced by organic solvents, but prolonged and/or severe exposures may leave permanent impairment.

6. GENOTOXICITY

6.1. Mutagenic Effects

Mutagenesis includes the induction of DNA damage and a wide variety of genetic alterations, which can range from simple gene mutations (DNA base-pair changes) to grossly visible changes in chromosome structure or number (clastogenesis). Some of these changes may cause genetic damage transmissible to subsequent generations, and/or some may cause cancer or other problems in the exposed generation (Hoffmann, 1991). Arsenic has long been known to cause chromosomal damage (Hindmarsh and McCurdy, 1986; Leonard and Lauwerys, 1980; Pershagen et al., 1981; Squibb and Fowler, 1983; Windhorst et al., 1977; Winship, 1984), but most investigators have been unable to induce direct gene mutations (Goyer, 1991; Squibb and Fowler, 1983). This apparent paradox, plus occasional poor correlation between arsenic exposure dose and resultant frequency of chromosomal aberrations, have been explained by the concept that arsenic promotes genetic damage in large part by inhibiting DNA repair (Bencko, 1977; Leonard and Lauwerys, 1980; Nordenson et al., 1978; Rossman et al., 1977). The repair inhibition may be a basic mechanism for the comutagenicity and presumably the cocarcinogenicity of arsenic (Okui and Fujiwara, 1986). Such DNA damage could have been produced by a wide variety of other mutagens or conceivably even by arsenic itself. In the absence of such DNA damage, inhibition of repair would not be manifest, which could account for different outcomes among epidemiologic or experimental investigations controlled for arsenic but not for other latent mutagens.

Comparisons of chromosome aberration frequencies induced by trivalent and pentavalent arsenic have indicated that the trivalent forms are far more potent and genotoxic than the pentavalent forms (Barrett et al., 1989; Nakamuro and Sayato, 1981; Nordenson et al., 1981). Enzymes such as superoxide dismutase and catalase that scavenge for oxygen free radicals seem to provide protection against arsenic-induced DNA damage, indicating a possible basis for the genotoxic effect of arsenic (Nordenson and Beckman, 1991). Follow-up studies of arsenic-exposed workers showed that diminished arsenic exposures led to lesser frequencies of chromosomal aberrations (Nordenson and Beckman, 1982).

6.2. Reproductive Effects

For over 50 years, we have known that inorganic arsenic readily crosses the placental barrier and affects fetal development (Squibb and Fowler, 1983). Organic arsenicals do not seem to cross the placenta so readily and are stored in the placenta instead (Leonard and Lauwerys, 1980). There is extensive documentation of experimental induction of malformations in a variety of species (Bencko, 1977; Ferm, 1977; Squibb and Fowler, 1983; Windhorst et al., 1977). The nature of the malformation or developmental arrest depends on the timing of the exposure.

At the Ronnskar smelter in northern Sweden, ores with a high arsenic content are handled. Women employed in the plant as well as those who live nearby delivered babies whose weight was significantly lower than those delivered by women who were not so exposed (Nordstrom et al., 1978a). Among those same women, the frequency of spontaneous abortion was generally higher with closer proximity of residence to the smelter (Nordstrom et al., 1978b). Although residential proximity to the Ronnskar smelter had no effect on the incidence of congenital malformations, pregnancies during which the mother had worked at the smelter were significantly more apt to yield babies with single or multiple malformations, particularly urogenital malformations or hip-joint dislocations (Nordstrom et al., 1979).

More recently, in eastern Massachusetts, women who consumed high levels of arsenic in water (1.4–1.9 mg/L) had 1.7 times [95% C.I. (confidence interval) 0.7–4.2] the frequency of spontaneous abortion (Aschengrau et al., 1989). In an elaborate epidemiologic study of congenital cardiac malformations in Massachusetts during 1980 to 1983, cases of coarctation of the aorta were 3.4 times (95% C.I. 1.3–8.9) more apt to have had any detectable level of arsenic in drinking water than a population sample of control births (Zierler et al., 1988). Arsenic had no relationship to the occurrence of other cardiac malformations, whereas selenium, a well-known antagonist of arsenic, was associated with lower frequencies of all types of cardiovascular anomaly.

6.3. Carcinogenic Effects

Over 100 years ago, it was observed that patients who received chronic treatment with arsenical medications had greatly increased incidence of both basal cell and squamous cell carcinomas of the skin (reviewed in Hindmarsh and McCurdy, 1986; Winship, 1984; more recent data reviewed in Neubauer, 1947). These arsenical skin cancers commonly occurred in the presence of other dermatologic manifestations of arsenicism, and some had internal neoplasms that were regarded as arsenical in origin.

Also a century ago, workers involved in the mining and industrial handling of arsenic were noted to have excessive rates of skin cancers and lung cancer (reviewed by Hamilton, 1925). More recently, Michigan workers involved in formulating and packaging arsenical insecticides were observed to have had a more than 3-fold greater than expected risk of respiratory cancer mortality compared to controls and a 7-fold excess risk for individuals exposed more than 8 years (Ott et al., 1974); however, subsequent follow-up has shown the longer term overall respiratory cancer mortality excess risk to be closer to 2-fold (Sobel et al., 1988). Arsenic pesticide formulators in Baltimore also had a significant excess lung cancer death ratio (Mabuchi et al., 1980). Among Swedish glass workers exposed to arsenic and other metals, the standardized mortality ratios were significantly doubled for lung and gastric cancers (Wingren and Axelson, 1985). Likewise, among workers at an arsenic calcining plant attached to an English tin mine, there was excessive mortality from lung and gastric cancers (Hodgson and Jones, 1990).

A mortality analysis by the company physician for workers and retirees at the copper smelter in Tacoma, Washington, concluded there was no evidence of arsenic-induced cancer in the cohort, despite figures that suggested that lung cancer deaths had occurred at about twice the number expected (Pinto and Bennett, 1963). Subsequent analysis by state health department epidemiologists indicated that the excess respiratory cancer mortality among these smelter workers was highly significant (Milham and Strong, 1974). The next mortality analysis by the company physician was performed with a different statistician, it focused on pensioners, and it clearly identified a 3-fold excess of respiratory cancer deaths (Pinto et al., 1978). A subsequent analysis for all workers from 1940 to 1964, who were followed until 1976, showed a significant 2-fold excess lung cancer mortality experience overall, and a rising risk gradient with longer duration of employment (Enterline and Marsh, 1982). Re-analysis of those data in conjunction with improved air arsenic data showed there had been a significant augmentation of lung cancer risk by arsenic levels that were lower than previously reported, indicating that arsenic was a more potent carcinogen than previously estimated (Enterline et al., 1987). Compared to controls, female lung cancer cases in the community were slightly more apt to live closer to the Tacoma smelter, but the difference was not significant (Frost et al., 1987). Fitting the employee mortality data into a multistage model indicated that the arsenic effect occurred primarily at a late stage in the carcinogenesis process (Mazumdar et al., 1989), which agreed with the cocarcinogenesis concept discussed in Section 6.1. Thus, arsenic would seem to be a cancer promoter rather than a cancer initiator, the risk of cancer that it posed would indeed seem to be dose-dependent, and that cancer risk would be expected to decline again when the arsenic exposure ceased and the substance was cleared from the body.

The initial 1938 to 1963 mortality analysis of workers at the copper smelter at Anaconda, Montana, demonstrated a more than 3-fold excess respiratory cancer ratio, with an excess risk as high as 8-fold among heavily exposed men who had worked there 8 years or more (Lee and Fraumeni, 1969). Re-analysis of those data showed a clear dose–response relationship between arsenic exposure and respiratory cancer risk, with men in the high-exposure category having a 7-fold excess risk (Welch et al., 1982). Fitting these data into a multistage model also indicated that the arsenic effect seemed to have occurred at a late stage of carcinogenesis (Brown and Chu, 1983), which was later substantiated by the Tacoma data (Mazumdar et al., 1989). An analysis of the 1938 to 1977 mortality experience of the Anaconda smelter workers showed that the calendar years of employment had been an important risk determinant (Lee-Feldstein, 1983). Respiratory cancer deaths were about 5 times greater than expected for men employed 15 or more years starting before 1925, while others had had only about 2.5 times the mortality experience from that cause. When further subdivided by heavy, medium, or light estimated exposure to As_2O_3, the group first employed prior to 1925 had an excess respiratory cancer mortality 7 to 8 times greater than expected. Lee-Feldstein (1986), using a numerical scale for estimation of cumulative arsenic exposure, found that respiratory cancer mortality increased linearly with increasing cumulative arsenic exposure. When conditional logistic regression analysis was applied to a matched case-control study of Anaconda workers employed prior to 1957 for at least 12 months, significant predictions of respiratory cancer mortality could be made by three estimates of arsenic exposure: cumulative exposure, time-weighted average exposure, and maximum category of exposure (Lee-Feldstein, 1989).

With respect to the Ronnskar copper smelter in northern Sweden, a case-control study of 1960 to 1976 lung cancer deaths showed they were 4 to 5 times more apt to have occurred among arsenic-exposed workers (Axelson et al., 1978). A cohort follow-up study of workers employed during 1928 to 1967 who were followed until 1981 showed a positive dose–response relationship and an overall 3.7-fold excess mortality experience (Jarup et al., 1989). A case-control investigation from the same data set showed that the arsenic augmentation of lung cancer mortality risk was greater among light and medium smokers than among heavy smokers (Jarup and Pershagen, 1991).

Like other exposures, excessive arsenic in drinking water will tend to concentrate in the skin, where there will be visible direct toxic effects as well as measurable chromosomal changes and increased rates of skin cancers. An Oregon case of multiple basal cell carcinomas manifested after 14 years of using water containing 1.2 ppm arsenic (Wagner et al., 1979). Skin cancers were described among the other manifestations of arsenicism present among the residents of Antofagasta, Chile, where the arsenic level in the drinking water ranged from 0.05 to 0.96 ppm (mean 0.60 ppm) (Zaldivar, 1974). In southwestern Taiwan, where the arsenic level ranged from 0.01 to 1.82 ppm (mean 0.50 ppm) in artesian well water, the prevalence rate of skin cancer among persons surveyed was directly correlated with arsenic concentration in the household water supply and with duration of intake of arsenical water (Tseng, 1977). In the same Taiwan region, subsequent investigators have demonstrated significant elevations of standardized 1968 to 1982 mortality ratios for both sexes for bladder cancer, renal cancer, skin cancer, lung cancer, liver cancer, and colon cancer, and for males for leukemia (Chen et al., 1985). When arsenic dose was estimated by years of use of high-arsenic water, both adjusted odds ratio and multiple logistic regression analyses showed consistent risk increases for 1980 to 1982 fatal cancers of the bladder, lung, and liver (Chen et al., 1986). Data from 42 southwestern Taiwan villages analyzed by age-standardized mortality rates showed significant effects of well-water arsenic on 1973 to 1986 rates for cancers of the bladder, kidney, skin, lung, and liver in both sexes, and on rates for cancer of the prostate in males; rates for leukemia were unaffected (Wu et al., 1989). It has been suggested that the high melanoma incidence rates in southwestern Britain are caused by the region's higher arsenic levels in soil and water (Clough, 1980; Phillip et al., 1983).

The use of arsenical insecticides by German vintners prior to 1942 was linked to subsequent observations of higher mortality from lung cancers and from hepatic angiosarcomas than expected (Luchtrath, 1983; Roth, 1957). Medicinal and industrial contact with arsenic has been reported as etiologic for angiosarcoma of the liver in United States as well (Falk et al., 1981; Kasper et al., 1984; Popper et al., 1978; Vianna et al., 1981). Although a British cohort that had been treated with potassium arsenite showed no particular risk of subsequent internal neoplasia (Cuzick et al., 1982), a more recent and broader analysis of data arranged by type of exposure has shown that ingested arsenic probably does cause cancers of the bladder, kidney, lung, and liver (Bates et al., 1992). A cohort follow-up study of former workers and neighbors exposed to As_2O_3 in Toroku, Japan, showed significant excess mortality from cancers of the bladder, kidney, and respiratory tract (Tsuda et al., 1990).

7. SUMMARY

Arsenic is a common toxic substance with exceedingly diverse manifestations of poisoning. Different species of arsenic have different degrees of toxicity, with arsine and trivalent arsenic causing the most damage. The body's toxic response depends on the route and dose of exposure plus individual and local tissue susceptibilities. We have reviewed the effects of arsenic on the major target organ systems. The evidence for human carcinogenicity seems quite strong, although arsenic probably serves as a promoter rather than an initiator of neoplasia.

REFERENCES

Adams, R. M. (1990). *Occupational Skin Disease*. Saunders, Philadelphia, 2nd ed., pp. 147–151, 350–351.

Aschengrau, A., Zierler, S., and Cohen, A. (1989). Quality of community drinking water and the occurrence of spontaneous abortion. *Arch. Environ. Health* **44**, 283–290.

Axelson, O., Dahlgren, E., Jansson, C. D., and Rehnlund, S. O. (1978). Arsenic exposure and mortality: A case-referent study from a Swedish copper smelter. *Br. J. Ind. Med.* **35**, 8–15.

Barrett, J. C., Lamb, P. W., Wang, T. C., and Lee, T. C. (1989). Mechanisms of arsenic-induced cell transformation. *Biol. Trace Elem. Res.* **21**, 421–429.

Bates, M. N., Smith, A. H., and Hopenhayn-Rich, C. (1992). Arsenic ingestion and internal cancers: A review. *Am. J. Epidemiol.* **135**, 462–476.

Bencko, V. (1977). Carcinogenic, teratogenic and mutagenic effects of arsenic. *Environ. Health Perspect.* **19**, 179–182.

Bencko, V., and Symon, K. (1977). Health aspects of burning coal with a high arsenic content. I. Arsenic in hair, urine and blood in children residing in a polluted area. *Environ. Res.* **13**, 378–385.

Bencko, V., Symon, K., Chladek, V., and Pihrt, J. (1977). Health aspects of burning coal with a high arsenic contact. II. Hearing changes in exposed children. *Environ. Res.* **13**, 386–395.

Brown, C. C., and Chu, K. C. (1983). Implications of the multistage theory of carcinogenesis applied to occupational arsenic exposure. *J. Natl. Cancer Inst.* **70**, 455–463.

Brown, C. E., and McNamara, D. H. (1940). Acute interstitial myocarditis following administration of arsphenamines. *Arch. Dermatol. Syph.* **42**, 312–321.

Buchanan, W. D. (1962). *Toxicity of Arsenic Compounds.* Elsevier, New York, pp. 15–66.

Cannon, A. B. (1936). Chronic arsenical poisoning. *N.Y. State J. Med.* **36**, 219–241.

Chen, C. J., Chuang, Y. C., Lin, T. M., and Wu, H. Y. (1985). Malignant neoplasms among residents of a blackfoot disease endemic area in Taiwan: High-arsenic artesian well water and cancers. *Cancer Res.* **45**, 5895–5899.

Chen, C. J., Chuang, Y. C., You, S. L., Lin, T. M., and Wu, H. Y. (1986). A retrospective study on malignant neoplasms of bladder, lung, and liver in blackfoot disease endemic area in Taiwan. *Br. J. Cancer* **53**, 399–405.

Chen, C. J., Wu, M. M., Lee, S. S., Wang, J. D., Cheng, S. H., and Wu, H. Y. (1988). Atherogenicity and carcinogenicity of high-arsenic artesian well water. *Arteriosclerosis* **8**, 452–460.

Chi, I. C., and Blackwell, R. Q. (1968). A controlled retrospective study of blackfoot disease, an endemic peripheral gangrene disease in Taiwan. *Am. J. Epidemiol.* **88**, 7–24.

Clarkson, T. W. (1991). Inorganic and organometal pesticides. In W. J. Hayes, Jr. and E. R. Laws, Jr. (Eds.), *Handbook of Pesticide Toxicology.* Academic Press, San Diego, pp. 545–552.

Clough, P. (1980). Incidence of malignant melanoma of the skin in England and Wales. *Br. Med. J.* **280**, 112.

Cuzick, J., Evans, S., Gillman, M., and Price-Evans, D. A. (1982). Medicinal arsenic and internal malignancies. *Br. J. Cancer* **45**, 904–911.

Edge, J. R. (1946). Myocardial fibrosis following arsenical therapy. *Lancet* **2**, 675–677.

Ellenhorn, M. J., and Barceloux, D. G. (1988). *Medical Toxicology: Diagnosis and Treatment of Human Poisoning.* Elsevier, New York, pp. 1012–1016.

Enterline, P. E., and Marsh, G. M. (1982). Cancer among workers exposed to arsenic and other substances in a copper smelter. *Am. J. Epidemiol.* **116**, 895–911.

Enterline, P. E., Henderson, V. L., and Marsh, G. M. (1987). Exposure to arsenic and resipratory cancer; a reanalysis. *Am. J. Epidemiol.* **125**, 929–938.

Falk, H., Caldwell, G. G., Ishak, K. G., Thomas, L. B., and Popper, H. (1981). Arsenic-related hepatic angiosarcoma. *Am. J. Ind. Med.* **2**, 43–50.

Ferm, V. H. (1977). Arsenic as a teratogenic agent. *Environ. Health Perspect.* **19**, 215–217.

Feussner, J. R., Shelburne, J. D., Bredehoeft, S., and Cohen, H. J. (1979). Arsenic-induced bone marrow toxicity. *Blood* **53**, 820–827.

Fowler, B. A., and Weissberg, J. B. (1974). Arsine poisoning. *N. Engl. J. Med.* **291**, 1171–1174.

Fowler, B. A., Oskarsson, A., and Woods, J. S. (1987). Metal- and metalloid-induced porphyrinurias. *Ann. N. Y. Acad. Sci.* **514**, 172–182.

Frost, F., Harter, L., Milham, S., Royce R., Smith, A. H., Hartley, J., and Enterline, P. (1987). Lung cancer among women residing close to an arsenic emitting copper smelter. *Arch. Environ. Helath* **42**, 148–152.

Fuortes, L. (1988). Arsenic poisoning. *Postgrad. Med.* **83**(1), 233–234, 241–244.

Garcia-Vargas, G. G., Garcia-Rangel, A., Aguilar-Romo, M., and Garcia-Salcedo, J. (1991). Pilot study on the urinary excretion of porphyrins in human populations chronically exposed to arsenic in Mexico. *Hum. Exp. Toxicol.* **10**, 189–193.

Gerhardsson, L., Dahlgren, E., Eriksson, A., Lagerkvist, B. E. A., Lundstrom, J., and Nordberg, G. F. (1988). Fatal arsenic poisoning—a case report. *Scand. J. Work Environ. Health* **14**, 130–133.

Gerhardt, R. E., Hudson, J. B., Rao, R. N., and Sobel, R. E. (1978). Chronic renal insufficiency from cortical necrosis induced by arsenic poisoning. *Arch. Intern. Med.* **138**, 1267–1269.

Giberson, A., Vaziri, N. D., Mirahamadi, K., and Rosen, S. M. (1976). Hemodialysis of acute arsenic intoxication with transient renal failure. *Arch. Intern. Med.* **136**, 1303–1304.

Goldsmith, S., and From. A. H. L. (1980). Arsenic-induced atypical ventricular tachycardia. *N. Engl. J. Med.* **303**, 1096–1098.

Gorby, M. S. (1988). Arsenic poisoning. *West. J. Med.* **149**, 308–315.

Goyer, R. A. (1991). Toxic effects of metals. In M. O. Amdur, J. Doull, and C. D. Klaassen (Eds.), *Toxicology*. Pergamon, New York, 4th ed., pp. 629–633.

Hamilton, A. (1925). *Industrial Poisons in the United States*. Macmillan, New York, pp. 206–233.

Hindmarsh, J. T., and McCurdy, R. F. (1986). Clinical and environmental aspects of arsenic toxicity. *CRC Crit. Rev. Clin. Lab. Sci.* **23**, 315–347.

Hodgson, J. T., and Jones, R. D. (1990). Mortality of a cohort of tin miners 1941–86. *Br. J. Ind. Med.* **47**, 665–676.

Hoffmann, G. R. (1991). Genetic toxicology. In M. O. Amdur, J. Doull, and C. D. Klaassen, (Eds.), *Toxicology*. Pergamon, New York, 4th ed., pp. 201–225.

Hotta, N. (1989). Clinical aspects of chronic arsenic poisoning due to environmental and occupational pollution in and around a small refining spot. *Jpn. J. Const. Med.* **53**, 49–70.

Hughes, J. P., Polissar, L., and Van Belle, G. (1988). Evaluation and synthesis of health effects studies of communities surrounding arsenic producing industries. *Int. J. Epidemiol.* **17**, 407–413.

Jarup, L., and Pershagen, G. (1991). Arsenic exposure, smoking, and lung cancer in smelter workers—a case-control study. *Am. J. Epidemiol.* **134**, 545–551.

Jarup, L., Pershagen, G., and Wall, S. (1989). Cumulative arsenic exposure and lung cancer in smelter workers: A dose-response study. *Am. J. Ind. Med.* **15**, 31–41.

Jenkins, R. B. (1966). Inorganic arsenic and the nervous system. *Brain* **89**, 479–498.

Kasper, M. L., Schoenfeld, L., Strom, R. L., and Theologides, A. (1984). Hepatic angiosarcoma and bronchoalveolar carcinoma induced by Fowler's solution. *J. Am. Med. Assoc.* **252**, 3407–3408

Kjeldsberg, C. R., and Ward, H. P. (1972). Leukemia in arsenic poisoning. *Ann. Intern. Med.* **77**, 935–937.

Kraetzer, A. F. (1930). Raynaud's disease associated with chronic arsenical retention. *J. Am. Med. Assoc.* **94**, 1035–1037.

Kyle, R. A., and Pease, G. L. (1965). Hematologic aspects of arsenic intoxication. *N. Engl. J. Med.* **273**, 18–23.

Lagerkvist, B., Linderholm, H., and Nordberg, G. F. (1986). Vasospastic tendency and Raynaud's phenomenon in smelter workers exposed to arsenic. *Environ. Res.* **39**, 465–474.

Landrigan, D. J., Costello, R. J., and Stringer, W. T. (1982). Occupational exposure to arsine: An epidemiological reappraisal of current standards. *Scand. J. Work Environ. Health* **8**, 169–177.

Lee, A. M., and Fraumeni, J. F., Jr. (1969). Arsenic and respiratory cancer in man: An occupational study. *J. Natl. Cancer Inst. (U.S.)* **42**, 1045–1052.

Lee-Feldstein, A. (1983). Arsenic and respiratory cancer in humans: Followup of copper smelter employees in Montana. *J. Natl. Cancer Inst.* **70**, 601–610.

Lee-Feldstein, A. (1986). Cumulative exposure to arsenic and its relationship to respiratory cancer among copper smelter employees. *J. Occup. Med.* **28**, 296–302.

Lee-Feldstein, A. (1989). A comparison of several measures of exposure to arsenic. *Am. J. Epidemiol.* **129**, 112–124.

Leonard, A., and Lauwerys, R. R. (1980). Carcinogenicity, teratogenicity, and mutagenicity of arsenic. *Mutat. Res.* **75**, 49–62.

Luchtrath, H. (1983). The consequences of chronic arsenic poisoning among Moselle wine growers. *J. Cancer Res. Clin. Oncol.* **105**, 173–182.

Mabuchi, K., Lilienfeld, A. M., and Snell, L. M. (1980). Cancer and occupational exposure to arsenic: A study of pesticide workers. *Prev. Med.* **9**, 51–77.

Mazumdar, S., Redmond, C K., Enterline, P. E., Marsh, G. M., Costantino, J. P., Zhou, S. Y., and Patwardhan, R. N. (1989). Multistage modeling of lung cancer mortality among arsenic exposed copper smelter workers. *Risk Anal.* **9**, 551–563.

Milham, S., Jr. (1977). Studies of morbidity near a copper smelter. *Environ. Health Perspect.* **19**, 131–132.

Milham, S., Jr., and Strong, T. (1974). Human arsenic exposure in relation to a copper smelter. *Environ. Res.* **7**, 176–182.

Morton, W. E., and Caron, G. A. (1989). Encephalopathy: An uncommon manifestation of workplace arsenic poisoning? *Am. J. Ind. Med.* **15**, 1–5.

Nakamuro, K., and Sayato, Y. (1981). Comparative studies of chromosomal aberration induced by trivalent and pentavalent arsenic. *Mutat. Res.* **88**, 73–80.

Nelson, R. L. (1934). Acute diffuse myocarditis following exfoliative dermatitis. *Am. Heart J.* **9**, 813–816.

Neubauer, O. (1947). Arsenical cancer: A review. *Br. J. Cancer* **1**, 192–251.

Nordenson, I., and Beckman, L. (1982). Occupational and environmental risks in and around a smelter in northern Sweden. VII. Reanalysis and followup of chromosomal aberrations in workers exposed to arsenic. *Hereditas* **96**, 175–181.

Nordenson, I., and Beckman, L. (1991). Is the genotoxic effect of arsenic mediated by oxygen free radicals? *Hum. Hered.* **41**, 71–73.

Nordenson, I., Beckman, G., Beckman, L., and Nordstrom, S. (1978). Occupational and environmental risks in and around a smelter in northern Sweden. II. Chromosomal aberrations in workers exposed to arsenic. *Hereditas* **88**, 47–50.

Nordenson, I., Sweins, A., and Beckman, L. (1981). Chromosome aberrations in cultured human lymphocytes exposed to trivalent and pentavalent arsenic. *Scand. J. Work Environ. Health* **7**, 277–281.

Nordstrom, S., Beckman, L., and Nordenson, I. (1978a). Occupational and environmental risks in and around a smelter in northern Sweden. I. Variations in birth weight. *Hereditas* **88**, 43–46.

Nordstrom, S., Beckman, L., and Nordenson, I. (1978b). Occupational and environmental risks in and around a smelter in northern Sweden. III. Frequencies of spontaneous abortion. *Hereditas* **88**, 51–54.

Nordstrom, S., Beckman, L., and Nordenson, I. (1979). Occupational and environmental risks in and around a smelter in northern Sweden. VI. Congenital malformations. *Hereditas* **90**, 297–302.

Nriagu, J. O., and Azcue, J. M. (1990). Food contamination with arsenic in the environment. *Adv. Environ. Sci. Technol.* **23**, 121–143.

Okui, T., and Fujiwara, Y. (1986). Inhibition of human excision DNA repair by inorganic arsenic and the co-mutagenic effect in V 79 Chinese hamster cells. *Mutat. Res.* **172**, 69–76.

Ott, M. D., Holden, B. B., and Gordon, H. L. (1974). Respiratory cancer and occupational exposure to arsenicals. *Arch. Environ. Health* **29**, 250–255.

Pershagen, G. (1983). The epidemiology of human arsenic exposure. In B. A. Fowler (Ed.), *Biological and Environmental Effects of Arsenic.* Elsevier, New York, pp. 199–232.

Pershagen, G. Braman, R. S., and Vahter, M. (1981). In *Environmental Health Criteria 18: Arsenic.* World Health Organ., Geneva, pp. 76–146.

Phillip, R., Hughes, A. O., Robertson, M. C., and Mitchell, T. F. (1983). Malignant melanoma incidence and association with arsenic. *Bristol Med.-Chir. J.* **98**, 165–169.

Pinto, S. S., and Bennett, B. M. (1963). Effect of arsenic trioxide exposure on mortality. *Arch. Environ. Health* **7**, 583–591.

Pinto, S. S., and McGill, C. M. (1953). Arsenic trioxide exposure in industry. *Ind. Med. Surg.* **22**, 281–287.

Pinto, S. S., Henderson, V., and Enterline, P. E. (1978). Mortality experience of arsenic-exposed workers. *Arch. Environ. Health* **33**, 325–331.

Polissar, L., Lowry-Coble, K., Kalman, D. A., Hughes, J. P. van Belle, G., Covert, D. S., Burbacher, T. M., Bolgiano, D., and Mottet, N. K. (1990). Pathways of human exposure to arsenic in a community surrounding a copper smelter. *Environ. Res.* **53**, 29–47.

Politis, M. J., Schaumburg, H. H., and Spencer, P. S. (1980). Neurotoxicity of selected chemicals. In P. S. Spencer, and H. H. Schaumburg (Eds.), *Experimental and Clinical Neurotoxicology*. Williams & Wilkins, Baltimore, MD, pp. 613–615.

Popper, H., Thomas, L. B., Telles, N. C., Falk, H., and Selikoff, I. J. (1978). Development of hepatic angiosarcoma in man induced by vinyl chloride, thorotrast and arsenic. *Am. J. Pathol.* **92**, 349–369.

Reynolds, E. S. (1901). An account of the epidemic outbreak of arsenical poisoning occurring in beer-drinkers in the north of England and the midland counties in 1900. *Lancet* **1**, 166–170.

Rezuke, W. N. (1991). Arsenic intoxication presenting as a myelodysplastic syndrome. *Am. J. Hematol.* **36**, 291–293.

Rosenberg, H. G. (1974). Systemic arterial disease and chronic arsenicism in infants. *Arch. Pathol.* **97**, 360–365.

Rossman, T. G., Meyn, M. S., and Troll, W. (1977). Effects of arsenite on DNA repair in *Escherichia coli*. *Environ. Health Perspect.* **19**, 229–233.

Roth, F. (1957). The sequelae of chronic arsenic poisoning in Moselle vintners. *Ger. Med. Mon.* **82**, 172–175.

Schoolmeester, W. L., and White, D. R. (1980). Arsenic poisoning. *South. Med. J.* **73**, 198–208.

Shannon, R. L., and Strayer, D. S. (1989). Arsenic-induced skin toxicity. *Hum. Toxicol.* **8**, 99–104.

Sobel, W., Bond, G. G., Baldwin, C. L., and Ducommun, D. J. (1988). An update of respiratory cancer and occupational exposure to arsenicals. *Am. J. Ind. Med.* **13**, 263–270.

Squibb, K. S., and Fowler, B. A. (1983). The toxicity of arsenic and its compounds. In B. A. Fowler (Ed.), *Biological and Environmental Effects of Arsenic*. Elsevier, New York, pp. 233–269.

Stone, O. J., and Willis, C. J. (1968). The effect of arsenic on inflammation. *Arch. Environ. Health* **16**, 801–804.

Teitelbaum, D. T., and Kier, L. C. (1969). Arsine poisoning. *Arch. Environ. Health* **19**, 133–143.

Tseng, W. P. (1977). Effects and dose-response relationships of skin cancer and blackfoot disease with arsenic. *Environ. Health Perspect.* **19**, 109–119.

Tseng, W. P. (1989). Blackfoot disease in Taiwan, a 30-year followup study. *Angiology* **40**, 547–558.

Tsuda, T., Nagira, T., Yamamoto, M., and Kume, Y. (1990). An epidemiological study on cancer in certified arsenic poisoning patients in Toroku. *Ind. Health* **28**, 53–62.

Tsuda, T., Babazono, A., Ogawa, T., Hamada, H., Aoyama, H., Kurumatani, N., Nagira, T., Hotta, N., Harada, M., and Inomata, S. (1992). Inorganic arsenic: A dangerous enigma for mankind. *Appl. Organomet. Chem.* **6**, (in press).

Vahter, M. (1983). Metabolism of arsenic. In B. A. Fowler (Ed.), *Biological and Environmental Effects of Arsenic*. Elsevier, New York, pp. 171–198.

Vahter, M. E. (1988). Arsenic. In T. W. Clarkson, L. Friberg, G. F. Nordberg, and P. R. Sager (Eds.), *Biological Monitoring of Toxic Metals*. Plenum, New York, pp. 303–321.

Vianna, N. J., Brady, J. A., and Cardamone, A. T. (1981). Epidemiology of angiosarcoma of liver in New York State. *N.Y. State Med. J.* **81**, 895–899.

Wagner, S. L., Maliner, J. S., Morton, W. E., and Braman, R. S. (1979). Skin cancer and arsenical intoxication from well water. *Arch. Dermatol.* **115**, 1205–1207.

Weinberg, S. L. (1960). The electrocardiogram in acute arsenic poisoning. *Am. Heart J.* **60**, 971–975.

Welch, K., Higgins, I., Oh, M., and Burchfiel, C. (1982). Arsenic exposure, smoking and respiratory cancer in copper smelter workers. *Arch. Environ. Health* **37**, 325–335.

Westhoff, D. D., Samaha, R. T., and Barnes, A. (1975). Arsenic intoxication in a case of megaloblastic anemia. *Blood* **45**, 241–246.

Windhorst, D. B., Albert, R. E., and Boutwell, R. K. (1977). Biologic effects of arsenic on man. In *NRC Committee on Medical and Biological Effects of Environmental Pollutants: Arsenic*. National Academy of Sciences, Washington, DC, pp. 173–215.

Wingren, G., and Axelson, O. (1985). Mortality pattern in a glass-producing area in SE Sweden *Br. J. Ind. Med.* **42**, 411–414.

Winship, K. A. (1984). Toxicity of inorganic arsenic salts. *Adv. Drug React. Acute Poisoning Rev.* **3**, 129–160.

Wu, M. M., Kuo, T. L., Hwang, Y. H., and Chen, C. J. (1989). Dose-response relation between arsenic concentration in well water and mortality from cancers and vascular diseases. *Am. J. Epidemiol.* **130**, 1123–1132.

Zaldivar, R. (1974). Arsenic contamination of drinking water and foodstuffs causing endemic chronic poisoning. *Beitr. Pathot.* **151**, 384–400.

Zaldivar, R. (1980). A morbid condition involving cardiovascular, bronchopulmonary, digestive, and neural lesions in children and young adults after dietary arsenic exposure. *Zentralbl. Bakteriol., Abt. I, Orig. B* **170**, 44–56.

Zierler, S., Theodore, M., Cohen, A., and Rothman, K. J. (1988). Chemical quality of maternal drinking water and congenital heart disease. *Int. J. Epidemiol.* **17**, 589–594.

3

TOXICITY AND METABOLISM OF INORGANIC AND METHYLATED ARSENICALS

Hiroshi Yamauchi and Bruce A. Fowler

University of Maryland Baltimore County, The University Program In Toxicology, Baltimore, Maryland 21227

Arsenic in the Environment, Part II: Human Health and Ecosystem Effects,
Edited by Jerome O. Nriagu.
ISBN 0-471-30436-0

---

1. INTRODUCTION

The arsenic compounds discussed in this chapter may be classed into two major categories: (1) inorganic arsenic (Asi) and the three methylated arsenic compounds: methylated arsenic (MA), dimethylated arsenic (DMA), and trimethylated arsenic (TMA), which are detected in biologic samples of human origin; and (2) arsenic-based semiconductor crystals, such as, gallium arsenide (GaAs) and indium arsenide (lnAs), for which there are new demands in the electronics industry for the production of the III-V semiconductors. It is expected that the use of III-V semiconductors will increase in electronic devices used for optical communications and as semiconductor elements in supercomputers. Arsine gas is currently in use as a dopant in the production of arsenic semiconductors, but because of its high toxicity, there is active research in search of adequate substitutes for the gas. Trimethylarsine (TMAs) and trisdimethylaminoarsine (TDAAs, $As[N(CH_3)_2]_3$) have been introduced as new substitute materials. There will be increased use of III-V semiconductors, methylarsenic compounds, and alkylarsines in the electronic industry, and it may therefore be presumed that occupational and environmental exposures may also increase as these compounds are discarded as industrial wastes into the environment. At present, little attention has been paid to the potential for ecologic contamination from these compounds.

Because there are excellent books and chapters on the basic toxicology of inorganic arsenic (Squibb and Fowler, 1983; Vahter, 1983; lshinishi et al., 1986) and on the symptomatology of inorganic arsenic poisoning [World Health Organization (WHO), 1981b; Pershagen, 1983; lshinishi et al., 1986], these subjects are omitted from this chapter. On the other hand, there are few reviews on arsenic metabolism and toxicity, and on the toxicologic aspects of arsenicals used in relatively new fields of industry (e.g., III-V semiconductors). The methylarsenic compounds and alkylarsines are examples of such substances and they are the focus of the current review.

2. METABOLISM, DISTRIBUTION, AND EXCRETION OF ARSENIC

2.1. Metabolism of Inorganic Arsenic

It has been postulated based on the form of urinary excretion of various arsenic species following ingestion by human beings of trivalent arsenic (arsenite)

(Yamauchi and Yamamura, 1979a) and pentavalent arsenic (arsenate) (Yamauchi and Yamamura, 1979b) that these species of inorganic arsenic are oxidized and reduced in vivo. It has also been found that dosing marmoset monkeys with arsenate resulted in the secretion of arsenite in the urine (Vahter and Marafante, 1985). Although it is not easy to determine the valence of inorganic arsenic compounds present in tissues, it is possible to determine the valence of such compounds in the urine by chromatography (Morita et al., 1981; Reay and Asher, 1977) and also by the following method.

Our method utilizes a fresh urine sample without treatment with NaOH under heating; the reaction solution is adjusted to a pH of 3.5 to 4 for arsenite and 1.5 for total inorganic arsenic; the respective compounds are reduced with sodium borohydride before determination by atomic absorption spectrophotometry. The amount of arsenate is calculated by deducting the amount of arsenite from the amount of total arsenic. The advent of arsenic speciation chemistry has proven critical in studies of metabolic pathways of inorganic arsenic compounds. The metabolism of inorganic arsenic to MA and DMA was demonstrated in experimental animals (Yamauchi et al., 1980; Vahter and Norin, 1980) as well as in humans (Crecelius, 1977; Yamauchi and Yamamura, 1979a, b; Tam et al., 1979; Buchet et al., 1981). The actual steps in the process by which inorganic arsenic compounds undergo biotransformation to MA and DMA (that is, the pathway that inorganic arsenic takes to MA and then to DMA, or that it takes to MA or to DMA) remains to be elucidated. Methylation of inorganic arsenic compounds is a phenomenon common to mammals (Vahter and Marafante, 1988), but there are several reports on species differences in the methylating capability in marmoset monkeys, which indicates that this species is very poor at methylating inorganic arsenic (Vahter et al., 1982; Vahter and Marafante, 1985).

Studies of possible methyl group donors in the methylation of inorganic arsenic compounds have been made both in vitro and in vivo, and these results indicate that *S*-adenosylmethionine is regarded as the most likely candidate donor (Marafante and Vahter, 1984; Buchet and Lauwerys, 1985, 1987; Takahashi et al., 1990). Although the liver is the main site for methylation of inorganic arsenic compounds, it has been demonstrated in humans (Buchet et al., 1984) as well as in experimental animals (Takahashi et al., 1988) that the methylation of inorganic arsenic compounds is not affected by impaired hepatic function. These studies also demonstrated the important role of liver-reduced glutathione (GSH) in the methylation of inorganic arsenic in vivo (Buchet and Lauwerys, 1987, 1988; Hirata et al., 1990).

2.2. Metabolism, Distribution, and Excretion of III-V Semiconductors

Some studies on the metabolism of GaAs in rats (Webb et al., 1984; Schwetz et al., 1993) and in hamsters (Yamauchi et al., 1986a; Rosner and Carter, 1987) have been reported, and studies on the metabolism of InAs in hamsters have been reported (Yamauchi et al., 1992a, 1993). It has been shown that GaAs and InAs are slowly dissolved in vivo into inorganic arsenic and gallium and inorganic arsenic and indium, respectively. Daily urinary arsenic excretion rates following adminstration of GaAs or InAs are generally low (less than about 0.1% of total dose), although the rates vary slightly with the route of administration, and tend to be reduced with the passage of time (Fig. 1). Because of the low in vivo solubility of GaAs and InAs, a trace of gray crystals may still remain visible at the site of administration even a few months after subcutaneous administration of the GaAs or InAs suspension. Both GaAs and InAs liberate inorganic arsenic in vivo, and the inorganic arsenic so liberated is methylated chiefly to form DMA (Yamauchi et al., 1986b, 1992a). In a study of InAs, it was observed that increasing amounts of inorganic arsenic liberated in vivo from InAs and its metabolite DMA were detected in the hair with the passage of time. Because studies on this aspect of arsenic disposition following exposure to GaAs or InAs have only recently been initiated, a great deal of future research will be needed to assess the toxicological importance of arsenical species released following dissolution of InAs or GaAs particles.

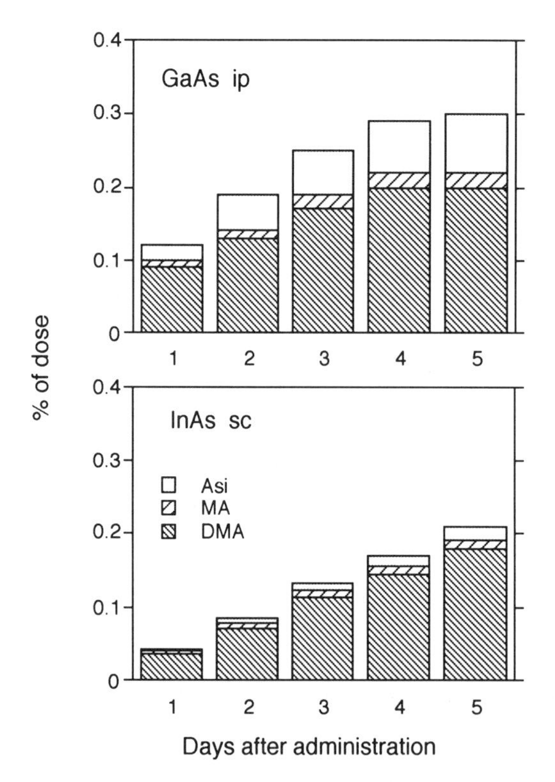

Figure 1. Urinary excretion rate of chemical species of arsenic up to 5 days following a single intraperitoneal (ip) administration of 100 mg/kg body weight of GaAs and a single subcutaneous (sc) administration of 100 mg/kg body weight of InAs to the hamster. Data represent the mean of 5 hamsters. (Data from Yamauchi et al., 1986a, 1992a.)

2.3. Metabolism and Excretion of Methylated Arsenic Compounds

In studies on the metabolism and excretion of MA (e.g., monomethylarsonate) and DMA (dimethylarsinate) in humans (Buchet et al., 1981), it has been shown that MA is slightly (13% of the ingested dose) methylated into DMA in vivo. DMA is not converted to other chemical species of arsenic in vivo. On the other hand, comparison of human data with data from animal studies showed relatively few differences. Only a very small fraction of MA is converted to DMA and TMA in hamsters, with 4.5% converted to DMA and no more than a trace to TMA (Yamauchi et al., 1988). The major route of elimination of MA varies with the route of administration. When MA is administered orally, it is mainly eliminated into the feces, and when administered intraperitoneally the major fraction is excreted into the urine. DMA is transformed in vivo to TMA, with a conversion rate of 15.7% (within 5 days after administration) as reported by Yamauchi and Yamamura (1984a) and 6.7% (within 2 days after administration) as reported by Marafante et al. (1987). The chemical structure of the TMA produced in vivo is not thought to be trimethylarsine oxide (TMAO) (Marafante et al., 1987).

No arsenobetaine is produced in mammals, and the arsenobetaine detected in human urine represents the direct excretion of that compound ingested from seafood. In studies on the metabolism and excretion of arsenobetaine in humans (Yamauchi and Yamamura, 1984b; Brown et al., 1990) as well as in experimental animals (Vahter et al., 1983; Yamauchi et al., 1986b), it has been shown that arsenobetaine is extremely stable in vivo and is not transformed to other species of arsenic. Arsenobetaine is rapidly eliminated from the body, but it has been reported that the compound took about 20 days to be excreted at a rate of less than 1% of the administered dose (Brown et al., 1990).

It is characteristic of the methylated arsenic compounds discussed in this chapter that the compounds are neither demethylated nor converted in vivo to form arsenobetaine in mammals.

2.4. Distribution of Inorganic and Methylated Arsenic Compounds

When arsenic compounds are administered to experimental animals, the compounds tend to be found in high concentrations in the liver, kidneys, lungs, and spleen. This is a phenomenon common to both inorganic (Yamauchi et al., 1980; Lindgren et al., 1982; Yamauchi and Yamamura, 1985) and methylated arsenic compounds (Vahter et al., 1983, 1984; Yamauchi and Yamamura, 1984a; Yamauchi et al., 1986b, 1988, 1989a, 1990). Neither inorganic trivalent nor methylated arsenic compounds are accumulated in the bone tissue, but inorganic pentavalent arsenic compounds will accumulate in this tissue. This appears to result from the substitution of the administered arsenate for the phosphorus present in the bone, since arsenate has a chemistry similar to that of phosphate (Lindgren et al., 1982). Trivalent inorganic arsenic compounds tightly bind to the SH groups contained in the keratin of hair. As a result, arsenicals are accumulated in high concentrations in hair following in vivo exposure. On the other hand, the accumulation of methylated arsenic compounds in hair depends on the presence of DMA (Yamauchi and Yamamura, 1984a), but this phenomenon does not hold true of arsenobetaine, MA, or other methylated arsenic compounds. TDAAs is hydrolyzed in vivo to produce inorganic arsenic, and this inorganic arsenical has likewise been shown to accumulate in hair (unpublished data).

There is a common trend in the distribution of arsenical compounds in the blood. These compounds occur in higher concentrations in the plasma in the early stages after administration, and they appear in higher concentrations in the blood cells with the passage of time. However, the distribution pattern of arsenical compounds in the blood of rats is different. Rat red blood cells have a high affinity for inorganic arsenic and DMA (Yamauchi et al., 1980; Vahter et al., 1984), and these arsenic compounds are retained by the red cells of this species' cells over long periods of time (usually over the life span of the red blood cells); hence, it is necessary to prudently interpret the metabolic data regarding arsenic in rats.

2.5. Metabolism and Excretion of Alkylarsines

TMAs is new material that is being used in the electronics industry. TMAs occurs as a colorless liquid having a garlicky odor, and it boils at 70 °C. When administered orally to hamsters, TMAs is oxidized in vivo to produce TMAO (Fig. 2); 80% of the administered amount is excreted in the urine and only a trace in the feces during the first 5 days (Yamauchi et al., 1990). The characteristic elimination pattern of TMAs is that the compound is eliminated in expired air as the administered chemical. On the other hand, when synthetic TMAO is administered orally in hamsters, a major portion is excreted in the urine as the parent compound, but an extremely small portion is reduced in vivo to form TMAs

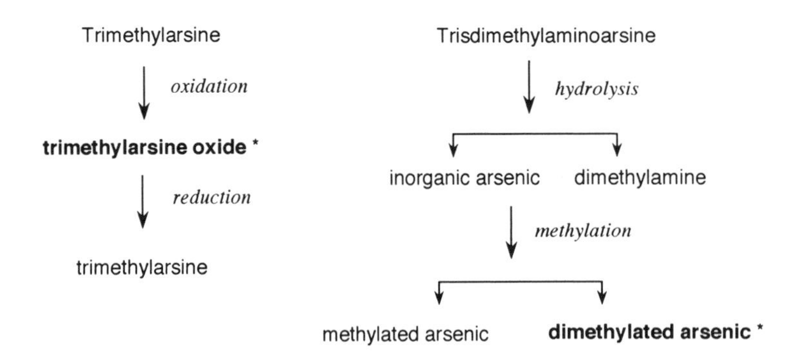

Figure 2. Metabolism of trimethylarsine and trisdimethylaminoarsine in vivo.

before it is eliminated in the expired air (a few minutes after administration) (Yamauchi et al., 1989a).

TDAAs occurs as a colorless, odorless liquid, boiling at 170 °C. When administered subcutaneously in hamsters, it is hydrolyzed in vivo to inorganic arsenic and dimethylamine (Fig. 1). Like other inorganic arsenic compounds, the inorganic arsenic so produced in vivo is methylated mainly to form DMA, which in turn is excreted into the urine. TDAAs belongs to the class of arsines, but it characteristically acts as inorganic arsenic when absorbed in vivo. Total excretion rates in the urine and feces have been shown to be lower than those of methylated arsenic compounds (Yamamura et al., 1993).

2.6. Biological Half-Life

Many arsenic compounds are eliminated with a three-exponential function pattern. In this section, their elimination rates are compared in terms of the half-life in the first phase. The half-life of arsenite in humans has been reported to be 30 hr by Buchet et al. (1980) and 28 hr by Yamauchi and Yamamura (1979a, recalculated value). The half-life of arsenate has been reported to be 2.1 days by Pomroy et al. (1980). Our calculations of the half-lives of inorganic and methylated arsenic compounds administered orally in hamsters gave a half-life of 28.6 hr for arsenic trioxide, 7.4 hr for MA, 5.6 hr for DMA, 6.1 hr for arsenobetaine, 3.7 hr for TMAs, and 5.3 hr for TMAO (Yamauchi et al., 1990). According to Vahter et al. (1984), the first-phase half-life of DMA in mice is 2.5 hr, the second phase is 10 hr, and the third phase is 20 days. The first-phase half-life of arsenobetaine likewise calculated is 12 hr (Vahter et al., 1983).

3. INTAKE OF DIETARY ARSENIC

The intake of dietary arsenic depends almost solely on the intake of seafood, and the intake of dietary arsenic by the Japanese, whose diet has a large seafood component, is far higher than that of Europeans and Americans. The reported average intakes are about 60 μg/day for Americans (Gartrell et al., 1985), 45 ± 95 μg/day for Belgians (Buchet et al., 1983), and 126 μg/day (Ishizaki, 1979) or 273 ± 263 μg/day (Ikebe et al., 1988) for the Japanese. The arsenic intake for the Japanese by chemical species has been calculated by Mohri et al. (1990) at a mean of 182 ± 114 μg/day (5.7% Asi, 3.6% MA, 27.4% DMA, and 47.9% arsenobetaine) in 21 specimens, and by Yamauchi et al. (1992b) at a mean of 195 ± 235 μg/day (17.1% Asi, 1.1% MA, 6.6% DMA, and 75.2% arsenobetaine) in 35 male and female subjects. These studies demonstrated that the principal chemical species of arsenic compounds ingested from food is arsenobetaine. According to Yamauchi et al. (1992b), the intakes of MA, DMA, and arsenobetaine vary with diet components, while inorganic arsenic is characteristically ingested from all meals. Nothing is known about the daily dietary intake of trimethylated arsenic compounds (TMAO and arsenocholine), but it is probably extremely small.

4. ARSENIC CONCENTRATIONS IN HUMAN BIOLOGICAL SAMPLES

4.1. Urine and Feces

Urinary arsenic concentrations reflect more sensitively the influence of seafood ingestion than do arsenic concentrations in any other biological matrix. Because diets in Japan and Sweden are high in seafood, arsenobetaine concentrations in urine samples also tend to be high in people living in these countries. It has been shown that all four chemical species of arsenic compounds that are detected in the urine originate from meals (Vahter, 1986; Vahter and Lind, 1986; Yamato, 1988; Yamauchi et al., 1992b). However, urinary MA and DMA are derived from the diet and are also the principal metabolites of inorganic arsenic compounds. It appears that the trimethylated arsenic compounds detected in the urine consist mostly of arsenobetaine, but TMAO may also be present. Evidence for this does not exist.

There are only a few reports on arsenic elimination in the feces of healthy subjects. Mohri et al. (1990) reported the elimination of 46 ± 35 μg As/day (16.9% Asi, 36.2% MA, 26.3% DMA, and 20.6% arsenobetaine). This finding indicates that more MA is excreted into the feces than into the urine, and this is consistent with the finding from an animal experiment (Yamauchi et al., 1988).

Urinary arsenic concentrations in workers occupationally exposed to inorganic arsenic compounds have been studied in workers following exposure to arsenic trioxide (Yamamura and Yamauchi, 1980; Vahter et al., 1986) and following exposure to other inorganic arsenic compounds (Buchet et al., 1980; Yamauchi et al., 1989b). It has been observed in workers with high-level exposure to inorganic arsenic compounds that the elevation in urinary inorganic arsenic concentrations was associated with variations in the concentrations of MA and DMA, the major metabolites of inorganic arsenic compounds (Vahter et al., 1986; Yamauchi et al., 1989b). On the other hand, in workers with low-level exposure to inorganic arsenic compounds (i.e., those engaged in the production of GaAs), urinary concentrations of inorganic arsenic metabolites remained unchanged, but urinary inorganic arsenic concentrations were slightly elevated (Yamauchi et al., 1989b). Determination of urinary arsenic compounds by chemical species in workers following exposure to inorganic arsenic compounds allows arsenic originating from exposure in the work environment to be distinguished from dietary arsenic (arsenobetaine) ingested from food. Hence it makes possible the biological monitoring of inorganic arsenic exposure with extremely high precision.

4.2. Blood

There are only a few reports on arsenic concentrations in the blood of healthy subjects (WHO, 1981a). Yamauchi et al. (1992b) reported a concentration of 0.73 ± 0.57 μg/100 mL (9.6% Asi, 14% DMA, and 73% TMA). Although the chemical species of TMA in the blood is not known yet, the compound is probably arsenobetaine. There is a significant correlation between the concentrations of blood and urinary TMA ($r = 0.407$, $p < 0.01$) but not between the concentrations of blood and urinary inorganic arsenic or DMA.

4.3. Hair

Arsenic concentrations in the hair play an important role in evaluating arsenic poisoning from oral ingestion of arsenic. However, determined arsenic concentrations in the hair are not usually considered reliable for biologic monitoring due to external contamination from arsenic exposure. Although there are various methods for washing hair, including one officially set forth, it is impossible to completely remove the arsenic as an external contaminant. Based on the idea that arsenic concentrations in the hair offer a method suitable for assessing the levels of arsenic contamination in working environments, studies have used unwashed hair specimens from healthy subjects. The mean arsenic concentration in the unwashed hair from 100 subjects was $0.075 \pm 0.043\ \mu g/g$ (75% Asi and 25% DMA) (Yamato, 1988). Neither MA nor arsenobetaine has been detected in the hair. On the other hand, inorganic arsenic concentrations in the hair of workers following exposure to inorganic arsenic were as high as $20\ \mu g/g$ (Yamamura and Yamauchi, 1980; Yamauchi et al., 1989b), but it is impossible to determine whether the inorganic arsenic is endogenous or from external contamination. DMA in the hair does not result from external contamination, but some workers show a tendency for elevated DMA concentrations. The assessment of DMA involves problems yet to be solved on a biological basis.

5. TOXICITY OF ARSENIC COMPOUNDS

5.1. Inorganic Arsenic

The incidence patterns of chronic inorganic arsenic poisoning in the past and in the present show that arsenic contamination of drinking water is the most frequent cause. Arsenic poisoning has occurred in Taiwan (Tseng, 1979; Chen et al., 1986), Chile (Zaldivar and Guilier, 1977), and Mexico (Cebrián et al., 1983; Garcia-Vargas et al., 1991) through consumption of contaminated well water. In contrast, such arsenic poisoning due to exposure to arsenic through inhalation in the smelting industry is rare owing to safety equipment. Groundwater containing high concentrations of inorganic arsenic is ubiquitous in nature; hence, it should be borne in mind that there is a risk of inorganic arsenic poisoning at any time. It is possible today to determine the chemical species of arsenic contained in groundwater, and according to some of the reports, arsenate (As^{5+}) accounts for the major portion of the chemical species of inorganic arsenic detected, while the arsenite content tends to be low. Arsenite-rich samples of hot-spring water have also been reported (Yamauchi et al., 1984). It has generally been accepted that among the chemical species of inorganic arsenic, arsenite is more toxic

Table 1 A Comparison of Arsenic Compounds of LD_{50} in Animals

Arsenic Compound	LD^{50} (mg/kg)	Animal/Mode of Administration
Arsenite: arsenic trioxide[a]	34.5	Mouse/oral
Arsenite: sodium arsenite[b]	4.5	Rat/intraperitoneal
Arsenate: sodium arsenate[b]	14–18	Rat/intraperitoneal
MA: monomethylarsonic acid[c]	1,800	Mouse/oral
DMA: dimethylarsinic acid[c]	1,200	Mouse/oral
TMA: arsenobetaine[a]	10,000	Mouse/oral
Trimethylarsine oxide[c]	10,600	Mouse/oral
Trimethylarsine[d]	8,000	Mouse/subcutaneous
Trisdimethylaminoarsine[d]	15	Mouse/subcutaneous

[a] Kaise et al. (1985).
[b] Franke and Moxon, (1936).
[c] Kaise et al. (1989).
[d] Yamamura et al. (1993).

than arsenate. However, this conclusion is derived from the median (50%) lethal doses in animals (Table 1), and there is, in reality, no great difference in toxicity between the chemical species of inorganic arsenic. Both inorganic species should be understood to be rather toxic arsenic compounds.

5.2. Methylated Arsenic Compounds

The methylated arsenic compounds are far less acutely toxic than the inorganic arsenic compounds. The trimethylated compounds appear to be least toxic. Arsenic poisoning with aromatic organic arsenic compounds (e.g., salvarsan) has long been known, whereas there appear to be no reports on arsenic poisoning from the methylated arsenic compounds. Derivatives of MA and DMA compounds are used as agrochemicals and herbicides, but at present these compounds do not seem to have caused any particularly problematic environmental contamination or damage to health.

MA and DMA are derived from inorganic arsenic compounds in vivo, and DMA tends to be the second most abundant biological form in relation to inorganic arsenic compounds detected in human tissues (but not in the urine) (Yamauchi and Yamamura, 1983). A conclusion has been drawn that the methylation of inorganic arsenic in mammals is a detoxification mechanism. On the other hand, there are reports on toxicologic problems with DMA, such as damage to DNA (Yamanaka et al., 1989a, 1991) and mutagenicity (Yamanaka et al., 1989b). In this regard, these genetic studies indicate that the major metabolites of inorganic arsenic are not innocuous.

The main chemical species of arsenic that human beings ingest from foods, especially people who eat large quantities of fish and shellfish, is arsenobetaine. Arsenobetaine is practically nontoxic (Table 1), and this may be the reason why there is no arsenic poisoning reported in such people. Among the trimethylated arsenic compounds, trimethylarsine oxide is rarely detected in fish. However, it is not ingested in such amounts as to cause concern over its toxic effect on the body, and it is as nontoxic as arsenobetaine (Table 1).

5.3. III-V Semiconductors

GaAs is presently the major representative of the III-V semiconductors, and InAs is a well-known semiconductor now in the research and development stage. There has been no report on arsenic poisoning with arsenic-based semiconductors as yet. However, low-level exposure to inorganic arsenic (dust of GaAs crystals) does exist at plants manufacturing ingots of GaAs crystals (Yamauchi et al., 1989b). Webb et al. (1984) were the first to report the animal toxicity of GaAs in 1984; hence, the reports so far available concern only relatively new studies. GaAs is a poorly soluble substance. It dissolves slowly in the body, so that not only its acute toxicity but also its chronic toxicity create problems. In assessing the toxicity of GaAs and lnAs, not only the compounds themselves but also their decomposition products (i.e., inorganic arsenic, gallium, and indium) have to be evaluated for toxicity. The toxicological assessment of these compounds is experimentally more complex than that of single metallic elements.

In a study by intratracheal instillation of GaAs (a single dose of 100 mg/kg) in rats, Webb et al. (1986) observed that intrapulmonary contents of lipid, protein, collagen, and DNA increased, and that these increases were similar to those observed with administration of arsenic trioxide (a single dose of 17 mg/kg). It has been reported that intracellular changes with GaAs include inflammatory manifestations, such as lymphoid hyperplasia, pneumonocyte hyperplasia, and macrophages. On the other hand, Ohyama et al. (1988) made a long-term (2.5-year) toxicity study by continuous intratracheal instillation of GaAs (0.25 mg/body weight/week × 15 weeks) in hamsters, and reported that weight loss and increased mortality rates were seen in the hamsters of the treated groups compared with controls. Pathology of the lungs revealed proliferation of alveolar epithelial cells, pneumonitis, piling up of macrophages, and congestion associated with hemorrhage. However, neither carcinogenicity nor oncogenicity of GaAs was observed in the respiratory organs. In a study of inhalation exposure of rats and mice to GaAs dust (37 and 75 mg/kg) for 13 weeks, Schwetz et al. (1993) reported that the GaAs toxicity produced a regenerative anemia in both rats and mice, and respiratory tract changes consisting of pulmonary epithelial hyperplasia in mice, pulmonary inflammatory lesions in both rats and mice, tracheobronchial lymph node hyperplasia in rats, and laryngeal squamous metaplasia in both species; hemosiderosis was observed in the spleen and liver of mice; bone marrow hyperplasia was observed in male rats; and testicular atrophy was observed in both rats and mice. They further reported that developmental toxicity was more prominent in GaAs-treated animals than in those treated with Ga_2O_3. There are other reports available on the toxicity of GaAs. Goering et al. (1988) reported that red blood cell δ-aminolevulinic acid dehydratase (ALAD) activity decreased with intratracheal instillation of GaAs (a single dose of 50–200 mg/kg), showing a dose–response relationship, and that urinary δ-aminolevulinic acid (ALA) tended to increase. There is a report that intratracheal instillation of GaAs at doses of 100 and 200 mg/kg was immunologically active (Sikorski et al., 1989, 1991a, b), and in vitro findings suggested that this immunologic effect might have derived from arsenic (Burns et al., 1991).

Because InAs is similar in its physicochemical properties to GaAs, its chronic toxicity also poses potential health problems. Our group has studied the effects of subcutaneous InAs (a single dose of 1000 mg/kg) on the heme biosynthetic pathway in hamsters (Conner et al., 1994). It was observed that the administration of InAs was followed by significant decreases in blood ALAD activity, and that urinary porphyrin excretion increased. Comparison of the increased excretion of urinary porphyrins following a subcutaneous dose of InAs by porphyrin species showed marked changes in the excretions of hepta-, hexa-, penta-, copro I-, and copro III-porphyrin. Of these chemical species, copro I- and copro III-porphyrin tended to increase with the passage of time after the dose of InAs (Fig. 3), and it was postulated that this tendency was related to intracellular accumulation of indium in major target organ systems, (chiefly in the liver and kidneys (Yamauchi et al., 1992a). It is known that total urinary porphyrin increases with exposure to inorganic arsenic (Woods and Fowler, 1978) and that indium may also inhibit enzymes in this pathway (Woods and Fowler, 1982). Urinary ALA concentrations also increase following administration of InAs, but the increases tend to be smaller than those in urinary porphyrin excretion. On the other hand, the biochemistry of plasma samples taken 30 days after subcutaneous

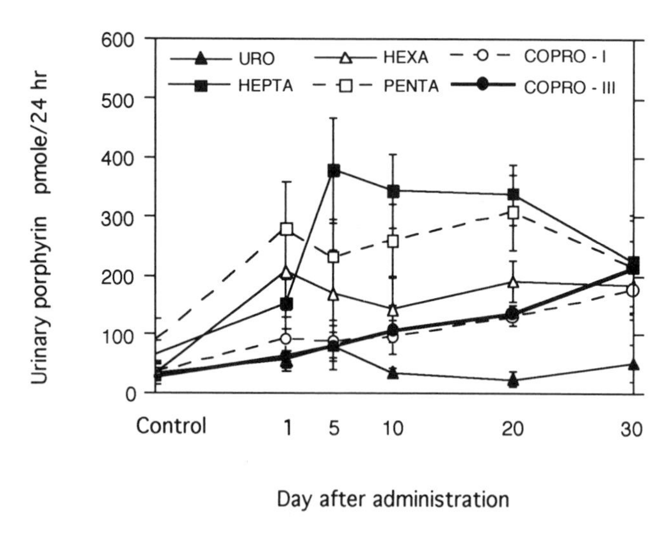

Figure 3. Changes with time in the urine concentration of porphyrins for periods up to 30 days following a single subcutaneous injection of 1000 mg/kg body weight of InAs in hamsters. The results are presented as the mean ± SD of 5 hamsters. (Data from Conner et al., 1994.)

injection of InAs at a dose of 1000 mg/kg showed that leucine aminopeptidase, lactate dehydrogenase (LDH), and γ-glutamyltranspeptidase (γ-GTP) tended to increase; hematology revealed a tendency for white blood cells to increase (Yamauchi et al., 1993).

It has been determined by the International Agency for Research on Cancer (1987) that arsenic is causally related to skin cancer in persons consuming arsenic-contaminated drinking water on a chronic basis and to respiratory cancers in workers on high-level exposure to inorganic arsenic. Because GaAs and InAs are inorganic arsenic compounds, the study of their carcinogenic potential will require the most serious attention in the future. However, the assessment of the carcinogenicity of arsenic semiconductors in animal experiments involves a technical problem. Since arsenic-based semiconductors occur as pulverized crystals in the workplace, it may be difficult to assess whether any carcinogenicity observed derives from their chemical actions or from their physical properties (i.e., shape, size, etc.).

5.4. Alkylarsines

TMAs and TDAAs are being used as substitutes for arsine gas in the production of III-V semiconductors, but practically no toxicologic studies have been conducted on these compounds. Very useful information on the acute toxicity of arsine gas is available from poisoning cases in the past and from animal experiments. It is known that its main toxic effect is hemolysis, and that renal failure is a secondary effect (Pinto et al., 1950; Fowler and Weissburg, 1974). In a recent report on the toxicity of arsine gas by the National Institute of Environmental Health Sciences, it was shown that exposure to arsine gas at concentrations of 2.5 ppm and more by inhalation had immunologic effects (Hong et al., 1989; Rosenthal et al., 1989; Blair et al., 1990a, b; Morrissey et al., 1990).

It has also been shown that the acute toxicity of TMAs is far lower than that of arsine gas, and that the toxicity varies greatly with metabolites (Yamauchi et al., 1990). TMAs is metabolized to form TMAO, and as shown in Table 1, the acute toxicity of the metabolite is very low. However, because TDAAs is hydrolyzed in vivo to produce inorganic arsenic, the toxicity of inorganic arsenic should also be considered. The median lethal dose of TDAAs is comparable to that of inorganic arsenic; hence, the acute toxicity of TDAAs is higher than that of TMAs. Because both TMAs and TDAAs belong to the arsine class, their hemolytic potential should be considered. In a study utilizing a single-dose, subcutaneous administration of these compounds (median lethal dose: TMAs, 8000 mg/kg; TDAAs, 15 mg/kg) in hamsters, we observed the occurrence of mild, transient hemolysis, whereas in a study by subacute exposure (to half of the median lethal dose for 10 days), neither of the compounds was found to be hemolytic (Yamamura et al., 1993). TMAs tends to be slightly more hemolytic than TDAAs. Hematologic studies of these compounds showed that acute exposure to either compound tended to increase hemoglobin concentrations, but that subacute exposure to either compound was not associated with any appreciable changes. Biochemical studies of plasma samples showed that acute exposure to either TMAs or TDAAs resulted in damage in hepatocytes as manifested by elevations of glutamic oxaloacetic transaminase activity, glutamic pyruvate transaminase activity, and LDH activity, and that subacute exposure to either compound elevated γ-GTP but caused no renal damage. The toxicity of alkylarsines is chiefly manifested as hemolysis, and tends to be far lower than that of arsine gas. From the toxicologic point of view, alkylarsines are desirable substitutes for arsine gas.

6. CONCLUSION

Excellent results have been achieved in the elucidation of the chemical species and structures of environmental arsenic compounds and in studies of the methylation mechanism of inorganic arsenic that were initiated in the latter half of the 1970s. With the emergence of III-V semiconductors as new materials, the general toxicity, metabolism, genetic toxicity, and carcinogenicity of these compounds will be new subjects for arsenical research, since the toxicological potential of arsenic will be modified by gallium and indium. The application of stress-protein induction patterns is a new method for studying the toxicity of arsenic compounds with regard to the mechanisms of III-V semiconductor toxicity. In vitro (Aoki et al., 1990) and in vivo (Conner et al., 1993) stress-protein induction studies have been reported for GaAs and InAs using renal proximal tubule cells. The induction of specific stress proteins following exposure of GaAs or InAs is characterized by the appearance of this response early in the cell injury process. From a basic study of the induction of stress proteins by inorganic arsenic, Taketani et al. (1990) showed that the 32-kDa stress protein is heme oxygenase. Further studies of the stress-protein induction response by III-V semiconductors in relation to alterations in the heme pathway are in progress (Conner et al., 1993, 1994), and a linkage between these two early cellular responses should be established. Such data should yield new insights into the relationship between these possible biomakers by III-V semiconductor toxicity and the cell injury process.

ACKNOWLEDGMENT

Supported in grant by NIH ESO 4979.

REFERENCES

Aoki, Y., Lipsky, M. M., and Fowler, B. A. (1990). Alteration in protein synthesis in primary cultures of rats kidney proximal tubule epithelial cells by exposure to gallium, indium, and arsenite. *Toxicol. Appl. Pharmacol.* **106**, 462–468.

Blair, P. C., Thompson, M. B., Morrissey, R. E., Moorman, M. P., Sloane, R. A., and Fowler, B. A. (1990a). Comparative toxicity of arsine gas in B6C3F mice, Fischer 344 rats, and Syrian golden hamsters. System organ studies and comparison of clinical indices of exposure. *Fundam. Appl. Toxicol.* **14**, 776–787.

Blair, P. C., Thompson, M. B., Bechtold, M., Wilson, R. E., Moorman, M. P., and Fowler, B. A. (1990b). Evidence for oxidative damage to red blood cells in mice induced by arsine gas. *Toxicology* **63**, 25–34.

Brown, R. M., Newton, D., Pickford, C. J., and Sherlock, J. C. (1990). Human metabolism of arsenobetaine ingested with fish. *Hum. Exp. Toxicol.* **9**, 41–46.

Buchet, J. P., and Lauwerys, R. (1985). Study of inorganic arsenic methylation by rat liver in vitro: Relevance for the interpretation of observations in man. *Arch. Toxicol.* **57**, 125–129.

Buchet, J. P., and Lauwerys, R. (1987). Study of factors influencing the in vivo methylation of inorganic arsenic in rats. *Toxicol. Appl. Pharmacol.* **91**, 65–74.

Buchet, J. P., and Lauwerys, R. (1988). Role of thiols in the in vitro methylation of inorganic arsenic by rat liver cytosol. *Biochem. Pharmacol.* **37**, 3149–3153.

Buchet, J. P., Lauwerys, R., and Roels, H. (1980). Comparison of several methods for the determination of arsenic compounds in water and in urine. Their application for the study of arsenic metabolism and for the monitoring of workers exposed to arsenic. *Int. Arch. Occup. Environ. Health* **46**, 11–29.

Buchet, J. P., Lauwerys, R., and Roels, H. (1981). Comparison of the urinary excretion of arsenic metabolites after a single oral dose of sodium arsenite, monomethylarsonate, or dimethylarsinate in man. *Int. Arch. Occup. Environ. Health* **48**, 71–79.

Buchet, J. P., Lauwerys, R., Vandevoorde, A., and Pycke, J. M. (1983). Oral daily intake of cadmium, lead, manganese, copper, chromium, mercury, calcium, zinc and arsenic in Belgium: A duplicate meal study, *Food Chem. Toxicol.* **21**, 19–24.

Buchet, J. P., Geubel, A., Pauwels, S., Mahieu, P., and Lauwerys, R. (1984). The influence of liver disease on the methylation of arsenite in humans. *Arch. Toxicol.* **55**, 151–154.

Burns, L. A., Sikorski, E. E., Saady, J. J., and Munson, A. E. (1991). Evidence for arsenic as the immunosuppressive component of gallium arsenide *Toxicol. Appl. Pharmacol.* **110**, 157–169.

Cebrián, M. E., Albores, A., Aguilar, M., and Blakely, E. (1983) . Chronic arsenic poisoning in the north of Mexico. *Hum. Toxicol.* **2**, 121–133.

Chen, C. J., Chuang, Y. C., You, S. L., Lin, T. M., and Wu, H. Y. (1986). A retrospective study on malignant neoplasma of bladder, lung and liver in blackfoot disease endemic area in Taiwan. *Br. J. Cancer* **53**, 399–405.

Conner, E. A., Yamauchi, H., and Fowler, B. A. (1994). Alterations in the heme biosynthetic pathway from III-V semiconductors (submitted for publication).

Conner, E. A., Yamauchi, H., Fowler, B. A., and Akkerman, M. (1993). Biological indicators for monitoring exposure/toxicity from III-V semiconductors. *J. Exposure Anal. Environ. Epidemiol.* **3** (in press).

Crecelius, E A. (1977). Changes in the chemical speciation of arsenic following ingestion by man. *Environ. Health Perspect.* **19**, 147–150.

Fowler, B. A., and Weissburg, J. B. (1974). Arsine poisoning. *N. Engl. J. Med.* **291**, 1171–1174.

Franke, K. W., and Moxon, A. L. (1936). A comparison of the minimum fatal doses of selenium, tellurium, arsenic, and vanadium. *J. Pharmacol. Exp. Ther.*, **58**, 454–459.

Garcia-Vargas, G. G., Garcia-Rangel, A., Aguilar-Romo, M., Garcia-Salcedo, J., Razo, L. M., Ostrosky-Wegman, P, Nava, C. C., and Cebrián, M. E. (1991). A pilot study on the urinary excretion of porphyrins in human populations chronically exposed to arsenic in Mexico. *Hum. Exp. Toxicol.* **10**, 189–193.

Gartrell, M. J., Craun, J. C., Podrebarac, D. S., and Gunderson, E. L. (1985). Pesticides, selected elements, and other chemicals in adult total diet samples, October 1978–September 1979. *J. Assoc. Off. Anal. Chem.* **68**, 862–875.

Goering, P., Maronpot, R., and Fowler, B. A. (1988). Effect of intratracheal gallium arsenide administration on δ-aminolevulinic acid dehydratase in rats: Relationship to urinary excretion of aminolevulinic acid. *Toxicol. Appl. Pharmacol.* **92**, 179–193.

Hirata, M., Tanaka, A., Hisanaga, A., and Ishinishi, N. (1990). Effects of glutathione depletion on the acute nephrotoxic potential of arsenite and on arsenic metabolism in hamsters. *Toxicol. Appl. Pharmacol.* **106**, 469–481.

Hong, H. L., Fowler, B. A., and Boorman, G. A. (1989). Hematopoietic effects in mice exposed to arsine gas. *Toxicol. Appl. Pharmacol.* **97**, 173–182.

Ikebe, K., Tanaka, Y., and Tanaka, R. (1988). Daily intake of 15 metals according to the duplicate portion studies. *J. Food Hyg. Soc. Jpn.* **29**, 52–57.

International Agency for Research on Cancer (IARC) (1987). *IARC Monograph on the Evaluation of Carcinogenic Risk of Chemicals to Humans. Suppl. 7. Overall Evaluations of Carcinogenicity.* Updating of IARC Monographs, Vol. 1–42. World Health Organization, IARC, Lyon, France, pp. 29–33.

Ishinishi, N., Tsuchiya, K., Vahter, M., and Fowler, B. A. (1986). Arsenic. In L. Friberg (Ed.), *Handbook on the Toxicology of Metals.* Elsevier, Amsterdam, pp. 47–48.

Ishizaki, M. (1979). Arsenic content in foods on the market and its average daily intake. *Jpn. J. Hyg.* **34**, 605–611.

Johnson, D L., and Braman, R. S. (1975). Alkyl- and inorganic arsenic in air samples. *Chemosphere* **6**, 333–338.

Kaise, T., Watanabe, S., and Ito, K. (1985). The acute toxicity of arsenobetaine. *Chemosphere* **14**, 1327–1332.

Kaise, T., Yamauchi, H., Horiguchi, Y., Tani, T., Watanabe, S., Hirayama, T., and Fukui, S. (1989). A comparative study on acute toxicity of methylarsonic acid, dimethylarsinic acid and trimethylarsine oxide in mice. *Appl. Organomet. Chem.* **3**, 237-277.

Lindgren, A., Vahter, M., and Dencker, L. (1982). Autoradiographic studies on the distribution of arsenic in mice and hamsters administered ^{74}As-arsenite or -arsenate. *Acta Pharmacol. Toxicol.* **51**, 253–265.

Marafante, E., and Vahter, M. (1984). The effect of methyltransferase inhibition on the metabolism of [^{74}As]arsenite in mice and rabbits. *Chem. -Biol. Interact.* **50**, 49–57.

Marafante, E., Vahter, M., Norin, H., Envall, J., Sandström, S., Christakopoulos, A., and Ryhage, R. (1987). Biotransformation of dimethylarsinic acid in mouse, hamster and man. *J. Appl. Toxicol.* **7**, 111–117.

Mohri, T., Hisanaga, A., and Ishinishi, N. (1990). Arsenic intake and excretion by Japanese adults: A 7-day duplicate diet study. *Food Chem. Toxicol.* **28**, 521–529.

Morita, M., Uehiro, T., and Fuwa, K. (1981). Determination of arsenic compounds in biological samples by liquid chromatography with inductively coupled argon plasma-atomic emission spectrometric detection. *Anal. Chem.* **53**, 1806–1808.

Morrissey, R. E., Fowler, B. A., Harris M. W., Moorman, M. P., Jameson, C. W. and Schwetz, B. A. (1990). Arsine, absence of developmental toxicity in rats and mice. *Fundam. Appl. Toxicol.* **15**, 350–356 (1990).

Ohyama, S., Ishinishi, N., Hisanaga, A., and Tanaka, A. (1988). Comparative chronic toxicity, including tumorigenicity of gallium arsenide and arsenic trioxide intratracheally instilled into hamsters. *Appl. Organomet. Chem.* **2**, 333–337.

Pershagen, G. (1983). The epidemiology of human arsenic exposure. In B. A. Fowler (Ed.), *Biological and Environmental Effects of Arsenic.* Elsevier, Amsterdam, pp. 199–232.

Pinto, S. S., Petronella, S. J., Johns, D. R., and Arnold, M. F. (1950). Arsine poisoning: A study of thirteen cases. *Arch. Ind. Hyg. Occup. Med.* **1**, 437–451.

Pomroy, C., Charbonneau, S. M., McCullough, R. S., and Tam, G. K. H. (1980). Human retention studies with ^{74}As. *Toxicol. Appl. Pharmacol.* **53**, 550–556.

Reay, P. F., and Asher, C. J. (1977). Preparation and purification of ^{74}As-labeled arsenate and arsenite for use in biological experiments. *Anal. Biochem.* **78**, 557–560.

Rosenthal, G. J., Fort, M. M., Germolec, D. R., Ackermann, M. F., Lamm, K. R., Blair, P. C., Fowler, B. A., and Luster, M. I. (1989). Effect of subchronic arsine inhalation on immune function and host resistance. *Inhalation Toxicol.* **1**, 113–127.

Rosner, M. H., and Carter, D. E. (1987). Metabolism and excretion of gallium arsenide and arsenic oxides by hamsters following intratracheal instillation. *Fundam Appl. Toxicol.* **9**, 730–737.

Schwetz, B. A., Roycroft, J. H., Moorman, M. P., Rosenthal., G. J., Chou, B. J., Mast, T. J., Morrissey, R. E., and Fowler, B. A. (1993). Toxicity studies on inhaled arsine, gallium arsenide and gallium oxide. *Environ. Health Perspect.* (in press).

Sikorski, E. E. McCay, J. A., White, K. L., Jr., Bradley, S. G., and Munson, A. E. (1989). Immunotoxicity of the semiconductor gallium arsenide in female B6C3F1 mice. *Fundam. Appl. Toxicol.* **13**, 843–858.

Sikorski, E. E., Burns, L. A., Stern, M. L., Luster, M. I., and Munson, A. E. (1991a). Splenic cell targets in gallium arsenide-induced suppression of the primary antibody response. *Toxicol. Appl. Pharmacol.* **110**, 129–142.

Sikorski, E. E., Burns, L. A., McCoy, K. L., Stern, M. L., and Munson, A. E. (1991b). Suppression of splenic accessory cell function in mice exposed of gallium arsenide. *Toxicol. Appl. Pharmacol.* **110**, 143–156.

Squibb, K., and Fowler, S. A. (1983). The toxicity of arsenic and its compounds. In B. A. Fowler (Ed.), *Biological and Environmental Effects of Arsenic.* Elsevier, Amsterdam, pp. 233–269.

Takahashi, K., Yamauchi, H., Yamato, N., and Yamamura, Y. (1988). Methylation of arsenic trioxide in hamsters with liver damage induced by long-term administration of carbon tetrachloride. *Appl. Organomet. Chem.* **2**, 309–314.

Takahashi, K., Yamauchi, H., Mashiko, M., and Yamamura, Y. (1990). Effect of S-adenosylmethionine on methylation of inorganic arsenic. *Jpn. J. Hyg.* **45**, 613–618.

Taketani, S., Sato, H., Yoshinaga, T., Tokunaga, R., Ishii, T., Bannai, S. (1990). Induction in mouse peritoneal macrophages of 34 KDa stress protein and heme oxygenase by sulfhydrl-reactive agents. *J. Biochem. (Tokyo)* **108**, 28–32.

Tam, G. K. H., Charbonneau, S. M., Bryce, F., Pomroy, C., and Sandi, E. (1979). Metabolism of inorganic arsenic (^{74}As) in humans following oral ingestion. *Toxicol. Appl. Pharmacol.* **50**, 319–322.

Tseng, W. P. (1979). Effects and dose-response relationships of skin cancer and blackfoot disease with arsenic. *Environ. Health Perspect.* **19**, 109–119.

Vahter, M. (1983). Metabolism of arsenic. In B. A. Fowler (Ed.), *Biological and Environmental Effects of Arsenic.* Elsevier, Amsterdam, pp. 171–198.

Vahter, M. (1986). Environmental and occupational exposure to inorganic arsenic. *Acta Pharmacol. Toxicol.* **59**, Suppl. 7, 31–34.

Vahter, M., and Lind, B. (1986). Concentrations of arsenic in urine of the general population in Sweden. *Sci. Total Environ.* **54**, 1–12.

Vahter, M., and Marafante, E. (1985). Reduction and binding of arsenate in marmoset monkeys. *Arch. Toxicol.* **57**, 119–124.

Vahter, M., and Marafante, E. (1988). *In vivo* methylation and detoxication of arsenic. In P. J. Craig (Ed.), *The Biological Alkylation of Heavy Elements.* Royal Society of Chemistry, London, pp. 105–119.

Vahter, M., and Norin, H. (1980). Metabolism of ^{74}As-labeled trivalent and pentavalent inorganic arsenic in mice. *Environ. Res.* **21**, 446–457.

Vahter, M., Marafante, E., Lindgren, A., and Dencker, L. (1982). Tissue distribution and subcellular binding of arsenic in marmoset monkeys after injection of ^{74}As-arsenite. *Arch. Toxicol.* **51**, 65–77.

Vahter, M., Marafante, E., and Dencker, L. (1983). Metabolism of arsenobetaine in mice, rats and rabbits. *Sci. Total Environ.* **30**, 197–211.

Vahter, M., Marafante, E., and Dencker, L. (1984). Tissue distribution and retention of ^{74}As-dimethylarsinic acid in mice and rats. *Arch. Environ. Contam. Toxicol.* **13**, 259–264.

Vahter, M., Friberg, L., Rahnster, B., Nygren, Å., and Nolinder, P. (1986). Airborne arsenic and urinary excretion of metabolites of inorganic arsenic among smelter workers. *Int. Arch. Occup. Environ. Health* **57**, 79–91.

Webb, D. R., Sipes, I. G., and Carter, D. E. (1984). *In vitro* solubility and *in vivo* toxicity of gallium arsenide. *Toxicol. Appl. Pharmacol.* **76**, 96–104.

Webb, D. R., Wilson, S. E., and Carter, D. E. (1986). Comparative pulmonary toxicity of gallium arsenide, gallium(III) oxide, or arsenic(III) oxide intratracheally instilled into rats. *Toxicol. Appl. Pharmacol.* **82**, 405–416.

Woods, J. S., and Fowler, B. A. (1978). Altered regulation of mammalian hepatic heme biosynthesis and urinary porphyrin excretion during prolonged exposed to sodium arsenate. *Toxicol. Appl. Pharmacol.* **43**, 361–371.

Woods, J. S., and Fowler, B. A. (1982). Selective inhibition of δ-aminolevulinic acid dehydratase by indium chloride in rat kidney: Biochemical and ultrastructural studies. *Exp. Mol. Pathol.* **36**, 306–315.

World Health Organization (WHO) (1981a). *Environmental Health Criteria 18: Arsenic.* WHO, Geneva, pp. 77–78.

World Health Organization (WHO) (1981b). *Environmental Health Criteria 18: Arsenic.* WHO, Geneva, pp. 87–146.

Yamamura, Y., and Yamauchi, H. (1980). Arsenic metabolites in hair, blood and urine in workers exposed to arsenic trioxide. *Ind. Health* **18**, 203–210.

Yamamura, Y., Takahashi, K., Kaise, T., Fowler, B. A., and Yamauchi, H. (1993). Relations of the toxicity of alkylarsines to their metabolism. *Environ. Health Perspect.* (in press).

Yamanaka, K., Hasegawa, A., Sawamura, R., and Okada, S. (1989a). Dimethylated arsenics induce DNA strand breaks in lung via the production of active oxygen in mice. *Biochem. Biophys. Res. Commun.* **165**, 43–50.

Yamanaka, K., Ohba, H., Hasegawa, A., Sawamura, R., and Okada, S. (1989b). Mutagenicity of dimethylated metabolites of inorganic arsenics. *Chem. Pharm. Bull.* **37**, 2753–2756.

Yamanaka, K., Ohba, H., Hasegawa, A., Sawamura, R., and Okada, S. (1991). Cellular response to oxidative damage in lung induced by the administration of dimethylarsinic acid, a major metabolite or inorganic arsenic, in mice. *Toxicol. Appl. Pharmacol.* **108**, 205–213.

Yamato, N. (1988). Concentrations and chemical species of arsenic in human urine and hair. *Bull. Environ. Contam. Toxicol.* **40**, 633–640.

Yamauchi, H., and Yamamura, Y. (1979a). Dynamic change of inorganic arsenic and methylarsenic compounds in human urine after oral intake as arsenic trioxide. *Ind. Health* **17**, 79–83.

Yamauchi, H., and Yamamura, Y. (1979b). Urinary inorganic arsenic and methylarsenic excretion following arsenate-rich seaweed ingestion. *Jpn. J. Ind. Health* **21**, 47–54.

Yamauchi, H., and Yamamura, Y. (1983). Concentration and chemical species of arsenic in human tissue. *Bull. Environ. Contam. Toxicol.* **31**, 267–277.

Yamauchi, H., and Yamamura, Y. (1984a). Metabolism and excretion of orally administered dimethylarsinic acid in the hamsters. *Toxicol. Appl. Pharmacol.* **74**, 134–140.

Yamauchi, H., and Yamamura, Y. (1984b). Metabolism and excretion of orally ingested trimethylarsenic in man. *Bull. Environ. Contam. Toxicol.* **32**, 682–687.

Yamauchi, H., and Yamamura, Y. (1985). Metabolism and excretion of orally administered arsenic trioxide in the hamsters. *Toxicology* **34**, 113–121.

Yamauchi, H., Iwata, M., and Yamamura, Y. (1980). Metabolism and excretion of arsenic trioxide in rats. *Jpn. J. Ind. Health* **22**, 111–121.

Yamauchi, H., Yamamura, Y., and Harako, A. (1984). Chemical species of arsenic in the waters of land sources. *Jpn. J. Public Health* **31**, 357–362.

Yamauchi, H., Takahashi, K., and Yamamura, Y. (1986a). Metabolism and excretion of orally and intraperitoneally administered gallium arsenide in the hamster. *Toxicology* **40**, 237–246.

Yamauchi, H., Jaise, T., and Yamamura, Y. (1986b). Metabolism and excretion of orally administered arsenobetaine in the hamster. *Bull. Environ. Contam. Toxicol.* **36**, 350–355.

Yamauchi, H., Yamato, N., and Yamamura, Y. (1988). Metabolism and excretion of orally and intraperitoneally administered methylarsonic acid in the hamster. *Bull. Environ. Contam. Toxicol.* **40**, 280–286.

Yamauchi, H., Kaise, T., and Yamamura, Y. (1989a). Metabolism and excretion of orally and intraperitoneally administered trimethylarsine oxide in the hamster. *Toxicol. Environ. Chem.* **22**, 69–76.

Yamauchi, H., Takahashi, K., Mashiko, M., and Yamamura, Y. (1989b). Biological monitoring of arsenic exposure of gallium arsenide and inorganic arsenic-exposed workers by determination of inorganic arsenic and its metabolites in urine and hair. *Am. Ind. Hyg. Assoc. J.* **50**, 606–612.

Yamauchi, H., Kaise, T., Takahashi, K., and Yamamura, Y. (1990). Toxicity and metabolism of trimethylarsine in mice and hamster. *Fundam. Appl. Toxicol.* **14**, 399–407.

Yamauchi, H., Takahashi, K., Yamamura, Y., and Fowler, B. A., (1992a). Metabolism of subcutaneous administered indium arsenide in the hamster. *Toxicol. Appl. Pharmacol.* **116**, 66–70.

Yamauchi, H., Takahashi, K., Mashiko, M., Saitoh, J., and Yamamura, Y. (1992b). Intake of different chemical species of dietary arsenic by the Japanese, and their blood and urinary arsenic levels. *Appl. Organomet. Chem.* **6**, 383–388.

Yamauchi, H., Takahashi, K., Yamamura, Y., and Fowler B. A. (1993). Toxicity and metabolism of gallium arsenide and indium arsenide. *Environ. Health Perspect.* (in press).

Zaldivar, R. and Guilier, A. (1977). Environmental and clinical investigations on endemic chronic arsenic poisoning in infants and children. *Zentralbl. Bakteriol., Parasitenkd Infektionskr. Hyg., Abt. 1. Orig.* **165**, 226–234.

4

TOXICITY AND METABOLISM OF ARSENIC IN VERTEBRATES

Syed M. Naqvi

Department of Biology & Health Research Center, Southern University, Baton Rouge, Louisiana 70813

Chetana Vaishnavi

Department of Experimental Medicine, Postgraduate Institute of Medical Education & Research, Chandigarh-16001, India

Harpal Singh

Department of Biology, Savannah State College, Savannah, Georgia 31404

Arsenic in the Environment, Part II: Human Health and Ecosystem Effects,
Edited by Jerome O. Nriagu.
ISBN 0-471-30436-0

---

1. INTRODUCTION

Arsenic is a naturally occurring element that ranks twentieth in abundance in the earth's crust, fourteenth in the seawater, and twelfth in the human body (Woolson, 1975). It is relatively common in environmental sources such as air, water, and soil. It is also found in plants and other organisms. Plants and animals collected from naturally arseniferous areas or from anthropogenic sources generally contain a high level of arsenic residues in their tissues (Bagatto and Alikhan, 1987). Pure arsenic is a steel-gray-colored brittle metal that is uncommon in the environment. It usually occurs in combination with one or more elements such as oxygen, chlorine, and sulfur and is referred to as inorganic arsenic. It occurs as arsenides and arsenates along with arsenic trioxide (a weathering product or arsenides) in sulfidic ores. When arsenic occurs in combination with carbon and hydrogen it is known as organic arsenic. It is important to maintain a distinction between inorganic and organic arsenic, since compounds in the former form are usually more toxic than those in the latter group. Moreover, the inorganic form is more mobile than the organic form and poses greater problems by leaching into surface waters and groundwaters [National Research Council of Canada (NRCC), 1978].

As much as about 100,000 tons of arsenic are produced worldwide; the United States alone produces about 21,000 tons while using about 44,000 tons, most of which is "imported from Sweden" [Environmental Protection Agency (EPA), 1980]. Arsenic trioxide (As_2O_3), the chief commercially important compound, is produced primarily from flue dust generated at copper and lead smelters. Elemental arsenic is produced by reducing trioxide with carbon [Hazardous Substances Data Bank (HSDB), 1990]. Since 1985, the major U.S. producer of arsenic trioxide (the ASARCO smelter in Tacoma, Washington) ceased operation, and the compound is no longer produced in this country (U.S. Bureau of Mines, 1990). As much as 66 million pounds is now being imported into the United States annually (U.S. Bureau of Mines, 1990).

Arsenic trioxide is principally used for wood preservation (74%), and about 19% is used for the production of agricultural chemicals [insecticides, herbicides, algicides, and growth stimulants for plants and animals (Rosenberg et al., 1980)]. Smaller amounts are used in the production of glass and nonferrous alloys and in the electronics industry [National Toxicology Program (NTP) 1989b; U.S. Bureau of Mines, 1990]. The past use of arsenic in rodenticide formulations has been discontinued due to human health risk and accidental poisonings. Its use in medicines for asthma, psoriasis, syphilis, and trypanosomiasis has also been severely curtailed, but it is still used for severe parasitic diseases (Eisler, 1988).

Potential environmental contamination from arsenic occurs due to (1) disposal of industrial waste chemicals, (2) manufacturing of copper and other metals, (3) burning of fossil fuels, and (4) application of pesticides. Many reviews on the ecotoxicological aspects of arsenic in the environment are available. Readers are referred to those by Woolson (1975), National Academy of Sciences (NAS) (1977a, b), NRCC (1978), Pershagen and Vahter (1979), EPA (1980, 1985), Hood (1985), Andreae (1986), Nriagu (1988), and U.S. Department of Health and Human Services (USDHHS) (1992).

The toxic effects of arsenic are complicated by its existence in several different valence states and many different inorganic and organic compounds. Most cases of vertebrate toxicity have been associated with exposure to inorganic arsenic. The most common inorganic arsenical in air is arsenic trioxide, while a variety of inorganic arsenates and arsenites occur in water, soil, and food. Pentavalent inorganic compounds include arsenic acid, arsenic pentoxide, calcium arsenate, and lead arsenate. Trivalent compounds commonly encountered are arsenic trioxide and arsenic trichloride (oily liquid); the former is oxidized catalytically or by bacteria to arsenic pentoxide or orthoarsenic acid. Other trivalent inorganic compounds are calcium arsenite, cupric arsenite, lead arsenite, and sodium arsenite. Arsenic compounds are dangerous; when they are heated or come in contact with acids, acid fumes, or active metals (i.e., Fe, Al, or Zn) they emit highly toxic fumes. The solubility of arsenic compounds in water ranges from quite soluble (sodium arsenite and arsenic acid) to practically insoluble (arsenic trisulfide). Soluble compounds are more toxic than the nonsoluble forms because of their greater absorption potential.

A number of studies have noted that trivalent arsenites tend to be more toxic than pentavalent arsenates (Willhite, 1981; Sardana et al. 1981; Maitani et al., 1987a). However, the relative difference in toxicity is reasonably small (2- to 3-fold). Different forms of arsenic may be interconverted, both in the environment and in the vertebrate body. Several cases of accidental human poisoning by arsenic have been reported (Park and Currier, 1991; Zhuang et al., 1990; Sheabar et al., 1989; Mitchell-Heggs et al., 1990; Fernandez-Sola et al., 1991; Roses et al., 1991), and successful suicide attempts using arsenic have been documented (Joliffe et al., 1991; Martin et al., 1990; Di Napoli et al., 1989). Mossop (1989) reported that as little as 0.25 ppm inorganic arsenic can cause symptoms of poisoning in humans. Gallium arsenide is another inorganic arsenic posing a potential human health hazard, and it is widely used in the microelectronics industry. Although this compound is poorly soluble, it undergoes slow dissolution and oxidation to form gallium trioxide and arsenite (Webb et al., 1986). The toxicity of gallium trioxide is attributed to the arsenite that is liberated, plus the additional effects of the gallium species.

Although organic arsenicals are generally viewed as being less toxic than inorganics, some methyl and phenyl derivatives that are used in agriculture are of possible human health concern. Among these are monomethylarsonic acid (MMA) and its salts monosodium methanearsonate (MSMA) and disodium methanearsonate (DSMA), and dimethylarsenic acid (DMA, or cacodylic acid) and its sodium salts sodium dimethylarsenite and roxarsone (3-nitro-4-hydroxyphenylarsenic acid). Like the inorganic compounds, these differ toxicologically.

Arsine, or arsenous hydride (AsH_3), is a colorless gas, heavier than air. It is produced accidentally as a result of generation of nascent hydrogen in the presence of arsenic or by the action of water on metallic arsenide. Arsine and its methyl derivatives may be formed from other arsenic compounds by microbial action or inadvertent chemical reactions that generate strong reducing conditions. Biotransformation of arsenic results in the production of volatile arsenicals that are returned to the land by the natural arsenic cycle, resulting in soil adsorption, plant uptake, erosion, leaching, and reduction of arsines. Thus arsenic is constantly being oxidized, reduced, and metabolized. Arsenic in soil is taken up by plants or reduced by microorganisms and chemical processes.

2. TOXICITY

2.1. Inorganic Arsenicals

Animals are not as sensitive to arsenic as humans, which can be partially attributed to differences in gastrointestinal absorption. The LD_{50} values and toxicity of inorganic arsenic compounds vary greatly depending on the chemical form and oxidation state. In rats, the oral LD_{50} ranges from 15 to 293 mg/kg, whereas it ranges from 10 to 150 mg/kg in other vertebrates [International Agency for Research on Cancer (IARC), 1987].

Although some animals appear to absorb and metabolize inorganic arsenicals in a manner similar to humans, others such as hamsters have low gastrointestinal absorption. Rats incur excessive binding to red blood cells, and marmoset monkeys exhibit low methylating capacity. This suggests that caution should be exercised when applying toxicokinetic data from animals to humans.

In a subacute study, rats were given 1.5 or 7.6 mg/kg arsenic trioxide per day by gavage for 40 days. Rats receiving the higher dose showed hair loss and eczema, hyperplasia, and hyperkeratosis of the skin. However, no adverse effects were observed in the group receiving the lower dose. Dysfunction of the blood–brain barrier was noticed in rats fed 290 mg/kg arsenic as arsenite for 35 days. In rats given 40, 85, or 125 mg/kg As in drinking water for 6 weeks, an increase in relative kidney weights was found [World Health Organization (WHO), 1981].

Blair et al. (1990a) exposed mice subchronically to arsine gas (0–2.5 ppm) for 90 days and found that red blood cell damage was due to the oxidation of heme (ferrous to ferric), which resulted in the depletion of intracellular reduced glutathione. This in turn oxidized sulfhydryl groups in the hemoglobin and possibly in the red blood cell membranes. These investigators also reported (1990b) that arsine exposure increased the activity of δ-aminolevulinic acid dehydratase in mice, rats, and hamsters. Bannai et al. (1991) noticed that the glutathione (GSH) content of mouse peritoneal macrophages increased significantly when they were exposed to sodium arsenite. This increase was attributed to the induction of cysteine transport by arsenic.

Low arsine gas (potent hemolytic) exposure by female mice produced stress on the hematopoietic system, which was characterized by hemolysis that persisted for a long period (Fowler and Boorman, 1989). Several other effects on laboratory mammals have been documented for inorganic arsenic: (1) a decrease in the resistance to viral infections in mice exposed subcutaneously (2–4 mg/kg), orally (70–150 mg), and via intraperitoneal injections (1.8 mg); (2) impaired kidney function in rabbits exposed to 0.7 mg/kg for 2 to 12 weeks three times a week; (3) diminished hearing ability in guinea pigs exposed intraperitoneally to sodium arsenite (0.07 mg/kg) for two months; (4) diminished AcChE (acetylcholinesterase) activity in the brain and decreased blood AcChE levels in guinea pigs (WHO, 1981); (5) liver cirrhosis and dose-dependent proliferation in rabbit's bile duct fed 0.4–4.7 mg/kg/day arsenite; (6) changes in ECG of cats exposed to 1.5 mg/kg arsenite; (7) decrease in hematocrit and hemoglobin concentrations in rats exposed to aerosols containing 3.7 mg As/m^3 as arsenic trioxide during three months; and (8) decrease in cellular ATP levels in a dose-dependent manner in HeLa S-3 cells due to mitochondrial damage (Yih et al., 1991).

Vahter and Marafante (1989) reported that injection of ^{4}As-arsenate to rabbits resulted in 40% recovery of the arsenic in the nuclear fraction of kidney cells and 30% in liver cells after one day. They found marked accumulation of arsenic in the mitochondria of kidney cells. Reichl et al. (1991) suggested glucose treatment for improving the survival of mice poisoned with inorganic arsenic. They found the dead mice had leukocyte distruptions, increased fat, and decreased glycogen in the spleen. Neiger and Osweiler (1989) studied the effect of subacute dietary sodium arsenite on 7- to 8-month-old dogs by feeding them 0 to 10 mg/kg/day. They doubled the dosage from the fifty-ninth day until day 183. This resulted in significant weight loss and elevation of serum alanine aminotransferases and aspartate aminotransferases.

Hudson et al. (1984) observed signs of inorganic trivalent arsenite poisoning in birds within 1 hr, causing death in 1 to 6 days. Lethal effects were due to the destruction of blood vessels lining the gut, which resulted in decreased blood pressure and subsequent shock (Nystrom, 1984). Hepatic damage was observed in coturnix (*Coturnix coturnix*) exposed to acute oral doses of As(III), possibly caused by osmotic imbalance induced by indirect inhibition of the sodium pump

(Nystrom, 1984). No deleterious effects were noticed when western grasshoppers (*Melanophis* sp.) poisoned by As(III) were fed to nestling northern bobwhites (*Colinus virginianus*), mockingbirds (*Mimus polyglottos*), American robins (*Turdus migratorius*), and other songbirds (NAS, 1977a, b).

Chickens excrete arsenicals rapidly with only 2% of dietary sodium arsenite retention after 60 hr (NAS, 1977a, b), and arsanilic acid is excreted largely unchanged (Woolson, 1975). Excretion was more rapid when arsenic was administered through intramuscular injection than through the mouth (NRCC, 1978). Studies by Fullmer and Wasserman (1985) indicate that arsenates rapidly penetrate the mucosal and serosal surfaces of the epithelial membranes of chickens, and intestinal absorption is essentially completed within 1 hr at 370 mg As(V)/kg body weight, but it is only 50% complete at a 3700 mg/kg dosage. Additionally, vitamin D_3 enhances duodenal absorption in rachitic chicks, and As(V) and phosphate do not appear to share a common transport pathway in the avian duodenum. Brown et al. (1990) examined the ulcerative cholecystitis in a flock of 21-day-old turkey hens that were accidently given about 0.004% 3-nitro-4-hydroxyphenylarsenic acid (3-NITRO) in their water for 2 days. Turkeys 31 days old were given 0.004% 3-NITRO for 6 days and did not develop ulcers, exhibiting the age-dependent toxicity of 3-NITRO to turkeys. Hood (1985) reported that snowshoe hares (*Lepus* sp.) died after consuming plants heavily contaminated with MSMA as a result of careless silviculture practices. Accidental deaths of domestic animals are well documented. More than 100 cattle died due to arsenic poisoning caused by overdosing topical applications of As_2O_3 for lice control (Robertson et al., 1984). In Bangladesh, poisoned cattle died 12 to 36 hr after the onset of arsenic poisoning symptoms (Samad and Chowdhury, 1984). Weeds sprayed with arsenic compounds (being salty) were attractive to animals, resulting in accidental poisoning (Selby et al., 1977).

2.2. Organoarsenicals

Although organoarsenicals are believed to be less toxic than inorganics, Mathews and Porter (1989) documented a case of acute arsenic intoxication of a white-tailed deer (*Odocoileus virginianus*), which died in an intensely managed northern hardwood forest in northern New York. They hypothesized that the deer had licked trees that were treated with Silvisar 550® (containing MSMA herbicide). This is the first report of this type of poisoning in wild animals.

Toxicity of several organoarsenicals to livestock was investigated by El Bahri and Romdane (1991). They rated phenylarsonic compounds as the least toxic, followed by arsanilic acid and roxarsone. Phenylarsonic compounds are not well absorbed from the intestinal tract (NAS, 1977a), and phenylarsonates do not appear to undergo significant metabolism in the body. They are generally excreted in urine within 24 to 48 hr. Similarly, sodium arsanilate given to rats and guinea pigs was excreted unchanged (Christau et al., 1975). Chickens also do not metabolize arsanilic acid, acetylarsonic acid, or roxarsone (Moody and Williams,

1964a, b). Only slight increases in arsenic levels could be detected in the skin, muscle, liver, and kidney of chickens that were fed 50 ppm of roxarsone (Buck et al., 1973).

In pigs, 100 ppm arsanilic acid (about 4 mg/kg/day) in the diet for three generations showed no toxic effects (Frost et al., 1962), but in another study, Buck (1969) reported chronic poisoning of swine by 100 ppm roxarsone in the diet for 2 months. Similarly, Kennedy et al. (1986) observed nervous disorders after 10 days (of 187 ppm roxarsone in the diet) involving clonic convulsions and evidence of myelin and axonal degeneration. No such effects were noticed following similar doses of arsanilic acid (Rice et al., 1985). The mechanism of toxicity is unknown, but roxarsone decreases copper levels in the liver and other tissues of swine (Edmonds and Baker, 1986) and chickens (Czarnecki et al., 1984b). Roxarsone and copper interact synergistically at high doses, resulting in greater weight loss than by either agent alone (Czarnecki and Baker, 1984; Edmonds and Baker, 1986). Simultaneous feeding of roxarsone and cysteine also increases toxicity, possibly by reducing the pentavalent form to the more toxic trivalent form, resulting in "bioactivation" (Czarnecki et al., 1984a).

Several unusual organoarsenicals are found to accumulate in fish and shellfish. These derivatives (mainly arsenobetaine and arsenocholine, also called "fish arsenic") have been studied by several investigators and found to be essentially nontoxic (Brown et al., 1990; Cannon et al., 1983; Charbonneau et al., 1978a; Kaise et al., 1985; Luten et al., 1982; Siewicki, 1981; Tam et al., 1982; Yamauchi et al., 1986). Arsenobetaine, the principal organoarsenical in fish, shellfish, and crustanceans, has a very low toxicity to mice even in oral doses as high as 10,000 mg/kg (Kaise et al., 1985) or intraperitoneal injections of 500 mg/kg (Cannon et al., 1983).

The reported LD_{50} values for mono- and disodium methanearsonate (in rats) are 2800 and 700 mg/kg, respectively. Calves appear to be more sensitive than rats, having acute oral LD_{50} values ranging from 100 to 450 mg/kg (Hood, 1985). Data on methylarsonate toxicity are mostly derived from studies on cattle, sheep, and chicken. Palmer (1972) found that multiple doses of 5 to 10 mg/kg/day of DMA, NaMMA, and Na_2MMA in these species were without ill effects. Higher doses often resulted in diarrhea, anorexia, and weight loss. However, in chickens, up to 250 mg/kg/day did not affect normal weight gain. Siewicki (1981) found no effects on organ and body weight, hematology, or urinary excretion of amino-levulinic acid or coproporphyrins in rats that were fed diets containing 40 ppm (about 2 mg/kg/day) DMA for 42 days. Acute inhalation exposure to methane-arsonates usually induced mild signs of distress in rats; exposure to 4300 mg/m^3 DMA aerosols for 2 hours caused labored respiration, rhinorhhea, eye irritation, and diarrhea, and the estimated acute LC_{50} for these rats was 4300 mg/m^3 (Stevens et al., 1977). Recently, Blair et al. (1990a) observed weight loss in male rats exposed to 5.0 ppm arsine for 28 days. However, spleen weight increased in both sexes of mice, rats, and hamsters, but liver weight gain was of lesser magnitude.

3. TOXICITY OF ARSENIC TO HUMANS AND EXTRAPOLATION OF ANIMAL DATA

Most laboratory animals appear to be substantially less susceptible to arsenic than humans. It has been reported that chronic oral exposure to inorganic arsenic (0.05–0.1 mg/kg/day) causes neurological and hematological toxicity in humans but not in monkeys, dogs, and rats exposed to arsenite or arsenate at doses of 0.72 to 2.8 mg/kg/day (Byron et al., 1967; Heywood and Sortwell, 1979; Schaumburg, 1980). There is good evidence that arsenic is carcinogenic in humans if exposed orally or by inhalation, but not in animals. Therefore, quantitative dose-dependent data for animals should not be considered a reliable source to apply to humans.

3.1. Respiratory Effects

Humans exposed to arsenic dust in the air experience irritation of the mucous membranes, resulting in laryngitis, bronchitis, or rhinitis (Morton and Caron, 1989). Very high exposure, such as that experienced in the past by industrial workers, can cause perforation of the nasal septum (Pinto and McGill, 1953), but such effects are minor or absent at exposure levels of 0.1–1 mg/m^3 (Ide and Bullough, 1988). Intratracheal exposure to arsenic trioxide (13 mg As/kg) or gallium arsenide (1.5–52 mg As/kg) caused hyperplasia of rat and hamster lung (Goering et al., 1988; Ohyama et al., 1988). Similar results were obtained by Sikorski et al. (1989) and McKay et al. (1991) in mice exposed to 200 mg/kg GaAs intratracheally. Gallium arsenide dissociates in vivo to gallium and arsenic (Rosner and Carter, 1987). Webb et al. (1986) found that rat lung retained 44% of the gallium and 29% of the arsenic administered as gallium arsenide. Pierson et al. (1989) and Webb et al. (1987) reported the same findings to be time-dependent.

No reports exist on the respiratory effects of organoarsenicals in humans. Short-term exposure of rats and mice to high concentrations of DMA (2170 mg As/m^3) caused respiratory distress, and postmortem examination revealed bright-red lungs with dark spots (Stevens et al., 1979). These investigators obtained similar results by exposing the animals to 2480 mg As/m^3) of the disodium salt of MMA. It is not clear that such high exposure levels of organoarsenicals can occur in the human environment.

3.2. Cardiovascular Effects

It has been suggested by several epidemiological studies that chronic inhalation of arsenic trioxide can increase the risk of death in humans from cardiovascular disease (Axelson et al., 1978; Lee-Feldstein, 1983; Wall, 1980). However, Jarup et al. (1989) did not find such a correlation. Smelter workers exposed to arsenic trioxide dusts approximately 0.05–0.5 mg As/m^3 incurred a greater risk of Raynaud's disease and increased constriction of blood vessels in response to cold

(Lagerkvist et al., 1988). These researchers pointed out that long-term inhalation of inorganic arsenic could injure the blood vessels or the heart. Rosenberg (1974) reported that autopsies of five children exposed to arsenic (up to 0.8 mg/L) in drinking water (in Anto Fagasta, Chile) revealed evidence of myocardial infarction in two cases and arterial thickening in all cases. Zaldivar (1974) reported several cases of myocardial infarction and arterial thickening in children who consumed water containing about 0.6 mg/L arsenic.

Arsenic ingestion through food or water may have serious effects on the human cardiovascular system. Both acute and chronic arsenic exposure cause altered myocardial depolarization (prolonged Q-T interval and nonspecific ST segment changes) and cardiac arrhythmias that may lead to heart failure (Fennel and Stacy, 1981; Goldsmith and From, 1986; Little et al., 1990). Low-level arsenic exposure by humans may also cause vascular system damage, a classical example of which is blackfoot disease, which is endemic in an area of Taiwan where most drinking water contains 0.17 to 0.8 ppm arsenic (Tseng, 1977), corresponding to doses of about 0.01 to 0.5 mg As/kg/day (EPA, 1988).

Blackfoot disease is characterized by a progressive loss of circulation in the hands and feet, which ultimately leads to severely painful gangrene formation of the extremities (particularly the toes and feet), often necessitating amputation of the limb (C. J. Chen et al., 1988; Tseng, 1989; Lin and Yang, 1988). These authors pointed out that the onset of this disease was unusually insidious and began with numbness in one or more extremities, usually the feet. Ultimately it resulted in the formation of a gangrene. A dose-dependent relation between the disease and the duration of water intake was noticed. The 50% survival rate after the onset of the disease was 13.5 years. Hair arsenic content of the patients was elevated. Labadie et al. (1990) indicated that hepatic damage in arsenic poisoning is secondary to vascular endothelial injury. Other researchers have presented evidence that other factors, such as dietary deficiencies and other chemicals in the water, might also be responsible for the etiology of blackfoot disease (Ko, 1986; Lu et al., 1990). Chen et al. (1990) reported the results of a cell-growth assay study and indicated that trivalent arsenic inhibited endothelial cell proliferation and glycoprotein synthesis. Such damage may play an important role in the pathogenesis of blackfoot disease. Indubitably, arsenic is a contributing factor in this disease. However, Lu et al. (1990) demonstrated that the main cause of blackfoot disease is not arsenic but humic acid.

Effects of arsenic on the vascular system have also been reported in a number of other populations. In Chile, ingestion of 0.6 to 0.8 ppm arsenic in drinking water (equivalent to 0.02–0.06 mg As/kg/day) increased the incidence of Raynaud's disease and of cyanosis of fingers and toes (Zaldivar, 1977; Borgano and Greiber, 1972). Thickening of blood vessels and their occulsion were noticed in German vinters who were exposed to arsenical pesticides in wine (Roth, 1957; NAS, 1977b) and in patients from a region in Mexico where arsenic toxicity is endemic (Salcedo et al., 1984).

Vascular abnormalities have also been reported in rats exposed to 11 mg/kg/day arsenic trioxide for several weeks (Bekemeir and Hirschelmann,

1989), but no histological effects were detected in the hearts of rats or dogs exposed to arsenate or arsenites for two years (Byron et al., 1967). No data are available for cardiovascular effects due to organoarsenicals.

3.3. Gastrointestinal Effects

Workers exposed to high levels of arsenic dusts or fumes suffer from nausea, vomiting, and diarrhea (Morton and Caron, 1989). Buchet et al. (1981a) reported that gastrointestinal absorption of MMA or DMA is at least 75 to 80% in humans after organoarsenical ingestion through fish and seafood (Charbonneau et al., 1980). Clinical signs of gastrointestinal irritation from arsenic include burning lips, painful swallowing, thirst, nausea, and severe abdominal colic (EPA, 1984; Campbell and Alvarez, 1989; Goebel et al., 1990). These symptoms are usually not detectable at exposure levels below 0.01 mg As/kg/day (Valentine et al., 1985) and they decline within a short time after exposure ceases. The efficiency of absorption of inorganic arsenicals from the gastrointestinal tract is related to their water-solubility. In Australia, old stocks of lead arsenate that were used as pesticides prior to 1970 remain in sheds and cause chronic poisoning among workers; the substance is also used for suicides (Tallis, 1989). Chakraborty and Saha (1987) reported three deaths in India due to chronic arsenic poisoning by drinking water from tubewells having a mean arsenic content of 0.64 mg/L.

The most likely mechanism of gastrointestinal toxicity is damage to the epithelial cells, with resulting irritation. Armstrong et al. (1984) described an incident in which a family of eight suffered severe arsenic toxicity from ingestion of water containing 108 mg/L arsenic. Tay and Seah (1975) noted gastrointestinal involvement in 17 of 74 people ingesting arsenic at an estimated dose of 3 to 10 mg/day through an herbal preparation.

3.4. Hematological Effects

The hematopoietic system is also affected by both short- and long-term arsenic exposure. Anemia and leukopenia are common effects of poisoning and have been reported as resulting from acute (Armstrong et al., 1984), intermediate (Franzblau and Lilis, 1989), and chronic oral exposures (Glazner et al., 1968; Tay and Seah, 1975). These effects may be due to a direct hemolytic or cytotoxic effect on the blood cells (Lerman et al., 1980) and a suppression of erythropoiesis (Kyle and Pearse, 1965). No such effects were noticed in humans exposed chronically to 0.07 mg As/kg/day or less (Southwick et al., 1981; Huang et al., 1985). Relatively high doses of arsenic have been reported to cause bone marrow depression in humans (EPA, 1984). Hamamoto (1955) reported such effects in infants exposed to 3.5 mg As/day in contaminated milk, and Mizuta et al. (1956) reported anemia and leukopenia in adults ingesting 3 mg As/day in soy sauce.

High concentrations of arsine (10 ppm) cause death within hours [American Conference of Governmental Industrial Hygienists (ACGIH), 1986] due to red blood cell hemolysis (Sittig, 1985). Low levels of arsine (0.5–5.0 ppm) bring about these effects in a few weeks, and an average concentration of 0.5 ppm (0.2 mg/m^3) is considered acceptable in the workplace (ACGIH, 1986). Renal damage is secondary, and occurs due to clogging of nephrons with hemolytic debris (Sittig, 1985). Mono-, di-, and trimethylarsines are strong irritants but are less hemolytic than arsine, [National Institute for Occupational Safety and Health (NIOSH), 1979]. Arsine exposure by humans is usually fatal without proper therapy (Fowler and Weissburg, 1974). Arsine breaks down in the body to inorganic arsenic and methylated derivatives (less toxic than arsine). Laboratory animals exhibit similar hematological disorders (Peterson and Bhattacharya, 1985).

Rats exposed to arsenate (20–65 mg/L) for six weeks or more showed decreased activity of several enzymes (hepatic aminolevulinic acid synthetase and ferrochelatase) required for heme biosynthesis (Woods and Fowler, 1978), but the data did not indicate whether this resulted in anemia. Blair et al. (1990a) subchronically exposed mice to arsine (00.025–2500 ppm for 6 hr/day, 5 days/week for 90 days). It resulted in less anemia with an increase in mean corpuscular volume. They explained that the mechanism of hemolysis involved depletion of intracellular GSH, resulting in oxidation of sulfhydryl groups in the hemoglobin from ferrous to ferric.

The hematological effects of ingested arsenic have not been thoroughly studied in laboratory animals, but no adverse effects have been detected in monkeys exposed to arsenate for 2 weeks (Heywood and Sortwell, 1979). On the other hand, the effects of arsine gas exposure to mice reported by Blair et al. (1990b) included a decrease in packed cell volumes in mice, rats, and hamsters; bone marrow hyperplasia was also noticed. Inhibited growth was also observed in rainbow trout (*Salmo gairdneri*) exposed to 30 mg/kg arsenic, probably due to a decrease in hemoglobin content of the erythrocytes (Oladiemji et al., 1984). Naqvi and Flagge (1990) observed that chronic exposure by crayfish (*Procambarus clarkii*) to 15 ppm MSMA (an organoarsenical herbicide) reduced egg hatching in *P. clarkii* and led to some decrease in weight- and length-gain after hatching. They surmised that hemocyanin combines with arsenic, which reduces oxygen uptake by cells and thereby prevents hatching. However, their results need further confirmation.

No significant hematological effects have been reported from repeated administration of organoarsenicals to rats and mice, including 6- to 13-week exposures to MMA (Prukop and Savage, 1986), DMA (Siewicki, 1981), and roxarsone (NTP, 1989) at doses of 5 to 55 mg As/kg/day. Therefore, it is not certain that oral exposure to organoarsenicals has hematological effects.

3.5. Hepatic Effects

Arsenic was the first chemical agent to which liver disease was attributed in humans. Hepatic lesions that formed after prolonged ingestion of arsenic-containing medicines (Fowler's solution) have been described. Clinical examination often reveals that the liver is swollen and tender (Chakraborty and Saha, 1987; Mazumdar et al., 1988). The analysis of blood sometimes shows elevated levels of hepatic enzymes (Franzblau and Lilis, 1989). These effects are most often observed after chronic exposures to as little as 0.02 to 0.1 mg As/kg/day (Silver and Weinman, 1952; Mazumdar et al., 1988). The livers of chronically exposed persons show various degrees of fibrosis and expansion of the portal zone (Mazumdar et al., 1988; Morris et al., 1974; Piontek et al., 1989). Portal hypertension and bleeding from esophageal varices was also reported by Szuler et al. (1979). These hepatic effects have been considered to be secondary to hepatic blood vessel damage (Rosenberg, 1974; Labadie et al., 1990).

Hamamoto (1955) observed swollen livers in all of the victims of arsenic-contaminated milk. Franklin et al. (1950) noted hepatic fatty infiltration and cirrhosis of the liver in patients who used Fowler's solution. Further studies verified these results (Viallet et al., 1972; Morris et al., 1974; Datta, 1976). Significant histopathological changes were observed by Wang et al. (1990) in rat livers having increased microsomes. These rats were exposed to arsenic trioxide (45 mg/kg/day) for four days. Steinhelper and Olson (1988) reported that phenylarsine oxide and other trivalent organoarsenicals inhibited hepatic glycogenolysis. Histological examination of the livers of rabbits given repeated doses of MMA showed diffuse inflammation and hepatocellular degeneration (Jaghabir et al., 1988), but the lesions were not severe. No effects were noticed in rats exposed to DMA (Siewicki, 1981). These studies suggest that organoarsenicals may cause mild injury to the liver. No evidence of hepatic dysfunction was detected by clinical examination of several workers exposed to arsenic dusts by inhalation (Ide and Bullough, 1988), suggesting that liver injury is not likely to be of concern following inhalation exposure.

3.6. Renal Effects

Most researchers do not report significant renal injury by acute and chronic exposure to arsenic (Jenkins, 1966; Franzblau and Lilis, 1989; Kersjes et al., 1987). In some cases, elevated serum levels of creatinine or bilirubin have been noted (Levin-Scherz et al., 1987), and mild proteinuria has been reported (Armstrong et al., 1984; Tay and Seah, 1975). In a cross-sectional study, Foa et al. (1987) observed slight increases in albuminuria, β_2-microglobulin, and brush border antigen BB50 in occupational workers and matched controls. However, no definite conclusions can be drawn given the small sample size. In rare cases, renal failure has been reported, probably due to fluid imbalances or vascular injury (Fincher and Koerker, 1987).

Animal studies also indicate that kidneys are not the major target organs for arsenic injury (Woods and Southern, 1989; Schroeder and Balassa, 1967). Heywood and Sortwell (1979) noted mild histological changes in the renal tubules of monkeys exposed to arsenate for 2 weeks, and mild alterations in the renal mitochondria of rats were noticed after arsenate exposure for 6 weeks (Brown et al., 1976). In humans, the kidneys seem to be less sensitive to arsenic than most other organ systems. The effects of organoarsenicals on the human renal system have not been reported. In rats (but not mice), repeated oral doses of roxarsone caused tubular degeneration and necrosis (Abdo et al., 1989; NTP, 1989). Interstitial nephritis and tubular nephrosis occurred in rabbits after repeated oral doses of MMA (Jaghabir et al., 1989). Woods and Southern (1989) reported that prolonged oral exposure of rats to sodium arsenate at subtoxic levels decreased renal coproporphyrin oxidase activity. Most studies suggest that organoarsenicals can cause significant renal injury in laboratory animals, although the minimal dose is not well defined.

3.7. Dermal Effects

Skin disorders have been documented in several epidemiological studies in which people consumed drinking water that contained arsenic at levels of 0.01 to 0.1 mg As/kg/day or more. Characteristic effects of arsenic ingestion included generalized hyperkeratosis, warts or corns on the palms and soles, and areas of hyperpigmentation interspersed with small areas of hypopigmentation on the face, neck, and back (Bickley and Papa, 1989; Hindmarsh and McCurdy, 1986; Borgono et al., 1980; Cebrián et al., 1983; Huang et al., 1985; Zaldivar, 1977; Piontek et al., 1989). Several epidemiological studies involving 20 to 200 people detected no dermal or other effects as a result of exposure to chronic doses of 0.003 to 0.01 mg As/kg/day (Southwick et al., 1981; Valentine et al., 1985). However, these dose estimates did not include the arsenic contribution from the diet; therefore, the total intake may have been slightly higher. A chronic oral dose of 0.01 mg As/kg/day or less would pose little risk of noncancer effects in humans. Sodium arsenite induces a stress protein of about 32 kDa in human skin fibroblasts, identified as heme oxygenase, as a defense against oxidative damage (Keyse and Tyrell, 1989).

3.8. Neurological Effects

Several studies have indicated that ingestion of inorganic arsenic can result in neural injury. Acute high exposures (1 mg As/kg/day or more) often cause encephalopathy with such symptoms as headache, lethargy, mental confusion, hallucination, seizures, and coma (Dannan et al., 1984; Fincher and Koerker, 1987). Hindmarsh and McCurdy (1986) classified arsenic neuropathy as a distal axonopathy with axonal degeneration, especially of large myelinated fibers. Electromyographic technique (EMG) used to detect neuropathy showed decreased nerve conduction amplitude with little change in nerve conduction velocity (Donofrio and Wilbourn, 1985). Bansal et al. (1991) reported asymmetric bilateral phrenic nerve involvement in a patient who was poisoned by arsenic. Goebel et al. (1990) described polyneuropathy in a 41-year-old person who ingested 9 gm of arsenic, resulting in an acute Wallerian degeneration of myelinated fibers. Their findings were confirmed by laser microprobe analysis. Intermediate and chronic exposures (0.05–0.5 mg As/kg/day) cause symmetrical peripheral neuropathy, which begins as a numbness in the hands and feet but later may develop into a painful "pins and needles" sensation (Hindmarsh et al., 1977; Wagner et al., 1979; Mizuta et al., 1956). Both sensory and motor nerves are affected, accompanied by muscle weakness that sometimes leads to wrist- or ankle-drop (Chhuttani et al., 1967). Cessation of arsenic exposure leads to nerve function recovery, but it is often slow and incomplete (Fincher and Koerker, 1987; Murphy et al., 1981; EPA, 1984; LaQuesne and McLeod, 1977). No neurological effects could be detected in a chronically exposed population (0.001 mg As/kg/day or less) (Harrington et al., 1978; Valentine et al., 1985). Byron et al. (1967) and Heywood and Sortwell (1979) did not find adverse neurological effects in dogs or monkeys chronically exposed to arsenate or arsenite.

Inhalation of inorganic arsenic can cause neurological injury in humans that may include peripheral neuropathy of both sensory and motor neurons causing numbness, loss of reflexes, and muscle weakness (Feldman et al., 1979; Morton and Caron, 1989). Landon et al. (1977) reported a duration-dependent decrease in EMG measurements in smelter workers exposed to arsenic primarily through inhalation.

Pepin et al. (1989) reported that the use of melarsoprol, a trivalent arsenical derivative, induces a lethal encephalopathy in 2 to 10% of patients suffering from sleeping sickness. Veeken et al. (1989) reported a similar fatality percentage in 106 patients treated for trypanosomiasis. Schaumburg (1980) did not find any evidence of neuropathy in rats chronically exposed to 10 mg As/kg/day for 18 months, suggesting that rats are not the right model for arsenic toxicity in humans. Neurological effects have also not been reported for dogs and monkeys chronically treated by the oral route (Byron et al., 1967; Heywood and Sortwell, 1979). Repeated doses of roxarsone (0.87–5.8 mg As/kg/day) administered for a month orally to pigs caused significant neurotoxicity (Edmonds and Baker, 1986; Rice et al., 1985). It also caused a time-dependent degeneration of myelin and axons in the pigs' spinal cord (Kennedy et al., 1986). Such prominent signs of neurotoxicity were detected in rats and mice only at a high dose of 11.4 mg As/kg/day (NTP, 1989). These data suggest that organoarsenicals (the phenylarsenates) are neurotoxic to these animals at high doses.

3.9. Developmental Effects

It is not well established whether ingestion of inorganic arsenic can cause developmental abnormalities in humans. No overall association between arsenic in drinking water and congenital heart defects was found in a case-control study in Boston (Zierler et al., 1988), although an association with coarctation of the aorta was noted. Similarly, a marginal association was reported between detectable levels of arsenic in drinking water and the occurrence of spontaneous abortions (Aschengrau et al., 1989). However, the results were not significant statistically. Thus, neither of these studies provides convincing evidence that ingestion of the arsenic that is usually contained in drinking water is responsible for developmental toxicity in humans. However, Nordstrom et al. (1978, 1979b) found that babies born to women exposed to arsenic dusts during pregnancy had a higher than expected incidence of congenital malformations. The average birth weight of the babies was slightly below average (Nordstrom et al., 1978). The incidence of spontaneous abortion in women who lived near a copper smelter in Sweden tended to decrease as a function of distance (Nordstrom et al., 1979a). No data were presented regarding a correlation between exposure to the smelter and the presence of other toxicants (lead, cadmium, and sulfur dioxide). Therefore, the role of arsenic in the etiology of these effects is dubious. A couple of studies reported an increased number of miscarriages among women who worked in the semiconductor industry, which uses arsine (La Dou, 1983; Calabrese et al., 1987).

Arsenic is a known teratogen in other vertebrates. Studies suggest that high doses of ingested arsenic may be fetotoxic and weakly teratogenic. Chernoff and Rogers (1975) noticed that 40 to 100 mg/kg of DMA given orally to mice caused fetotoxicity (reduced weight, delayed ossification, and cleft palate) as well as maternal toxicity. Exposure of mice to 400 mg/kg/day of DMA and rats to 30 mg/kg/day had the same effects (Rogers et al., 1981). In hamsters, fetal malformations and maternal death were caused by the similar organoarsenicals NADMA (900–1000 mg/kg) and NA_2DMA (500–1500 mg/kg). These doses were 100 times larger than those that would have caused developmental toxicity in mice exposed to inorganic arsenic (Harrison et al., 1980; Hood, 1985; Hood and Bishop, 1972). Baxley et al. (1981) reported that a single oral dose of 20 mg/kg sodium arsenite to pregnant mice caused discernible teratogenic or maternal lethality. On the other hand, Fowler and Boorman (1989) and Morrissey et al. (1990) found no adverse effects of various concentrations of arsine in mice and rats. Mason et al. (1989) observed a synergistic effect when sodium arsenate, sodium dischromate, and copper sulfate were administered in combination, which resulted in increased fetal resorption and fetal abnormalities, as well as decreased fetal weights. When these compounds were administered separately, they were either non- or weakly teratogenic.

No reports exist concerning the developmental effects of organoarsenical compounds in humans. There have been five reported cases of human arsenic poisoning during pregnancy, in none of which the offspring exhibited adverse effects (Kantor and Levin, 1948). Possibly none was exposed prior to the second trimester. A case of neonatal death following arsenic poisoning has been reported (Lugo et al., 1969). Premature birth might have been responsible for the infant's death, although arsenic cannot be excluded with certainty. High doses of organoarsenical compounds (i.e., DMA) have caused developmental abnormalities in rats and mice (Rogers et al., 1981). Sodium arsenate and arsenite are both embryotoxic compounds in laboratory animals, the former requiring concentrations ten times greater than the latter to produce deformities (Chaineau et al., 1990), including hypoplasia of the prosencephalon, somite abnormalities, and failure of development of limb buds and sensory placodes.

3.10. Reproductive Effects

Hardly any published information exists regarding reproductive effects in humans and animals after inhalation exposure to inorganic arsenic or organoarsenicals. The same is true for studies of human oral exposure to these compounds. The only study related to this subject was published by Schroeder and Mitchener (1971). In this three-generation study, mice were given sodium arsenite in drinking water at an average dose of 1 mg As/kg/day; it had no effect on a number of parameters, although a small decrease in litter number and size and a slightly altered sex ratio (more males) were observed. It is difficult to judge whether these effects are either statistically or biologically significant, and whether they can be extrapolated to humans.

Data are very limited on the reproductive effects of organoarsenicals, and none exist on humans. Spermatogenesis might be impaired by organoarsenical exposure by mice, which were dosed with MMA (55 mg As/kg/day) prior to mating and during pregnancy (Prukop and Savage, 1986). Fewer than normal litters were produced, which could be attributable to decreased fertility of the males.

3.11. Genotoxic Effects

Inhalation exposure to arsenic trioxide increased the frequency of chromosomal aberrations in the peripheral lymphocytes of smelter workers (Beckman et al., 1977; Nordenson et al., 1978) and in fetal mouse livers of mothers exposed to 22 mg As/m^3 during the gestation period (days 9–12) (Nagymajtenyi et al., 1985). These data do not indicate that arsenic is mutagenic, but they do indicate that it is clastogenic. No published information exists on the genotoxic effects of organoarsenicals in humans or animals after inhalation exposure.

Burgdorf et al. (1977) reported an increase in sister-chromatid exchanges but no increase in chromosomal aberrations in humans who were exposed to Fowler's solution (which yields a usual dose of about 0.3 mg As/kg/day). Contrasting results were observed by Nordenson et al. (1979) in which chromosomal aberrations increased without an increase in sister chromatids. However, rats given oral doses of sodium arsenate (4 mg As/kg/day) for 2 to 3 weeks had an increased rate of chromosomal abnormalities (Datta et al., 1986). Poma et al. (1987) did not find a consistent increase in chromosomal aberrations in the bone marrow cells or spermatogonia of mice that were exposed to sodium arsenite (about 50 mg As/kg/day). Both arsenate and arsenite have been found *in vitro* to induce chromosomal aberrations and sister-chromatid exchanges in cultured animal and human cells (Jacobson-Kram and Montalbano, 1985; Barrett et al., 1988; Li and Rossman, 1989). Vig et al. (1984) were unable to detect any increase in chromosomal aberrations or sister chromatid exchange in residents of Fallon, Nevada, where drinking water contains about 0.1 mg/L arsenic.

There is no conclusive evidence that arsenic causes point mutations in any cellular system (Pershagen and Vahter, 1979; Belton et al., 1985; Lee et al., 1985; Deknudt et al., 1986). However, Li and Rossman (1989) have shown that arsenite causes inhibition of DNA repair after the incision step in Chinese hamster V79 cells. They also have found that DNA ligase II activity is inhibited after arsenite treatment, because it acts as a comutagen with a number of different types of mutagens. Nordenson and Beckman (1991) have shown that the oxygen radical-scavenging enzymes superoxide dismutase (SOD) and catalase have a protective effect against arsenic and can reduce the DNA damage in cultured human lymphocytes.

No studies in humans are reported to account for the genotoxic effect of organoarsenicals after oral exposure. Yamanaka et al. (1989a, b) exposed rats and mice to dimethylarsenic acid (DMA) and detected DNA single-strand breaks in the lung. They concluded that the damage was caused by dimethylarsine and active oxygen and was similar to damage caused by ionizing radiation. The same investigators later reported (1990) that the damage was caused by peroxy radical rather than active oxygen. These breaks were largely repaired within 24 hr, so the risk to human health is uncertain.

3.12. Immunologic Effects

The effect on the immune system of inhalation exposure to arsenic is not well studied. No abnormalities were detected in the serum levels of immunoglobulins of workers exposed to arsenic in a coal-burning power plant (Bencko et al., 1988). The levels of arsenic were not measured in this study, and they may have been too small to cause significant damage.

A number of studies have been conducted on animals. A single exposure of mice to 0.9 mg As/m^3 (arsenic trioxide) and then repeated exposures to 0.5 mg As/m^3 caused increased susceptibility to respiratory bacterial pathogens as a result of alveolar macrophage damage (Aranyi et al., 1985). Sikorski et al. (1989) exposed mice to 5.7 mg As/kg/day tracheally and noted a decrease in humoral response to antigens in several complement proteins, but no decrease occurred in resistance to bacterial or tumor cell challenges. These studies suggest that inhalation of inorganic arsenic can affect the immune system adversely. There is no published information concerning immunological effects in humans or animals exposed orally to organoarsenicals. Inorganic arsenate up to 100 ppm (20 mg As/kg/day) did not cause detectable immunosuppression in mice (Kerkvliet et al., 1980).

Sikorski et al. (1989) reported that a single intratracheal exposure to GaAs (200 mg/kg) suppresses in vitro the IgM antibody-forming cells. Higher doses decreased the thymus weight and spleen cellularity with no effect on the body weight. GaAs exposure also results in impaired ability of adherent cells to stimulate proliferation of T-cells, suggesting a defect in T-helper cell activation (Sikorski et al., 1991). Sodium arsenite decreases interferon production and its activity in mice (Gainer, 1972). Blakely et al. (1980) demonstrated that 0.5 to 10 ppm sodium arsenite in drinking water suppressed the IgM and IgG antibody-forming cells response to sheep red blood cells in mice. Simons et al. (1989) investigated the toxicity of arsenite and cadmium (Cd^{2+}) to HTC cells and observed that 99% of cells treated with 100 μM arsenite were nonviable. The LD_{50} was about 40 μM arsenite, whereas 10 μM had little or no effect. Leikin et al. (1991) investigated the use of immunotherapy for the treatment of arsenite toxicity. They noted no change in the mortality of female mice when antisera were administered.

3.13. Carcinogenic Effects

A number of epidemiological studies have presented convincing evidence that inorganic arsenic compounds increase the risk of lung cancer when exposure occurs through inhalation (Taylor et al., 1989). Arsenic trioxide dust at copper smelters has been implicated as the primary toxicant for workers (Pinto et al., 1977, 1978; Axelson et al., 1978; Enterline and Marsh, 1982; Enterline et al., 1987a, b; Jarup et al., 1989; Lee-Feldstein, 1983, 1986; Wall, 1980; Welch et al., 1982). Increased lung cancers have also been reported for chemical plant workers who were primarily exposed to arsenate (Ott et al., 1974; Mabuchi et al., 1979; Sobel et al., 1988). Several other studies suggested that persons living in the vicinity of smelters or arsenical chemical plants may also have an increased lung cancer risk (Matanoski et al., 1981; Cordier et al., 1983; Brown and Chu, 1983; Pershagen, 1985), although the increase in cancer cases is small and not clearly detectable in all cases, as pointed out by Frost et al. (1987).

In addition to inhalation exposure, ingestion of inorganic arsenic also increases the risk of skin cancer, as suggested by several epidemiological studies (Sommers and McManus, 1953; Tseng et al., 1968; Zaldivar, 1974; Tay and Seah, 1975; Tseng, 1977; Zaldivar et al., 1981; Cebrián et al., 1983; Luchtrath, 1983; Bickley and Papa, 1989; Piontek et al., 1989). The most common lesions are multiple squamous cell carcinomas, which appear to develop from some of the hyperkeratotic warts or corns. Basal cell carcinomas may also occur, arising from cells not assoicated with hyperkeratinization. Mostly, skin cancer develops after long-term exposure, but several reports indicate that it can occur after exposure of less than a year (Reyman et al., 1978; Wagner et al., 1979). Although both types of skin cancer can be removed surgically, they may be fatal if left untreated to develop into painful lesions (Shannon and Strayer, 1989). Studies that provide dose-dependent data have been reviewed by the EPA (1988).

Tseng et al. (1968) reported the incidence of arsenic-induced cancer in over 40,000 people from Taiwan, where arsenic levels in drinking water ranged from 0.001 to 1.82 mg/L. The occurrence of skin cancers and other signs of arsenic intoxication was compared to the rates in the 7500 people used as controls who consumed water with low arsenic content (17 μg/L). Several U.S. studies did not reveal any correlation in a small population that consumed water containing 0.1 to 0.2 ppm arsenic (Goldsmith et al., 1972; Morton et al., 1976; Harrington et al., 1978; Southwick et al., 1981). Arsenic cannot be implicated in the increased risk of skin cancers in the United States, though its role can hardly be denied (Andelman and Barnett, 1983). The incidence of some types of internal malignancies may also be increased by chronic oral exposure to arsenic (EPA, 1988), but the data are inconclusive (EPA, 1984; Phillips, 1985). Sommers and McManus (1953) noticed that 10 out of 27 patients with skin cancer following arsenic exposure also had an internal cancer. Reyman et al. (1978) documented a similar increase in the incidence of internal cancers in patients suffering from arsenical keratoses. Dobson et al. (1965) reported similar cancers earlier. Many case studies have noted the occurrence of liver tumors and other cancers in persons with arsenic-induced skin cancer (Sommers and McManus, 1953; Regelson et al., 1968; Lander et al., 1975; Tay and Seah, 1975; Zaldivar et al., 1981; Falk et al., 1981; Kasper et al., 1984; Koh et al., 1989).

More recent epidemiological studies in Taiwan have reported tumors of the lung, liver, bladder, and kidney associated with arsenic-induced skin cancers (K.S. Chen et al., 1988; Chen and Wang, 1990; Wu et al., 1989). Tsuda et al. (1990) indicated an interaction between exposure to arsenic and smoking in Japanese patients.

There is not much evidence that arsenates or arsenites cause skin or other types of cancers in laboratory animals (Byron et al., 1967; Schroeder et al., 1968; Kroes et al., 1974). The reason for the lack of carcinogenicity of arsenicals could be related to their genetic disposition, their mode of action, and their metabolism. Paradoxically, arsenic ingestion in mice may reduce the incidence of some types of tumors, such as urethane-induced pulmonary tumors (Blakely, 1987), spontaneous mammary tumors (Schrauzer and Ishmael, 1974; Schrauzer et al., 1976), and tumors induced by injections of sarcoma cells (Kerkvliet et al., 1980). Other studies indicated that arsenic increased the growth rate of existing tumors in animals and reduced lifespan (Schrauzer and Ishmael, 1974; Kerklviet et al., 1980). Some researchers have suggested that arsenic is mainly a tumor promoter, since it affects different types of neoplastic cells differently.

The effects of organoarsenical ingestion on humans and their role in carcinogensis have not been studied or published. Data for animals are also scanty. Sabbioni et al. (1991) did not find cytotoxicity of arsenobetaine (seafood arsenic) in the BALB/3T3 cells. They reported that low retention efficiency, inability to interact with intracellular components, and absence of biotransformation were the reasons for the lack of cytotoxicity of this organoarsenical. Prier et al. (1963) did not find any increase in tumor frequency when dogs were subjected to oral exposure of 1.5 mg As/kg/day or mice to 3.8 mg As/kg/day for 2 years in the form of roxarsone. Recently, roxarsone was administered during the lifetime of mice and rats of both sexes at doses of up to 1.4 mg As/kg/day (roxarsone), but no evidence of carcinogenicity was noticed (NTP, 1989); however, a slight increase in pancreatic tumors in male rats was seen. Initial exposure of rats to diethylnitrosamine and subsequently to DMA caused an increase in precancerous lesions (basophilic foci) in the liver, suggesting that this compound may also act as a cancer promoter (Johansen et al., 1984). Perhaps organoarsenicals possess a potential to be carcinogenic in rats.

Computer modeling studies based on available epidemiological data suggest that arsenic acts mainly as a promoter and increases lung cancer by enhancing a late stage in the carcinogenic sequence, although arsenic may also act at an early stage (Enterline and Marsh, 1982; Brown and Chu, 1983; Mazumdar et al., 1989). The biomedical mechanism of arsenic-induced carcinogenicity is not well understood. Arsenic does not seem to damage DNA directly, but several studies contend that arsenic inhibits one or more enzymes necessary for DNA repair or replication (Nordberg and Andersen, 1981; Rossman, 1981; Okui and Fujiwara, 1986; Li and Rossman, 1989). Nordberg and Andersen (1981) suggested an alternate mechanism, which is that arsenate is incorporated in the DNA in place of phosphate. This explanation is consistent with obervations that arsenate must be present during DNA synthesis in order to be effective. This also explains why arsenic is clastogenic, since the arsenate–phosphate bond would be weaker than the normal phosphodiester bond (Jacobson-Kram and Montalbano, 1985). Based on the existing information, the EPA has placed inorganic arsenic in Group A (as a known human carcinogen).

3.14. Biochemical Effects

Arsenical compounds are known to inhibit a number of important enzymes in both animals and humans. Phenylarsine oxide (PAO) blocks glucose transport activity by inhibiting insulin activation of glucose uptake in rat soleus muscles (Wang et al., 1991) and in 3T3-L1 adipocytes (Frost and Schwalbe, 1990), in which vicinal thiols are implicated in signal transmission. These vicinal groups include —SH, —SH/—OH and —SH/—CO_2H, which take part in insulin-stimulated sugar transport (Douen and Jones, 1990; Henriksen and Holloszy, 1990). Frost et al. (1989) reported that PAO blocks the fluid phase of endocytosis, while Gould et al. (1989) explained the biphasic effect of PAO on glucose transport: at 10 μM it activates the transport to threefold in 3T3-L1 adipocytes, but at higher concentrations it inhibits it. Kulanthaivel et al. (1990) noted the catalytic action of vicinal thiol groups and that PAO inhibits Na^+ and H^+ efflux of human placental Na^+–H exchanger. Musch et al. (1990) demonstrated that 250 μM oxyphenyl arsine (OPA) inhibited completely the Na–K–ATPase activity in tissue homogenates of winter flounder (*Pseudopleuronectus americanus*), concluding that it was due to the inhibition of tyrosine. Garcia-Morales et al. (1990) reported that PAO can also inhibit tyrosine phosphatase activity while leaving tyrosine function intact in a murine T-cell hybridoma.

The adverse effects of arsenicals on enzyme systems have been endorsed by a number of authors. The in vitro effects of arsenite on both human and rat blood enzymes were investigated by Sheabar and Yanni (1989). There was a 70 to 80% inhibition of glutamylpyruvate transaminase, and glutathione peroxidase was also affected adversely by 0.8 μg/mL arsenic. Also, blood glucose-6-phosphatase and cholinesterase were completely inhibited. Inorganic arsenite inhibited the protein ubiquitin in both intact and lysate reticulocytes of rabbit in a concentration-dependent manner due to the inhibition of arginylaminoacyl-tRNA transferase (Klemperer and Pickart, 1989). Arsenite also inhibited ubiquitin-substrate conjugate turnover.

Other studies have shown that sodium arsenite causes a marked increase in the cellular heme oxygenase activity of human HeLa cells (Taketani et al., 1989). Keyse and Tyrell (1989) further confirmed that heme oxygenase is the major 32-kDa stress protein that is induced by UVA radiation and sodium arsenite in human skin fibroblasts. These researchers concluded that the induction of this protein may be a general response to oxidant damage and an important cellular defense mechanism.

Arsenite is rapidly and extensively accumulated in the liver, where it inhibits NAD-linked oxidation of pyruvate or α-ketoglutarate. This occurs by complexation of trivalent arsenic with vicinal thiols necessary for the oxidation of these substrates (Squibb and Fowler, 1983). Rats treated intraperitoneally with sodium arsenite (10 mg/kg) develop hyperglycemia (Ghafghazi et al., 1980), suggesting the possibility of glucagon involvement.

Reichl et al. (1989) indicated that a single injection of AS_2O_3 affected mitochondrial metabolism in guinea pig livers, resulting in a significant decrease in the hydroxybutyrate: acetoacetate ratio and in acetyl-CoA and adenosine 5′-monophosphate, but an increase in pyruvate. Repeated injections decreased glycogen, pyruvate α-ketoglutarate, acetyl-CoA, and acetoacetate. These researchers later reported (1991) that there was a significant decrease in the pyruvate and lactate efflux in the perfused livers of guinea pigs exposed to As_2O_3.

Other biochemical effects included (1) induction of at least four stress proteins of 80 to 84 kDa by sodium arsenite (Pipkin et al., 1987), (2) increase in the cells' resistance to a subsequent 2-hr hyperthermic treatment (44 °C) of Chinese hamster 023 cells with arsenite (Crete and Landry, 1990), (3) inhibition (dose-dependent) of hepatic glycogenolysis, Ca^{2+} mobilization, and stimulated O_2 consumption (Steinhelper and Olson, 1988), and (4) significant increases in transferrin, orosomucoid, and ceruloplasmin levels in 47 adults working in a coal-burning power plant containing 900 to 1500 g arsenic per ton dry weight (Bencko et al., 1988). Ulthus (1990) noted that hamsters given 12 μg/g arsenic in the diet showed a decrease in the plasma levels of several amino acids—alanine, glycine, phenylalanine, and taurine—suggesting a possible role of arsenic in methionine/methyl metabolism.

4. METABOLISM

The metabolism of inorganic arsenic has been studied extensively in both animals and humans, but information on organoarsenical compounds is limited. Two major metabolic pathways for arsenics have been identified: oxidation–reduction reactions for the interconversion of arsenates and arsenites in the body, and methylation reactions that ultimately convert these compounds to monomethylarsine and dimethylarsine as metabolic products. Essentially, these processes are similar whether the animals or humans are exposed orally or by the inhalation of arsenic.

Most of the evidence for the above-mentioned processes has been derived from the toxicokinetics of these toxicants (Charbonneau et al., 1978b). Human exposure to either arsenates or arsenites usually results in increased levels of inorganic As(III), As(V), MMA, and DMA in the urine, as reported by several investigators (Crecelius, 1977; Smith et al., 1977; Tam et al., 1979; Lovell and Farmer, 1985;

Buchet et al., 1981a, b). Similar results were obtained from laboratory mice (Vahter, 1981; Vahter and Envall, 1983), hamsters (Marafante and Vahter, 1987; Hirata et al., 1988; Takahashi et al., 1988), and rabbits (Vahter and Marafante, 1983; Maiorino and Aposhian, 1985; Marafante et al., 1985).

It has been observed that the relative portions of As(III), As(V), MMA, and DMA in urine varied according to the chemical administered, the amount of time in which samples were analyzed after the initial exposure, the dose, and the animal species. However, DMA was the principal metabolite, followed by the inorganic arsenics [As(III) and As(V)] and MMA. The usual percentages of these metabolites in humans have been reported to be approximately 40 to 60% DMA, 20 to 25% inorganic arsenic, and 15 to 25% MMA (Smith et al., 1977; Tam et al., 1979; Buchet et al., 1981a; Vahter, 1986). Maiorino and Aposhian (1985) found that rabbits have a similar ratio of these metabolites, suggesting that this may be the best animal model for toxicokinetics. In contrast, marmoset monkeys do not methylate inorganic arsenic and cannot be considered a good model for humans (Vahter et al., 1982; Vahter and Marafante, 1985).

In vitro studies have indicated that the substrate for methylation is As(III), since As(V) is not methylated unless it is first reduced to As(III) (Lerman and Clarkson, 1983; Lerman et al., 1983; Buchet and Lauwerys, 1985, 1988). The same researchers indicated that the main site of methylation appears to be the liver, where the methylation process is mediated by enzymes that utilize *S*-adenosylmethionine as a cosubstrate. Additionally, under normal conditions, the availability of methyl donors (i.e., methylene, choline, and cysteine) does not appear to be rate limiting for the methylation process either in humans (Buchet et al., 1982) or in animals (Buchet et al., 1981b; Buchet and Lauwerys, 1987). These investigators also pointed out that severe dietary limitations on methyl donor intake can result in significant decreases in methylating capacity.

A number of studies have confirmed that methyl derivatives of arsenic appear to be less toxic than the parent compound, and since methylation tends to reduce the retention of inorganic arsenic in tissues, the methylation process is the "detoxification" mechanism. The importance of the arsenic dose that will saturate the methylation capacity has been pointed out, since methylation is an enzymatic process. Studies on humans suggest that methylation may begin to be limiting at doses of about 0.2 to 1 mg/day (0.003–0.015 mg As/kg/day) (Buchet et al., 1981b; Marcus and Rispin, 1988). However, these observations were based on data from a small group of people and cannot be considered decisive, since the pattern of urinary excretion products in those who ingested a near-lethal dose of arsenic or were exposed to elevated levels of arsenic in the workplace is not much different from that of the general population (Lovell and Farmer, 1985; Vahter, 1986). Thus, the dose rate at which methylation capacity becomes saturated cannot be defined precisely, as several of the studies mentioned earlier pointed out.

Buchet and Lauwerys (1987) have demonstrated that the in vitro methylation of arsenic in the liver requires glutathione (GSH) and *S*-adenosylmethionine and can be stimulated by hydroxycobalamin. However, a significant amount of GSH resulted in a marked increase in inorganic arsenic excretion and some retention in the liver. Thus, the important role of GSH in the metabolism of arsenic in vivo has been demonstrated (Klaassen, 1974; Rowland and Davies, 1981); however, the mechanism by which hepatic GSH influences the methylation process is not well understood. Investigating the role of GSH in arsenic metabolism, Gyurasics et al. (1991) stated that the hepatobiliary transport of arsenic depends on the GSH complex, since GSH depletion elicits a nephrotoxic effect in hamster kidneys and manifests as arsenic poisoning (Hirata et al., 1988, 1990).

Organoarsenicals tend to be metabolized in tissues very little as indicated by Buchet et al. (1981a), who reported that humans who ingested a dose of MMA converted a small amount (about 13%) to DMA; several other studies of hamsters noted the formation of low levels of the trimethyl derivative trimethylarsine oxide $[(CH_3)_3AsO]$ (Yamauchi and Yamamura, 1984; Yamauchi et al., 1988). However, the methylarsenates are not demethylated to inorganic arsenic either in humans (Marafante et al., 1987) or in rats and hamsters (Stevens et al., 1977; Yamauchi and Yamamura, 1984). Recently, Yamauchi et al. (1990) reported that trimethylarsine is never demethylated in vivo, but is oxidized to form trimethylarsine oxide and is excreted as such in the urine. A part of the trimethylarsine is directly expired into the air.

REFERENCES

Abdo, K. M., Elwell, M. R., Montgomery, C. A., Thomson, M. B., Thompson, R. B., and Prejean, J. D. (1989). Toxic responses in F344 rats and B6C3F mice given roxorsane in their diets for up to 13 weeks. *Toxicol. Lett.* **45**, 55–56.

American Conference of Governmental Industrial Hygienists (ACGIH) 1986. *Documentation of the Threshold Limit Values and Biological Exposure Indices.* 5th ed. ACGIH, Cincinnati, OH.

Andelman, J. B., and Barnett, M. (1983). *Feasibility Study to Resolve Questions on the Relationship of Arsenic in Drinking Water to Skin Cancer,* Final Report. Report to U.S. Environmental Protection Agency, Office of Research and Development, Washington, DC by Center for Environmental Epidemiology, University of Pittsburgh, Pittsburgh.

Andreae, M. O. (1986). Organoarsenic compounds in the environment. In P.G. Craig (Ed.), *Organometallic Compounds in the Environment: Principles and Reactions.* Wiley, New York, pp. 198–228.

Aranyi, C., Bradof, J. N., O'Shea, W. J., Graham, J. A., and Miller, F. J. (1985). Effects of arsenic trioxide inhalation exposure on pulmonary antibacterial defenses in mice. *J. Toxicol. Environ. Health* **15**, 163–172.

Armstrong, C. W., Stroube, R. B., Rubio, T., and Beckett, W. S. (1984). Outbreaks of fatal arsenic poisoning caused by contaminated drinking water. *Arch. Environ. Health* **39**, 276–279.

Aschengrau, A., Zierler, S., and Cohen, A. (1989). Quality of community drinking water and the occurrence of spontaneous abortion. *Arch. Environ. Health* **44**, 283–290.

Axelson, O., Dahlgren, E., Jansson, C. D., and Rehnlund, S. O. (1978). Arsenic exposure and mortality. A case reference study from a Swedish copper smelter. *Br. J. Ind. Med.* **35**, 8–15.

Bagatto, G., and Ali khan, M. A. (1987). Copper, cadmium and nickel accumulation in crayfish populations near copper-nickel smelter at Sudbury, Ontario, Canad. *Bull. Environ. Contam. Toxicol.* **38**, 540–545.

Bannai, S., Sato, H., Ishii, T., and Taketani, S. (1991). Enhancement of glutathione levels in mouse peritoneal macrophages by sodium arsenite, cadmium chloride and glucose/glucose oxidase. *Biochm. Biophys. Acta* **1092**, 175–179.

Bansal, S. K., Haldar, N., Dhand, U. K., et al. (1991). Phrenic neuropathy in arsenic poisoning. *Int. J. Dermatol.* **30**, 304–306.

Barrett, J. C., Oshimura, M., Tsutsui, T., and Wang, T. C. (1988). Mutation and neoplastic transformation: Correlations and dissociations. *Ann. N. Y. Acad. Sci.* **534**, 95–98.

Baxley, M. N., Hood, R. D., Vedel, G. C., Harrison, W. P., and Szczech, G. M. (1981). Prenatal toxicity or orally administered sodium arsenite in mice. *Bull. Environ. Contam. Toxicol.* **26**, 749–756.

Beckman, G., Beckman, L., and Nordenson, I. (1977). Chromosomal aberrations in workers exposed to arsenic. *Environ. Health Perspect.* **19**, 145–146.

Bekemeir, H., and Hirschelmann, R. (1989). Reactivity of resistance in blood vessels *ex vivo* after administration of toxic chemicals to laboratory animals. *Arteriolotoxicity. Toxicol. Lett.* **49**, 49–54.

Belton, J. C., Benson, N. C., Hanna, M. L., and Taylor, R. T. (1985). Growth inhibitory and cytotoxic effects of three arsenic compounds on cultured Chinese hamster ovary cells. *J. Environ. Sci. Health* **20A**, 37–72.

Bencko, V., Wagner, V., Wagnerova, M., and Batora, J. (1988). Immunological profiles in workers of a power plant burning coal rich in arsenic content. *J. Hyg. Epidemiol. Microbiol. Immunol.* **32**, 137–147.

Bickley, L. K., and Papa, C. M. (1989). Chronic arsenicism with vitiligo, hyperthyroidism, and cancer. *N. J. Med.* **86**, 377–380.

Blair, P. C., Thompson, M. B., Bechtold, M., Wilson, R. E., Moorman, M. P., and Fowler, B. A. (1990a). Evidence for oxidative damage to red blood cells in mice induced by arsine gas. *Toxicology* **63**, 25–34.

Blair, P. C., Thompson, M. B., Morrissey, R. E., Moorman, M. P., Sloane, R. A., and Fowler, B. A. (1990b). Comparative toxicity of arsine gas in B6C3 F1 mice, Fischer 344 rats and Syrian golden hamster: System organ studies and comparison of clinical indices of exposure. *Natl. Fundam. Appl. Toxicol.* **14**, 776–787.

Blakely, B. R. (1987). Alterations in urethan-induced adenoma formation in mice exposed to selenium and arsenic. *Drug Nutr. Interact.* **5**, 97–102.

Blakely, B. R., Sisodia, C. S., and Mukkur, T. K. (1980). The effect of methyl mercury, tetraethyl lead and sodium arsenite on the humoral immune response in mice. *Toxicol. Appl. Pharmacol.* **52**, 245–254.

Borgono, J. M., and Greiber, R. (1972). Epidemiological study of arsenicism in the city of Antofagasta. *Trace Subst. Environ. Health* **5**, 13–24.

Borgono, J. M., Venturino, H., and Vincent, P. (1980). Clinical and epidemiological study of arsenism in northern Chile (1977). *Rev. Med. Chile* **108**, 1039–1048.

Brown, C. C., and Chu, K. C. (1983). Implications of multistage theory of carcinogenesis applied to occupational arsenic exposure. *JNCI, J. Natl. Cancer Inst.* **70**, 455–463.

Brown, M. M., Rhyne, B. C., and Goyer, R. A. (1976). Intracellular effects of arsenic administration on renal proximal tubule cells. *J. Toxicol. Environ. Health* **1**, 505–514.

Brown, R. M., Newton, D., Pickford, C. J., and Button, D. K. (1990). Human metabolism of arsenobetaine ingested with fish. *Hum. Exp. Toxicol.* **9**, 41–46.

Buchet, J. P., and Lauwerys, R. (1985). Study of inorganic arsenic methylation by rat liver *in vitro*: Relevance for the interpretation of observations in man. *Arch. Toxicol.* **57**, 125–129.

Buchet, J. P., and Lauwerys, R. (1987). Study of factors influencing the *in vivo* methylation of inorganic arsenic in rats. *Toxicol. Appl. Pharmacol.* **91**, 65–74.

Buchet, J. P., and Lauwerys, R. (1988). Role of thiols in the *in vitro* methylation of inorganic arsenic by rat liver cytosol. *Biochem. Pharmacol.* **37**, 3149–3153.

Buchet, J.P., Lauwerys, R., and Roels, H. (1981a). Comparison of the urinary excretion of arsenic metabolites after a single dose of sodium arsenite, monomethyl arsonate or dimethyl arsenite in man. *Int. Arch. Occup. Environ. Health* **48**, 71–79.

Buchet, J. P., Lauwerys, T., and Roels, H. (1981b). Urinary excretion of inorganic arsenic and all its metabolites after repeated ingestion of sodium meta arsenite by volunteers. *Int. Arch. Occup. Environ. Health* **48**, 111–118.

Buchet, J. P., Lauwerys, R., and Mahieu, P. (1982). Inorganic arsenic metabolism in man. *Arch. Toxicol., Suppl.* **5**, 326–327.

Buck, W. B. (1969). Untoward reactions encountered with medicated feeds. In *The Use of Drugs in Animal Feeds.* National Academy of Sciences, Washington, DC, NAS Publ. No. 1679, pp. 196–217.

Buck, W. B., Osweiler, G.D., and Van Gelder, G.A. (1973). *Clinical and Diagnostic Veterinary Toxicology.* Kendall-Hunt Publ. Co., Dubuque, IA.

Burgdorf, W., Kurvink, K., and Cervenka, J. (1977). Elevated sister chromatid exchange rate in lymphocytes of subjects treated with arsenic. *Hum. Genet.* **36**, 69–72.

Burns et al. (1991). *Toxicol Appl. Pharmacol.* **110**, 157–169.

Byron, W. R., Bierbower, G. W., Brouwer, J. B., and Hanse, W. H. (1967). Pathological changes in rats and dogs from two-year feeding of sodium arsenite or sodium arsenate. *Toxicol. Appl. Pharmacol.* **10**, 132–147.

Calabrese, E. J., Kostecki, P. T., Gilbert, C. E. (1987). How much soil do children eat? An emerging consideration for environmental health risk assessment. *Comments Toxicol.* **1**, 229–241.

Campbell, J. P., and Alvarez, J. A. (1989). Acute arsenic intoxication. *Am. Fam. Physician* **40**, 93–97.

Cannon, J. R., Sauders, J. B., and Toia, R. F. (1983). Isolation and preliminary toxicological evaluation of arsenobetaine, the water soluble arsenical constituent from the hepatopancreas of the western rock lobster. *Sci. Total Environ.* **31**, 181–185.

Cebrián, M. E., Albores, A., Anguilar, M., and Blakely, E. (1983). Chronic arsenic poisoning in the north of Mexico. *Hum. Toxicol.* **2**, 121–133.

Chaineau, E., Binet, S., Pol, D., Chatellier, G., and Meininger, V. (1990). Embryotoxic effects of sodium arsenite and sodium arsenate on mouse culture embryos in culture. *Teratology* **41**, 105–112.

Chakraborty, A. K., and Saha. (1987). Arsenical dermatosis from tube-well water in West Bengal. *Indian J. Med. Res.* **85**, 326–334.

Charbonneau, S. M., Spencer, K., Bryce, F., and Sandi, E. (1978a). Arsenic excretion by monkeys dosed with arsenic containing fish or with inorganic arsenic. *Bull. Environ. Contam. Toxicol.* **20**, 470–477.

Charbonneau, S. M., Tam, G. K. H., Bryce, F., and Collins, B. (1978b). Pharmacokinetics and metabolism of inorganic arsenic in the dog. *Trace Subst. Environ. Health* **12**, 276–283.

Charbonneau, S. M., Tam, G. K. H., Bryce, F., and Sandi, E. (1980). *Proc. 19th. Annu. Meet. Soc. Toxicol.*, Washington, DC, *1980*, Abstract No. 362, A121.

Chen, C. J., and Wang, C. J. (1990). Ecological correlation between arsenic level in well water and age-adjusted mortality from malignant neoplasms. *Cancer Res.* **50**, 547–574.

Chen, C. J., Wu, M. M. Lee, S. S., and Lin T. M. (1988). Atherogenicity and carcinogenicity of high arsenic artesian well water. Multiple risk factors and related malignant neoplasms of blackfoot disease. *Arteriosclerosis* **8**, 452–460.

Chen, G. S., Asai, T., Suzuki, Y., Nishioka, K., and Nishiyama, S. (1990). A possible pathogenesis for blackfoot-disease effects of trivalent arsenic (As_2O_3) on cultured human umbilical vein endothelial cells. *J. Dermatol.* **17**, 599–608.

Chen, K. S., Huang, C. C., Liaw, C. C., and Lin, T. M. (1988). Multiple primary cancers in blackfoot endemic areas (case study). *J. Formosan Med. Assoc.* **87**, 1125–1128 (in Chinese).

Chernoff, N., and Rogers, E. H. (1975). Effects of cacodylic acid on the prenatal development in rats and mice. Substitute chemical program: The first year of progress. In *Toxicological Methods and Genetic Effects Workshop.* U.S. Environmental Protection Agency, Washington, D.C., Vol. II pp. 197–204.

Chhuttani, P. N., Chawla, L. S., and Sharma, T. D. (1967). Arsenical neuropathy *Neurology* **17**, 269–274.

Christau, B., Chabas, M. E., and Placidie, M. (1975). Is *p*-arsinilic acid transformed in the rat and guinea pig? *Ann. Pharm. Fr.* **33**, 37–41.

Cordier, S., Theriault, G., and Iturra, H. (1983). Mortality patterns in a population living near a copper smelter. *Environ. Res.* **31**, 311–322.

Crecelius, E. A. (1977). Changes in the chemical speciation of arsenic following ingestion by man. *Environ. Health Perspect.* **19**, 147–150.

Crete, P., and Landry, J. (1990). Induction of HSP 27 phosphorylation and thermoresistance in Chinese hamster cells by arsenite, cycloheximide, A 23187 and EGTA. *Radiat. Res.* **121**, 320–327.

Czarnecki, G. L., and Baker, D. H. (1984). Feed additive interactions in the chicken: Reduction of tissue copper deposition by dietary roxarsone in healthy and in *Eimeria acervalina*-infected or *Eimeria teneela*-infected chicks. *Poult. Sci.* **63**, 1412–1418.

Czarnecki, G. L., Edmonds, M. S., Isquierdo, O. A., and Baker, D. H. (1984a). Effect of 3-nitro-4-hydroxylphenylarsonic acid on copper utilization by the pig, rat and chick. *J. Anim. Sci.* **59**, 997–1002.

Czarnecki, G. L., Baker, D. H., and Garst, J. E. (1984b). Arsenic-sulfur amino acid interactions in the chick. *J. Anim. Sci.* **59**, 1572–1581.

Dannan, M., Dally, S., and Conso, F. (1984). Arsenic induced encepahalopathy (Letter). *Neurology* **34**, 1524.

Datta, D. V. (1976). Arsenic and non-cirrhotic portal hypertension. *Lancet* **1**, 433.

Datta, S., Talukdar, G., and Sharma, A. (1986). Cytotoxic effects of arsenic in dietary oil primed rats. *Sci. Cult.* **52**, 196–198.

Dekundt, G., Leonard, A., Arany, J., Du Buisson, G. J., and Delavignetta, E. (1986). *In vivo* studies in male mice on the mutagenesis effects of inorganic arsenic. *Mutagenesis* **1**, 33–34.

Di Napoli, J., Hall, A. H., Drake, R., and Rumack, B. H. (1989). Cyanide and arsenic poisoning by intravenous injection. *Ann. Emerg. Med.* **18**, 308–311.

Dobson, R. L., Young, M. R., and Pinto, J. S. (1965). Palmar keratoses and cancer. *Arch. Dermatol.* **92**, 553–556.

Donofrio, P. D., and Wilbourn, A. J. (1985). Subacute arsenic toxicity presenting on clinical and electromyographic examination as Guillain-Barré syndrome. *Ann. Neurol.* **18**, 156–157.

Douen, A. G., and Jones, M. N. (1990). Phenylarsine oxide and the mechanism of insulin-stimulated sugar transport. *Biofactors* **2**, 153–161.

Edmonds, M. S., and Baker, D. H. (1986). Toxic effects of supplemental copper and roxarsone when fed alone or in combination to young pigs. *J. Anim. Sci.* **63**, 533–537.

Eisler, R. (1988). *Arsenic Hazard to Fish, Wildlife, and Invertebrates: A Synoptic Review*, NTIS No. PB88-169404. U.S. Fish and Wildlife Service, Patuxent Wildlife Research Center, Laurel, MD.

El Bahri, L., and Ben Romdane, S. (1991). Arsenic poisoning in livestock. *Vet. Hum. Toxicol.* **33**, 259–264.

Enterline, P. E., and Marsh, G. M. (1982). Cancer among workers exposed to arsenic and other substances in a copper smelter. *Am. J. Epidemiol.* **116**, 895–911.

Enterline, P. E., Henderson, V. L., and Marsh, G. M. (1987a). Exposure to arsenic and respiratory cancer. A reanalysis. *Am. J. Epidemiol.* **125**, 929–938.

Enterline, P. E., Marsh, G. M., Esmen, N. A., and Henderson, V. L. (1987b). Some effects of cigarette smoking, arsenic and SO_2 on mortality among U.S. copper smelter workers. *J. Occup. Med.* **29**, 831–838.

Environmental Protection Agency (EPA) (1980). *Ambient Water Quality Criteria for Arsenic*, EPA 440/5-80-021. USEPA, Office of Water Regulations and Standards, Washington, DC.

Environmental Protection Agency (EPA) (1984). *Health Assessment Document for Inorganic Arsenic*, Final Report, EPA 600/8-83-021F. USEPA, Environmental Criteria and Assessment Office, Research Triangle Park, NC.

Environmental Protection Agency (EPA) (1985). *Health Advisory for Arsenic*, Draft. USEPA, Office of Drinking Water, Washington, DC.

Environmental Protection Agency (EPA) (1988). *Risk Assessment Forum. Special Report on Ingested Inorganic Arsenic: Skin Cancer: Nutritional Essentiality*, Draft Report, EPA/625/3-87/013. USEPA, Washington, DC.

Falk, H., Herbert, J. T., Edmonds, L., Heath, C. W., Thomas, L. B., and Pipper, H. (1981). Review of four classes of childhood hepatic angiosarcoma—elevated environmental arsenic exposure in one case. *Cancer (Philadelphia)* **47**, 382–392.

Feldman, R. G., Niles, C. A., Kelley-Hayes, M., and Wilson, G. (1979). Peripheral neuropathy in arsenic smelter workers. *Neurology* **29**, 939–944.

Fennel, J. S., and Stacy, W. K. (1981). Electrocardiographic changes in acute arsenic poisoning. *Ir. J. Med. Sci.* **150**, 338–339.

Fernandez-Sola, J., Nogue, S., Grau, J. M., Casademont, J., and Munne, P. (1991). Acute arsenical myopathy: Morphological description. *J. Toxicol. Clin. Toxicol.* **29**, 131–136.

Fincher, R. M., and Koerker, R. M. (1987). Long-term survival in acute encephalopathy: Follow-up using newer measures of electrophysiological parameters. *Am. J. Med.* **82**, 549–552.

Foa, B., Colombi, A., Maroni, M., Barbieri, F., Franchini, I., Mutti, A., De Rosa, E., and Bartolucci, G. B. (1987). Study of kidney function of workers with chronic low level exposure to inorganic arsenic. In E. DeRosa, G. B. Bartolucci, and V. Foa (Eds.), *Occupational and Environmental Chemical Hazards*. Halstead Press, Horwood, NY, pp. 362–367.

Fowler, B. A., and Weissburg, J. B. (1974). Arsine poisoning. *N. Engl. J. Med.* **291**, 1171–1174.

Fowler, S. A., and Boorman, G. A. (1989). Hematopoietic effects in mice exposed to arsine gas. *Toxicol. Appl. Pharmacol.* **97**, 173–182.

Franklin, M., Bean, W., and Harden, R. C. (1950). Fowler's solution as an etiological agent in cirrhosis. *Am. J. Med. Sci.* **219**, 589–596.

Franzblau, A., and Lilis, R. (1989). Acute arsenic intoxication from environmental arsenic exposure. *Arch. Environ. Health* **44**, 385–390.

Frost, D. V., Purdue, H. S., and Main, B. T. (1962). Further considerations of the safety of arsinilic acid for feed use. *World's Poult. Congr., Proc. 12th*, Sydney, Australia, Sect. Pap., pp. 234–237.

Frost, F., Harter, L., and Milham, S. (1987). Lung cancer among women residing close to an arsenic emitting copper smelter. *Arch. Environ. Health* **42**, 148–152.

Frost, S. C., and Schwalbe, M. S. (1990). Uptake and binding of radio-labelled phenylarsine oxide in 3T3-L1 adipocytes. *Biochem. J.* **269**, 589–595.

Frost, S. C., Lane, M. D., and Gibbs, E. M. (1989). Effect of phenylarsine oxide on fluid phase endocytosis; further evidence of the activation of the glucose transport. *J. Cell. Physiol.* **141**, 467–474.

Fullmer, C. S., and Wasserman. (1985). Intestinal absorption of arsenate in the chick. *Environ. Res.* **36**, 206–217.

Gainer, J. H. (1972). Effects of arsenicals on viral infections in mice. *Am. J. Vet. Res.* **33**, 2299–2309.

Garcia-Morales, P., Minami, Y., Luong, E., Klausner, R. D., and Samelson, L. E. (1990). Tyrosine phosphorylation in T cells is regulated by phosphatase activity: Studies with phenylarsine oxide. *Proc. Natl. Acad. Sci. U.S.A.* **67**, 9255–9259.

Ghafghazi, T., Ridlington, J. W., and Fowler, B. A. (1980). The effects of acute and subacute sodium arsenite administration on carbohydrate metabolism. *Toxicol. Appl. Pharmacol.* **55**, 126–130.

Glazner, F. S., Ellis, J. G., and Johnson, P. K. (1968). Electrocardiographic findings with arsenic poisoning. *Calif. Med.* **109**, 158–162.

Goebel, H. H., Schmidt, P. F., Bohl, J., Tettenborn, B., Kramer, G., and Guttman, L. (1990). Polyneuropathy due to arsenic intoxication: Biopsy studies. *J. Neuropathol. Exp. Neurol* **49**, 137–149.

Goering, P. L., Maronpot, R. R., and Fowler, B. A. (1988). Effect of intratracheal gallium arsenide administration on delta-amino-levulinic acid dehydratase in rats: Relationship to urinary excretion of aminolevulenic acid. *Toxicol. Appl. Pharmacol.* **92**, 179–193.

Goldsmith, J. R., Deanne, M., Thom, J., and Gentry, G. (1972). Evaluation of health implications of elevated arsenic in well water. *Water Res.* **6**, 1133–1136.

Goldsmith, S., and From, A. H. (1986). Arsenic-induced atypical ventricular tachycardia. *N. Engl. J. Med.* **303**, 1096–1097.

Gould, G. W., Lienhard, G. E., Tanner, L. I., and Gibbs, E. M. (1989). Phenylarsine oxide stimulates hexose transport in 3T3-L1 adipocytes by a mechanism other than an increase in surface transporters. *Arch. Biochem. Biophys.* **268**, 264–275.

Gyurasics, A., Varga, F., and Gregus, Z. (1991). Effect of arsenicals on biliary excretion of endogenous glutathione-dependent hepatobilary transport. *Biochem. Pharmacol.* **41**, 937–944.

Hamamoto, E. (1955). Infant arsenic poisoning by powdered milk. *Jpn. J. Med. Sci. Biol.* **8**, 3–12 (in Japanese).

Harrington, J. M., Middaugh, J. P., Morse, D. L., and Hornsworth, J. (1978). A survey of a population exposed to high arsenic in well water in Fairbanks, Alaska, *Am. J. Epidemiol.* **108**, 377–385.

Harrison, W. P., Frazier, J. C., Mazzanti, E. M., and Hood, R. D. (1980). Teratogenicity of disodium methanearsonate and sodium dimethylarsinate (sodium cacodylate) in mice. *Teratology* **21**, 43A (abstr.).

Hazardous Substances Data Bank (HSDB) (1990). National Library of Medicine, National Toxicology Information Program, Bethesda, MD.

Henriksen, E. J., and Holloszy, J. O. (1990). Effects of phenylarsine oxide on stimulation of glucose transport in rat skeletal muscle. *Am. J. Physiol.* **258**, C648–C653.

Heywood, R., and Sortwell, R. J. (1979). Arsenic intoxication in the rhesus monkey. *Toxicol. Lett.* **3**, 137–144.

Hindmarsh, J. T., and McCurdy, R. F. (1986). Clinical and environmental aspects of arsenic toxicity. *CRC Crit. Rev. Clin. Lab. Sci.* **23**, 315–347.

Hindmarsh, J. T., McLetchie, O. R., Heffernan, L. P., and McCurdy, R. F. (1977). Electromyographic abnormalities in chronic environmental arsenicalism. *J. Anal. Toxicol.* **1**, 270–276.

Hirata, M., Tanaka, A., Hisanaga, A., and Ishinishi, N. (1988). Glutathione and methylation of inorganic arsenic in hamsters. *Appl. Organomet. Chem.* **2**, 315–321.

Hirata, M., Tanaka, A., Hisanaga, A., and Ishinishi, N. (1990). Effects of glutathione depletion on the acute neurotoxic potential of arsenite and on the arsenic metabolism in hamsters. *Toxicol. Appl. Pharmacol.* **106**(3), 469–481.

Hood, R. D. (1985). *Cacodylic Acid: Agricultural Uses, Biological Effects and Environmental Fate.* U.S. Govt. Printing Office, Washington, DC.

Hood, R. D., and Bishop, S. L. (1972). Teratogenic effects of sodium arsenate in mice. *Arch. Environ. Health.* **24**, 62–65.

Huang, Y. Z., Quan, X. C., and Wang, G. Q. (1985). Endemic chronic arsenism in Xinjiang. *Chin. Med. J. (Peking, Engl. Ed.)* **98**, 219–222.

Hudson, R. H., Tucker, R. K., and Haegele, M. A. (1984). The behavior of dissolved arsenic in the estuary of the River Beaulieu. *Estuarine, Coastal Shelf Sci.* **19**, 493–504.

Ide, C. W., and Bullough, G. R. (1988). Arsenic and old glass. *J. Soc. Occup. Med.* **38**, 85–88.

International Agency for Research on Cancer (IARC) (1987). *IARC Monograph on the Evaluation of carcinogenic Risk of Chemicals to Humans. Suppl.* 7. Overall Evaluations of Carcinogenicity, Updating of IARC Monographs, Vol. 1–42. World Health Organization, IARC, Lyon, France, pp. 29–33, 57.

Jacobson-Kram, D., and Montalbano, D. (1985). The Reproductive Effects Assessment Group's report on the mutagenicity of inorganic arsenic. *Environ. Mutagen.* 7, 787–804.

Jaghabir, M. T., Abdelghani, A., and Anderson, A. C. (1988). Oral and dermal toxicity of MSMA to New Zealand white rabbits, *Oryctalagus cuniculus. Bull. Environ. Contam. Toxicol.* **40**, 119–122.

Jaghabir, M. T., Abdelghani, A., and Anderson, A. C. (1989). Histopathological effects of monosodium methanearsonate (MSMA) on New Zealand white rabbits (*Oryctalgus cuniculus*). *Bull. Environ. Contam. Toxicol.* **42**, 289–293.

Jarup, L., Pershagen, G., and Walls, S. (1989). Cumulative arsenic exposure and lung cancer in smelter workers: A dose dependent study. *Am. J. Ind. Med.* **15**, 31–41.

Jenkins, R. B. (1966). Inorganic arsenic and the nervous system. *Brain* **89**, 479–494.

Johansen, M. G., McGowan, J. P., Tu, S. H., and Shirachi, D. Y. (1984). Tumorigenic effect of dimethylarsinic acid in the rat. *Proc. West. Pharmacol. Soc.* **27**, 289–291.

Joliffe, D. M., Budd, A. J., and Gwilt, D. J. (1991). Massive acute arsenic poisoning. *Anaesthesia* **46**, 288–290.

Kaise, T., Watanabe, S., and Itoh, K. (1985). The acute toxicity of arsenobetaine. *Chemosphere* **14**, 1327–1332.

Kantor, H. I., and Leniv, P. M. (1948). Arsenic encephalopathy in pregnancy with recovery. *Am. J. Obstet. Gynecol.* **56**, 370–374.

Kasper, M. C., Shoenfield, L., Strom, R. L., and Theologides, A. (1984). Hepatic angiosarcoma induced by Fowler's solution. *JAMA, J. Am. Med. Assoc.* **252**, 3407–3408.

Kennedy, S., Rice, D. A., and Cush, P. F. (1986). Neuropathology of experimental 3-nitro-4-hydroxyphenylarsonic acid toxicosis in pigs. *Vet. Pathol.* **23**, 454–461.

Kerkvliet, N. I., Steppan, L. B., Koller, L. D., and Exon, J. Y. (1980). Immunotoxicology studies of sodium arsenate, effects of exposure on tumor growth and cell-mediated tumor immunity. *J. Environ. Pathol. Toxicol.* **4**, 54–79.

Kersjes, M. P., Maurer, J. R., and Trestrail, J. H. (1987). An analysis of arsenic exposures referred to the Blodgett Regional Poison Center. *Vet. Hum. Toxicol.* **29**, 75–78.

Keyes, S. M., and Tyrell, R. M. (1989). Heme oxygenase is the major 32-kDa stress protein induced in human skin fibroblast by UVA radiation, hydrogen peroxide and sodium arsenite. *Proc. Natl. Acad. Sci. U.S.A.* **86**, 99–103.

Klaassen, C. D. (1974). Biliary excretion of arsenic in rats, rabbits and dogs. *Toxicol. Appl. Pharmacol.* **29**, 447–457.

Klemperer, N. S., and Pickart, C. M. (1989). Arsenite inhibits two steps in the ubiquitin-dependent proteolytic pathway. *J. Biol. Chem.* **15**, 19245–19252.

Ko, Y. C. (1986). A critical review of epidemiologic status on Blackfoot disease. *Sangyo Ika Daigaku Zasshi* **8**, 339–353.

Koh, E., Kondoh, N., Kaihara, H., Fujioka, H., and Kitamura, K. (1989). Ureteral tumor with multiple Bowen's disease forty-two years after exposure to arsenic. *Eur. Urol* **16**, 398–400.

Kroes, R., van Logten, M. J., and Berkvens, J. M. (1974). Study on the carcinogenicity of lead arsenate and sodium arsenate and on the possible synergistic effects of diethylnitosomine. *Toxicology* **12**, 671–679.

Kulanthaivel, P., Simon, B. J., and Leibach, F. H. (1990). An essential role of vicinal dithiol groups in the catalytic activity of human placenta $Na(+) - H^+$ exchanger. *Biochim. Biophys. Acta* **1024**, 385–389.

Kyle, R. A., and Pearse, G. L. (1965). Hematological aspects of arsenic intoxication. *N. Engl. J. Med.* **273**, 18–23.

Labadie, H. Stoessel, P., Calalrd, P., and Beaugrand, M. (1990). Hepatic venoocclusive disease and perisinusoidal fibrosis secondary to arsenic poisoning. *Gastroenterology* **99**, 1140–1143.

La Dou, J. (1983). Potential occupational health hazards in the micro-electronics industry. *Scand. J. For. Environ. Health* **9**, 42–46.

Lagerkvist, B., Linderholm, H., and Nordberg, G. F. (1988). Arsenic and Reynaud's phenomenon: Vasospastic tendency and excretion of arsenic in smelter workers before and after the summer vacation. *Int. Arch. Occup. Environ. Health* **60**, 361–364.

Lander, J. J., Stanley, R. J., Summer, H. W., Boswell, D. C., and Aach, R. D. (1975). Angiosarcoma of liver associated with Fowler's solution (potassium arsenite). *Gastroenterology* **68**, 1582–1586.

Landan, E. D., Thompson, D. J., Feldman, R. G., Goble G. J., and Dixon, W. J. (1977). *Selected Non-carcinogenic Effects of Industrial Exposure to Inorganic Arsenic.* EPA 569/6-77-018. U.S. Environmental Protection Agency, Washington, DC.

LaQuesne, P. M., and McLeod, J. G. (1977). Peripheral neuropathy following a single exposure to arsenic. *J. Neurol. Sci.* **32**, 437–451.

Lee, T.-C., Oshimura, M., and Barrett, J. C. (1985). Comparison of arsenic-induced transformation, cytotoxicity, mutation and cytogenic effects in Syrian hamster embryo cells in culture. *Carcinogenesis (London)* **6**, 1421–1426.

Lee-Feldstein, A. (1983). Arsenic and respiratory cancer in man: Follow-up of an occupational study. In W. Lederer and R. Festerheim (Eds.), *Arsenic: Industrial, Biomedical and Environmental Perspectives.* Van Nostrand-Reinhold, New York, pp. 245–265.

Lee-Feldstein, A. (1986). Cumulative exposure to arsenic. Matched case-control study of copper smelter employees. *J. Occup. Med.* **28**, 296–302.

Leikin, J. B., Goldman-Leikin, R. E., Evans, M. A., Wiener, S., and Hryhorczuk, D. O. (1991). Immunotherapy in poisoning. *J. Toxicol. Clin. Toxicol.* **29**, 59–70.

Lerman, B. B., Ali, N., and Green, D. (1980). Megaloblastic, dyserythropoietic anemia following arsenic ingestion. *Ann. Clin. Lab. Sci.* **10**, 515–517.

Lerman, S. A., and Clarkson, T. W. (1983). The metabolism of arsenite and arsenate by the rat. *Fundam. Appl. Toxicol.* **3**, 309–314.

Lerman, S. A., Clarkson, T. W., and Gerson, R. J. (1983). Arsenic uptake and metabolism by liver cells is dependent on arsenic oxidation state. *Chem.-Biol. Interact.* **45**, 401–406.

Levin-Scherz, J. K., Patrick, J. D., Weber, F. H., et al. (1987). Acute arsenic ingestion. *Ann. Emerg. Med.* **16**, 702–704.

Li, J. H., and Rossman, T. G. (1989). Inhibition of DNA ligase activity by arsenite: A possible mechanism of its comutagenesis. *Mol. Toxicol.* **2**, 1–9.

Lin, S. M., and Yang, M. H. (1988). Arsenic, selenium and zinc in patients with Blackfoot disease. *Biol. Trace Elem. Res.* **15**, 213–221.

Little, R. E., Kay, G. N., and Cavender, J. B. (1990). Torsade de pontes and T-U wave alternans associated with arsenic poisoning. *PACE* **13**, 164–170.

Lovell, M. A., and Farmer, J. G. (1985). Arsenic speciation in urine from humans intoxicated by inorganic arsenic compounds. *Hum. Toxicol.* **4**, 203–214.

Lu, F. J., Shih, S. R., and Liu, T. M. (1990). The effect of fluorescent humic substances existing in the well water of Blackfoot disease endemic areas in Taiwan on prothrombin time and activated partial thromplastin time *in vitro*. *Thromb Res.* **57**, 747–753.

Luchtrath, H. (1983). The consequences of chronic arsenic poisoning among Moselle wine growers: Pathoanatomical investigations of post-mortem examinations performed between 1960 and 1977. *J. Cancer Res. Clin. Oncol.* **105**, 173–182.

Lugo, G. (1969). Acute maternal arsenic intoxication with neonatal death. *Am. J. Dis. Child.* **117**, 328–330.

Luten, J. B., Riekwel-Booy, G., and Rauchbar, A. (1982). Occurrence of arsenic in plaice (*Pleuronectes platessa*), nature of organoarsenical compounds present in its excretion by man. *Environ. Health Perspect.* **45**, 165–170.

Mabuchi, L., Lilienfield, A. M., and Snell, L. M. (1979). Lung cancer among pesticide workers exposed to inorganic arsenicals. *Arch. Environ. Health* **34**, 312–320.

Maiorino, R. M., and Aposhian, H. V. (1985). Dimercaptan metal-binding agents influence the biotransformation of arsenite in the rabbit. *Toxicol. Appl. Pharmacol.* **77**, 240–250.

Maitani, T., Saito, N., and Abe, M. (1987a). Chemical form-dependent induction of hepatic zinc-thionein by arsenic administration and effect of co-administered selenium in mice. *Toxicol. Lett.* **39**, 63–70.

Marafante, E., and Vahter, M. (1987). Solubility, retention and metabolism of intratracheally and orally administered inorganic arsenic compounds in the hamster. *Environ. Res.* **42**, 72–82.

Marafante, E., Vahter, M., and Envall, J. (1985). The role of methylation in the detoxification of arsenate in the rabbit. *Chem.-Biol. Interact.* **56**, 225–238.

Marafante, E., Vahter, M., Norin, H., Envall, J., Sandström, S., Christakopoulos, A., and Ryhage, R. (1987). Biotransformation of dimethylarsinic acid in mouse, hamsters and man. *J. Appl. Toxicol.* **7**, 111–117.

Marcus, W. L., and Rispin, A. S. (1988). Threshold carcinogenicity using arsenic as an example. In C. R. Cothern, M. A. Mehlman, and W. L. Marcus (Eds.), *Risk Assessment and Risk Management of Industrial and Environmental Chemicals*. Princeton Scientific Publ. Co., Princeton, NJ, Vol. 15, pp. 133–159.

Martin, D. S., Willis, S. E., and Cline, D. M. (1990). *N*-Acetylcysteine in the treatment of human arsenic poisoning. *J. Am. Breed. Fam. Pract.* **3**, 293–296.

Mason, R. W., Edwards, I. R., and Fisher, L. C. (1989). Teratogenicity of combinations of sodium dichromate, sodium arsenate and copper sulphate in the rat. *Comp. Biochem. Physiol.* **93**, 407–411.

Matanoski, G., Landau, E., and Towascia, J. (1981). Cancer mortality in an industrialized area of Baltimore. *Environ. Res.* **25**, 8–28.

Mathews, N. E., and Porter, W. F. (1989). Acute arsenic toxication of a free-ranging white-tailed deer in New York. *J. Wildl. Dis.* **25**, 132–135.

Mazumdar, D. N. G., Chakraborty, A. K., Ghose, A., Gupta, J. D., Chakraborty, D. P., Day, S. B., and Chatoopadya, S. (1988). Chronic arsenic toxicity from drinking tubewell water in rural West Bengal. *Bull. W.H.O.* **66**, 499–506.

Mazumdar, S., Redmond, C. K., Enterline, P. E., Marsh, G. M., Costantino, J. P., Zuo, S. Y., and Patwardhan, R. N. (1989). Multistage modeling of lung cancer among arsenic-exposed copper-smelter workers. *Risk Anal.* **9**, 551–563.

McCay, J. A., Sikorski, E. E., White, K. L., Page, D. G., Lysy, H. H., Musgrove, D. L., Munson, A. E. (1991). The toxicology of gallium arsenide in female B6 C3F1 mice exposed by the intratracheal route. Cited by Burns et al. (1991).

Mitchell-Heggs, C. A., Conway, M., and Cassar, J. (1990). Herbal medicine as a cause of combined lead and arsenic poisoning. *Hum. Exp. Toxicol.* **9**, 195–196.

Mizuta, N., Mizuta, M., and Ita, F. (1956). An outbreak of acute arsenic poisoning caused by arsenic contaminated soysauce (shoyu): A clinical report of 220 cases. *Bull. Yamaguchi Med. Sch.* **4**, 131–150.

Moody, J. P., and Williams, R. T. (1964a). The fate of arsanilic acid and acetylarsanilic acid in hens. *Food Cosmet. Toxicol.* **2**, 687–693.

Moody, J. P., and Williams, R. T. (1964b). The fate of 4-nitophenylarsonic acid in hens. *Food Cosmet. Toxicol.* **2**, 695–706.

Morris, J. S., Schmid, M., Newman, S., Scheuer, P. J., and Sherlock, S. (1974). Arsenic and non-cirrhotic portal hypertension. *Gastroenterology* **64**, 86–94.

Morrisey, R. E., Fowler, B. A., Harris, M. W., Moorman, M. P., Jameson, C. W., and Schwetz, B. A. (1990). Arsine: Absence of developmental toxicity in rats and mice. *Fundam. Appl. Toxicol.* **15**, 350–356.

Morton, W., Starr, G., Pohl, D., Stoner, J., Wagner, S., and Weswig, P. (1976). Skin cancer and water arsenic in Lane County, Oregon. *Cancer (Philadelphia)* **37**, 2523–2532.

Morton, W. E., and Caron, G. A. (1989). Encephalopathy: An uncommon manifestation of workplace arsenic poisoning? *Am. J. Ind. Med.* **15**, 1–5.

Mossop, R. T. (1989). On living in an arsenical atmosphere. Part 2. Clinical observations, animal experiments and ecological problems. *Cent. Afr. J. Med.* **35**, 546–551.

Murphy, M. J., Lyon, L. W., and Taylor, J. W. (1981). Subacute arsenic neuropathy: Clinical and electrophysiological observations. *J. Neurol., Neurosurg. Psychiatry* **44**, 896–900.

Musch, M. W., Chauncey, B., Schmid, E.C., Kinneh, R. K. H., and Goldstein, L. (1990). Mechanism of mercurial and arsenical inhibition of tyrosine absorption in intestine of the winter flounder, *Pseudopleuronectus americanus. Toxicol. Appl. Pharmacol.* **104**, 59–66.

Nagymajtenyi, L., Selypes, A., and Berencsi, G. (1985). Chromosomal aberrations and fetotoxic effects of atmospheric arsenic exposure in mice. *J. Appl. Toxicol.* **5**, 61–63.

Naqvi, S. M., and Flagge, C. T. (1990). Chronic effects of arsenic on American Red Crayfish, *Procambarus clarkii*, exposed to monosodium methanearsonate (MSMA) herbicide. *Bull. Environ. Contam. Toxicol.* **45**, 101–106.

National Academy of Sciences (NAS) (1977a). *Medical and Biological Effects of Environmental Pollutants: Arsenic.* NAS, Washington, DC.

National Academy of Sciences (NAS) (1977b). *Drinking Water and Health.* NAS, Washington, DC, pp. 316–344, 428–430.

National Institute for Occupational Safety and Health (NIOSH) (1979). *Arsine (Arsenic Hydroxide) Poisoning in the Workplace,* Current Intelligence Bulletin 32. NIOSH, Cincinnati, OH.

National Research Council of Canada (NRCC) (1978). *Effects of Arsenic in the Canadian Environment,* NRCC Publ. No. NRCC 15391. NRCC, Ottawa.

National Toxicology Program (NTP) (1989). *Toxicology and Carcinogenesis Studies of Roxarsone (CAS No. 121-19-7) in F344/N Rats and B6C3F1 Mice (Feed Studies),* Tech. Rep. Ser. No. 345. U.S. Department of Health and Human Services, Public Health Service, Research Triangle Park, NC.

Neiger, R. D., and Osweiler, G. D. (1989). Effect of subacute low level dietary sodium arsenite on dogs. *Fundam. Appl. Toxicol.* **13**, 439–451.

Nordberg, G. F., and Anderson, O. (1981). Metal interactions in carciongenesis: Enhancement, inhibition. *Environ. Health Perspect.* **40**, 65–81.

Nordenson, I., and Beckman, L. (1991). Is the genotoxic effect of arsenic mediated by oxygen-free radicals? *Hum. Hered.* **41**, 71–73.

Nordenson, I., Salmonsson, S., Brun, E., and Bechman, G. (1978). Occupational and environmental risks in and around a smelter in northern Sweden. II. Chromosomal aberrations in workers exposed to arsenic. *Hereditas* **88**, 47–50.

Nordenson, I., Salmonsson, S., Brun, E., and Bechman, G. (1979). Chromosome aberrations in psoriatic patients treated with arsenic. *Hum. Genet.* **48**, 1–6.

Nordstrom, S., Beckman, L., and Nordenson, I. (1978). Occupational and environmental risks in and around a smelter in northern Sweden. I. Variations in birthweight. *Hereditas* **88**, 43–46.

Nordstrom, S., Beckman, L., and Nordenson, I. (1979a). Occupational and environmental risks in and around a smelter in northern Sweden. V. Spontaneous abortion among female employees and decreased birth weight in their offspring. *Hereditas* **90**, 291–296.

Nordstrom, S., Beckman, L., and Nordenson, I. (1979b). Occupational and environmental risks in and around a smelter in northern Sweden. VI. Congenital malformations. *Hereditas* **90**, 297–302.

Nriagu, J. O. (1988). A silent epidemic of environmental metal poisoning? *Environ. Pollut.* **50**, 139–161.

Nystrom, R. R. (1984). Cytological changes occurring in the liver of coturnix quail with an acute arsenic exposure. *Chem. Toxicol.* **7**, 587–594.

Ohyama, S., Ishinishi, N., and Hisanaga, A. (1988). Comparative chronic toxicity, including tumorogenicity, of gallium arsenide ànd arsenic trioxidé intratracheally instilled into hamsters. *Appl. Organomet. Chem.* **3**, 333–337.

Okui, T., and Fujiwara Y. (1986). Inhibition of human excision DNA repair by inorganic arsenic and the co-mutagenic effect in V79 Chinese hamster cells. *Mutat. Res.* **172**, 69–76.

Oladiemji, A. A., Qudri, Su., and De Friestas, A. S. W. (1984). Long-term effects of arsenic accumulation in rainbow trout, *Salmo gairdneri. Bull. Environ. Contam. Toxicol.* **32**, 732–742.

Ott, M. G., Holder, B. B., and Gordon, H. L. (1974). Respiratory cancer and occupational exposure to arsenic. *Arch. Environ. Health* **29**, 250–255.

Palmer, J. S. (1972). Toxicity of 45 organic herbicides to cattle, sheep and chickens. *U.S. Dep. Agric., Prod. Res. Rep.* **137**.

Park, M. J., and Currier, J. (1991). Arsenic exposures in Mississippi: A review of cases. *South. Med. J.* **84**, 461–464.

Pepin, J., Milford, F., Guern, C., Moia, B., Ethier, L., and Mansinsa, D. (1989). Trial of prednisolone for prevention of melarsoprol-induced encephalopathy in Gambiense sleeping sickness. *Lancet* No. 8649, 1246–1250.

Pershagen, G. (1985). Lung cancer mortality among men living near an arsenic-emitting smelter. *Am. J. Epidemiol.* **122**, 684–694.

Pershagen, G., and Vahter, M. (1979). *Arsenic: A Toxicological and Epidemiological Appraisal*, Naturvardsverket Rapp. SNV PM 1128. Liber Tryck, Stockholm.

Peterson, D. P., and Bhattacharya, M. H. (1985). Hematological responses to arsine exposure: Quantitation of exposure responses in mice. *Fundam. Appl. Toxicol.* **5**, 499–505.

Phillips, R. (1985). Arsenic exposure: Health effects and the risk of cancer. *Rev. Environ. Health* **5**, 27–57.

Pierson, B., Wagenen, S. V., Nebesny, K. W., Fernando, Q., Scott, N., and Carter, D. E. (1989). Dissolution of crystalline calcium arsenide in aqueous solutions containing complexing agents. *Am. Ind. Hyg. Assoc. J.* **50**, 455–459.

Pinto, S. S., and McGill, C. M. (1953). Arsenic trioxide exposure in industry. *Ind. Med. Surg.* **22**, 281–287.

Pinto, S. S., Enterline, P.E., Henderson, V., and Varner, M. O. (1977). Mortality experience in relation to a measured arsenic trioxide exposure. *Environ. Health Perspect.* **19**, 127–130.

Pinto, S. S., Henderson, V., and Enterline, P. E. (1978). Mortality experience of arsenic-exposed workers. *Arch. Environ. Health* **33**, 325–331.

Piontek, M., Hengels, K. J., and Borchard, F. (1989). Noncirrhotic liver fibrosis after chronic arsenic poisoning. *Dtsch. Med. Wochenschr.* **114**, 1653–1657.

Pipkin, J. L., Anson, J. F., Hinson, W. G., Burns, E. R., Casciano, D. A., Sheehan, D. M. (1987). Cell cycle specific effects of sodium arsenite and hyperthermic exposure on incorporation of radioactive leucine and phosphate by stress proteins from mouse lymphoma cell nuclei. *Biochim. Biophys. Acta* **927**, 334–344.

Poma, K., Degraeve, N., and Suzanne, C. (1987). Cytogenic effects in mice after chronic exposure to arsenic followed by a single dose of ethylmethane sulfonate. *Cytologia* **52**, 245–249.

Prier, R. F., Nees, P. O., and Derese, P. H. (1963). The toxicity of an organic arsenical, 3-nitro-4-hydroxyphenyl arsonic acid. II. Chronic toxicity. *Toxicol. Appl. Pharmacol.* **5**, 526–542.

Prukop, J. A., and Savage, N. L. (1986). Some effects of multiple, sublethal doses of monosodium methanearsonate (MSMA) herbicide on hematology, growth, and reproduction of laboratory mice. *Bull. Environ. Contam. Toxicol.* **36**, 337–341.

Regelson, W., Kim, U., Opsina, J., and Holland, J. F. (1967). Hemangioendothelial sarcoma of liver from chronic arsenic intoxication by Fowler's solution. *Cancer (Philadelphia)* **21**, 514–522.

Reichl, F. X., Szinicz, L., Kreppel, A., and Forth, W. (1989). Effects of mitochondrial metabolism in livers of guinea pigs after a single or repeated injection of As_2O_3. *Arch. Toxicol.* **63**, 419–422.

Reichl, F. X., Szinicz, L., Kreppel, H., Fichtl, B., and Forth, W. (1991). Effect of glucose treatment on carbohydrate content in various organs in mice after acute As_2O_3 poisoning. *Vet. Hum. Toxicol.* **33**, 230–236.

Reyman, F., Moller, R., and Nielsen, A. (1978). Relationship between arsenic intake and internal malignant neoplasms. *Arch. Dermatol.* **114**, 378–381.

Rice, D. A., Kennedy, S., McMurray, C. H., Blanchflower, W. J. (1985). Experimental 3-nitro-4-hydroxyphenylarsonic acid toxicosis in pigs. *Res. Vet. Sci.* **39**, 47–51.

Robertson, I. D., Harms, W. E., and Ketterer, P. J. (1984). Accidental arsenical toxicity of cattle. *Aust. Vet. J.* **61**, 366–367.

Rogers, E. H., Chernoff, N., and Kavlock, R. J. (1981). The teratogenic potential of cacodylic acid in the rat and mouse. *Drug Chem. Toxicol.* **4**, 49–61.

Rosenberg, H. G. (1974). Systemic arterial disease and chronic arsenicism in infants. *Arch. Pathol.* **97**, 360–365.

Rosenberg, M. J., Landrigan, P. J., and Crowley, S. (1980). Low-level arsenic exposure in wood processing plants. *Am. J. Ind. Med.* **1**, 99–107.

Roses, O. E., Garcia-Fernandez, J. C., Villaamil, E. C., Camussa, N., Minetti, S. A., Martinez de Marco, M., Quiroga, P. N., Rattay, P., Sassone, A., and Valle, G. (1991). Mass poisoning by sodium arsenite. *J. Toxicol. Clin. Toxicol.* **29**, 209–213.

Rosner, M. H., and Carter, D. E. (1987). Metabolism and excretion of gallium arsenide and arsenic oxides by hamsters following intratracheal instillation. *Fundam. Toxicol.* **9**, 730–737.

Rossman, T. G. (1981). Enhancement of UV-mutagenesis by low concentrations of arsenic in *E. coli. Mutat. Res.* **91**, 207–211.

Roth, F. (1957). The sequelae of chronic arsenic poisoning in Moselle vintners. *Ger. Med. Mon.* **2**, 172–175.

Rowland, I. R., and Davies, M. J. (1981). *In vitro* metabolism of inorganic arsenic by the gastro-intestinal microflora of the rat. *J. Appl. Toxicol.* **1**, 278–283.

Sabbioni, E., Fischbach, M., Pozzi, G., Pietra, R., Gallorini, M., and Pietta, J. L. (1991). Cellular retention, toxicity and carcinogenic potential of seafood arsenic. I. Lack of toxicity and transforming activity of arsenobetaine in the BALB/3T3 cell line. *Carcinogenesis (London)* **12**, 1287–1291.

Salcedo, J. C., Portales, A., Landecho, E. X., and Diaz, R. (1984). Transverse study of a group of patients with vasculopathy from chronic arsenic poisoning in communities of the Francisco de Madero and San Pedro Districts, Coahuila, Mexico. *Rev. Fac. Med. Torreon* **12**, 16.

Samad, M. A., and Chowdhury, A. (1984). Clinical cases of arsenic poisoning in cattle. *Indian J. Vet. Med.* **4**, 107–108.

Sardana, M. K., Drummond, G. S., and Sassa, S. (1981). The potent heme oxygenase inducing action of arsenic in parasiticidal arsenicals. *Pharmacology* **23**, 247–253.

Schaumburg, H. A. (1980). *Failure to Produce Arsenic Neurotoxicity in the Rat. An Experimental Study*, Contract No. EPA 560/11-80-022, NTIS No. PB80-209505. Report to the US EPA, Office of Toxic Substances, Washington, DC by the American Public Health Association, Philadelphia.

Schrauzer, G. N., and Ishmael, D. (1974). Effects of selenium and of arsenic on the genesis of spontaneous mammary tumors in C3H mice. *Ann. Clin. Lab. Sci.* **4**, 441–447.

Schrauzer, G. N., White, D. A., and Schneider, C. J. (1976). Inhibition of the genesis of spontaneous mammary tumors in C3H mice: Effects of selenium and selenium-antagonistic elements and their possible role in human breast cancer. *Bioinorg. Chem.* **6**, 265–270.

Schroeder, H. A., and Balassa, J. J. (1967). Arsenic, germanium, tin and vanadium in mice: Effects on growth, survival and tissue levels. *J. Nutr.* **92**, 245–262.

Schroeder, H. A., and Mitchener, M. (1971). Toxic effects of trace elements on the reproduction of mice and rats. *Arch. Environ. Health* **23**, 102–106.

Schroeder, H. A., Kanisawa, M., and Frost, D. V. (1968). Germanium, tin and arsenic in rats: Effects on growth, survival, pathological lesions and life span. *J. Nutr.* **96**, 37–45.

Selby, L. A., Case, A. A., Osweiler, G. D., Hages, H. M, Jr. (1977). Epidemiology and toxicology of arsenic poisoning in domestic animals. *Environ. Health Perspect.* **19**, 183–189.

Shannon, R. L., and Strayer, D. S. (1989). Arsenic-induced skin toxicity. *Hum. Toxicol.* **8**, 99–104.

Sheabar, F. Z., and Yanni, S. (1989). *In vitro* effects of cadmium and arsenite on glutathione peroxidase, aspartate and alanine aminotranferases, cholinesterase and glucose-6-phosphate dehydrogenase activities in blood. *Vet. Hum. Toxicol.* **31**, 528–531.

Sheabar, F. Z., Yanni, S., and Taitelman, U. (1989). Efficiency of arsenic clearance from human blood *in vitro* from dogs *in vivo* by extracorporeal complexing haemodialysis. *Pharmacol. Toxicol.* **64**, 329–333.

Siewicki, T. C. (1981). Tissue retention of arsenic in rats fed with flounder or cacodylic acid. *J. Nutr.* **111**, 602–609.

Sikorski, E. E., McKay, J. A., White, K. L., Jr., Bradley, S. G., and Munson, A. E. (1989). Immunotoxicity of the semiconductor gallium arsenide in female BC3F1 mice. *Fundam. Appl. Toxicol.* **13**, 843–858.

Sikorski, E. E., Burns, L. A., McCoy, K. L., Stern, M. L., and Munson, A. E. (1991). Suppression of splenic accessory cell function in mice exposed to gallium arsenide. *Toxicol. Appl. Pharmacol.* **110**, 143–156.

Silver, A. S., and Weinman, P. L. (1952). Chronic arsenic poisoning following use of an asthma remedy. *JAMA, J. Am. Med. Assoc.* **250**, 584–585.

Simons, S. S., Jr., Sistare, F. D., and Chakraborty, P. K. (1989). Steroid binding activity is retained in a 16-kDa fragment of the steroid binding domain of rat glucocorticoid receptor. *J. Biol. Chem.* **264**, 14493–14497.

Sittig, M. (1985). *Handbook of Toxic and Hazardous Chemicals and Carcinogens*, 2nd ed. Noyes Publications, Park Ridge, NJ.

Smith, T. J., Crecelius, E. A., and Reading, J. C. (1977). Airborne arsenic exposure and excretion of methylated arsenic compounds. *Environ. Health Perspect.* **19**, 89–93.

Sobel, W., Bond, G. G., and Baldwin, C. L. (1988). An update of respiratory cancer and occupational exposure to arsenicals. *Am. J. Ind. Med.* **13**, 263–270.

Sommers, S. C., and McManus, R. G. (1953). Multiple arsenical cancers of the skin and internal organs. *Cancer (Philadelphia)* **6**, 347–359.

Southwick, J. W., Western, A. E., and Beck, M. M. (1981). *Community Health Associated with Arsenic in Drinking Water in Millard County, Utah.* EPA-600/1-81-064, NTIS No. PB82–108374. U.S. Environmental Protection Agency, Health Effects Laboratory, Cincinnati, OH.

Squibb, K. S., and Fowler, B. A. (1983). The toxicity of arsenic and its compounds. In B. A. Fowler (Ed.), *Biological and Environmental Effects of Arsenic.* Elsevier, Amsterdam, pp. 233–269.

Steinhelper, M. E., and Olson, M. S. (1988). Effects of phenylarsine oxide on agonist-induced hepatic vasoconstriction and glyco-genolysis. *Biochem. Pharmacol.* **37**, 1167–69.

Stevens, J. T., Hall, L. L., and Farmer, J. D. (1977). Disposition of ^{14}C and/or ^{74}As-cacodylic acid in rats after intravenous, intratracheal or peroral administration. *Environ. Health Perspect.* **19**, 151–157.

Stevens, J. T., Di Pasquale, L. C., and Farmer, J. D. (1979). The acute inhalation toxicology of the technical grade organoarsenical herbicides, cacodylic acid and disodium methanearsonic acid, a route comparison. *Bull. Environ. Contam. Toxicol.* **21**, 304–311.

Szuler, I. M., Williams, C. N., and Hindmarsh, J. T. (1979). Massive varicosal hemorrhage secondary to presinusoidal portal hypertension due to arsenic poisoning. *Can. Med. Assoc. J.* **120**, 168–171.

Takahashi, K., Yamauchi, H., Yamato, N., and Yamamura, Y. (1988). Methylation of arsenic trioxide in hamsters with liver damage induced by long-term administration of carbon tetrachloride. *Appl. Organomet. Chem.* **2**, 309–314.

Taketani, S., Kohno, H., Yoshinaga, T., and Tokunaga, R. (1989). The human 32-kDa stress protein induced by exposure to arsenite and cadmium ions is heme oxygenase. *FEBS Lett.* **245**, 173–176.

Tallis, G. A. (1989). Acute lead arsenate poisoning. *Aust. N. Z. J. Med.* **19**, 730–732.

Tam, G. K., Charbonneau, S. M., Bryce, F., Pomroy, C., and Sandi, E. (1979). Metabolism of inorganic arsenic (^{74}As) in humans following oral ingestion. *Toxicol. Appl. Pharmacol.* **50**, 319–322.

Tam, G. K. H., Charbonneau, S. M., Bryce, E., and Lacroix, G. (1982). Excretion of a single dose of fish-arsenic in man. *Bull. Environ. Contam. Toxicol.* **28**, 669–673.

Tay, C.-H., and Seah, C.-S. (1975). Arsenic poisoning from anti-asthmatic herbal preparations. *Med. J. Aust.* **2**, 424–428.

Taylor, P. R., Qiao, Y. L., Schatzkin, A., Yao, S. X., Lubin, J., Mao, B. L., Rao, J. Y., McAdams, M., and Xuan, X. Z. (1989). Relation of arsenic exposure to lung cancer among tin miners in Yunnan Province, China. *Br. J. Ind. Med.* **46**, 881–886.

Tseng, W. P. (1977). Effects of dose-response relationships of skin cancer and blackfoot disease with arsenic. *Environ. Health Perspect.* **19**, 109–119.

Tseng, W. P., (1989). Blackfoot disease in Taiwan: A 30-year follow-up study. *Angiology* **40**, 547–558.

Tseng, W. P., Chu, H. M., How, S. W., Fong, J. M., Lin, C. S., and Yen, S. (1968). Prevalence of skin cancer in an endemic area of chronic arsenicism in Taiwan. *J. Nat. Cancer Inst.* (*U.S.*) **40**, 453–463.

Tsuda, T., Nagira, T., Yamamoto, M., and Kume, Y. (1990). An epidemiological study on cancer in certified arsenic poisoning patients in Toruku. *Ind. Health* **28**, 53–62.

Ulthus, E. O. (1990). Effects of arsenic deprivation in hamsters. *Magnesium Trace Elem.* **9**, 227–232.

U.S. Bureau of Mines (1990). *Mineral Commodity Summaries.* U.S. Bureau of Mines, Washington, DC, pp. 22–23.

U.S. Department of Health and Human Services (USDHHS) (1992). *Arsenic. Toxicological Profile.* USDHHS, Public Health Service, Washington, DC, p. 175.

Vahter, M. (1981). Biotransformation of trivalent and pentavalent inorganic arsenic in mice and rats. *Environ. Res.* **25**, 286–293.

Vahter, M. (1986). Environmental and occupational exposure to inorganic arsenic. *Acta Pharmacol. Toxicol.* **59**, 31–34.

Vahter, M., and Envall, J. (1983). *In vivo* reduction of arsenate in mice and rabbits. *Environ. Res.* **32**, 14–24.

Vahter, M., and Marafante, E. (1983). Intracellular interaction and metabolic fate of arsenite and arsenate in mice and rabbits. *Chem. Biol. Interact.* **47**, 29–44.

Vahter, M., and Marafante, E. (1985). Reduction and binding of arsenate in marmoset monkeys. *Arch. Toxicol.* **57**, 119–124.

Vahter, M., and Marafante, E. (1989). Intracellular distribution and chemical forms of arsenic in rabbits exposed to arsenate. *Biol. Trace Elem. Res.* **21**, 233–239.

Vahter, M., Marafante, E., Lindgren, A., and Dencker, L. (1982). Tissue distribution and retention of ^{74}As-dimethylarsinic acid in mice and rats. *Arch. Environ. Contam. Toxicol.* **13**, 259–264.

Valentine, J. L., Reisbord, L. S., Kang, H. K., and Schluchter, M. D. (1985). Arsenic effects of population health histories. In C. F. Mills and K. J. Bremner Im Chesters (Eds.), *Trace Elements in Man and Animals.* Commonwealth Agricultural Bureau, Slough, UK, pp. 289–294.

Veeken, H. J. G. M., Ebeling, M. C. A., and Dolmans, W. M. V. (1989). Trypanosomiasis in a rural hospital in Tanzania. *Trop. Geogr. Med.* **41**, 113–117.

Viallet, A. Guillaume, I., Cote, J., Legare, A., and Lavoie, P. (1972). Presinusoidal hypertension following chronic arsenic intoxication. *Gastroenterology* **62**, 177.

Vig, B. K., Figueroa, M. L., Conforth, M. N., and Jenkins, S. H. (1984). Chromosome studies in human subjects chronically exposed to arsenic in drinking water. *Am. J. Ind. Med.* **6**, 325–338.

Wagner, S. L. Maliner, J. S., Morton, W. E., and Braman, R. S. (1979). Skin cancer and arsenical intoxication from well water. *Arch. Dermatol.* **115**, 1205–1207.

Wall, S. (1980). Survival and mortality pattern among Swedish smelter workers. *Int. J. Epidemiol.* **9**, 73–87.

Wang, C., Hsieh, C. H., and Wu, W. G. (1991). Phenylarsine oxide inhibits insulin-dependent glucose transport activity in rat soleus muscles. *Biochem. Biophys. Res. Commun.* **176**, 201–206.

Wang, S., Huang, J., and Chen, B. (1990). Comparative studies on acute hepatotoxicity of arsenic trioxide, phosphorus and carbon tetrachloride. *Weisheng Dulixue Zazhi* **4**, 4–6.

Webb, D. R., Wilson, S. E., and Carter, D. E. (1986). Comparative pulmonary toxicity of gallium arsenide, gallium (III) oxide, or arsenic (III) oxide intratracheally instilled into rats. *Toxicol. Appl. Pharmacol.* **82**, 405–416

Webb, D. R., Wilson, S. E., and Carter, D. E. (1987). Pulmonary clearances and toxicity of respirable gallium arsenide particulates intratracheally instilled into rats. *Am. Ind. Hyg. Assoc. J.* **48**, 660–667.

Welch, K., Higgins, I., Oh, M., and Burchfield, C. (1982). Arsenic exposure, smoking and respiratory cancer in copper smelter workers. *Arch. Environ. Health* **37**, 325–335.

Willhite, C. C. (1981). Arsenic-induced axial skeletal (dysraphic) disorders. *Exp. Mol. Pathol.* **34**, 145–158.

Woods, E. A., and Southern, M. R. (1989). Studies on the etiology of trace metal-induced porphyria: Effects of porphyrinogenic metals on coproporphyrinogen oxidase in rat liver and kidney. *Toxicol. Appl. Pharmacol.* **97**, 183–190.

Woods, J. S., and Fowler, B. A. (1978). Altered regulation of mammalian hepatic heme biosynthesis and urinary porphorin excretion during prolonged exposure to sodium arsenate. *Toxicol. Appl. Pharmacol.* **97**, 183–190.

Woolson, E. A. (1975). Arsenical pesticides. *ACS Symp. Ser.* **7**, 176.

World Health Organization (WHO) (1981). *Environmental Health Criteria 18: Arsenic.* WHO, Geneva.

Wu, M. M., Kuo, T. L., Hwang, Y. H., and Chen, C. J. (1989). Dose-dependent relation between arsenic concentration in well water and mortality from cancers and vascular diseases. *Am. J. Epidemiol.* **130**, 1123–1132.

Yamanaka, K., Hasegawa, A., Sawamura, R., and Okada, S. (1989a). Dimethylated arsenics induce DNA strand breaks in lung via the production of active oxygen in mice. *Biochem. Biophys. Res. Commun.* **165**, 43–50.

Yamanaka, K., Ohba, H., Hasegawa, A., Sawamura, R., and Okada, S. (1989b). Mutagenicity of dimethylated metabolites of inorganic arsenics. *Chem. Pharm. Bull.* **37**, 2753–2756.

Yamanaka, K., Hoshino, M., Okamota, M., Sawamura, R., Hasehawa, A., and Okada, S. (1990). Induction of DNA damage by dimethylarsine, a metabolite of inorganic arsenics, is for the major part likely due to its peroxy radical. *Biochem. Biophys. Res. Commun.* **168**, 58–64.

Yamauchi, H., and Yamamura, Y. (1984). Metabolism of excretion of orally administered dimethylarsinic acid in the hamster. *Toxicol. Appl. Pharmacol.* **74**, 134–140.

Yamauchi, H., Kaise, T., and Yamamura, Y. (1986). Metabolism and excretion of orally administered arsenobetaine in the hamster. *Bull. Environ. Contam. Toxicol.* **36**, 350–355.

Yamauchi, H., Yamamoto, N., and Yamamura, Y. (1988). Metabolism and excretion of orally and intraperitoneally administered methylarsonic acid in the hamster. *Bull. Environ. Contam. Toxicol.* **40**, 280–286.

Yamauchi, H., Kaise, T., Takahashi, K., and Yamamura, Y. (1990). Toxicity and metabolism of trimethylarsine in mice and hasmsters. *Fundam. Appl. Toxicol.* **14**, 399–404.

Yih, L. H., Huang, H. M., Jan, K., and Lee T. C. (1991). Sodium arsenite induced ATP depletion and mitochondrial damage in HeLa cells. *Cell Biol. Int. Rep.* **15**, 253–264.

Zaldivar, R. (1974). Arsenic contamination of drinking water and foodstuffs causing endemic chronic arsenic poisoning. *Beitr. Pathol.* **151**, 384–400.

Zaldivar, R. (1977). Ecological investigations on arsenic dietary intake and endemic chronic poisoning in man; Dose-dependent curve. *Zentralbl Bakteriol., Parasitenkd. Infektionskr. Hyg., Abt. I: Orig., Reihe B* **164**, 481–484.

Zaldivar, R., Prunes, L., and Ghai, G. (1981). Arsenic dose in patients with cutaneous carcinomata and hepatic hemangio-endothelioma after environmental and occupational exposure. *Arch. Toxicol.* **47**, 145–154.

Zhuang, G. S., Wang, Y. S., Tan, M. G., Zhi, M., Pan, W. Q., and Cheng, Y. D. (1990). Preliminary study of the distribution of the toxic elements As, Cd, and Hg in human hair and tissues by RNA. *Biol. Trace Elem. Res.* **26–27**, 729–736.

Zierler, S., Theodore, M., and Cohen, A. (1988). Chemical quality maternal drinking water and congenital heart disease. *Int. J. Epidemiol.* **17**, 589–594.

5

CHRONIC ARSENIC POISONING IN HUMANS: THE CASE OF MEXICO

M. E. Cebrián, A. Albores, G. García-Vargas, and L. M. Del Razo

Sección de Toxicología Ambiental, CINVESTAV-IPN, México, D.F., 07000 Mexico

Patricia Ostrosky-Wegman

Instituto de Investigaciones Biomedicas UNAM, México, D.F., 04510 Mexico

Arsenic in the Environment, Part II: Human Health and Ecosystem Effects,
Edited by Jerome O. Nriagu.
ISBN 0-471-30436-0

1. INTRODUCTION

Arsenic contamination of well water in Taiwan (Tseng et al., 1968), Minnesota, United States (Feinglass, 1973), and Canada (Grantham and Jones, 1977) and of public water supplies in Argentina (Astolfi, 1971) and Chile (Borgoño and Greiber, 1971; Zaldivar, 1974) has resulted in signs and symptoms of arsenic poisoning. Classically, these include cutaneous manifestations (skin pigmentation changes, keratosis, and skin cancer) and peripheral vascular disease (blackfoot disease and Raynaud's syndrome). Recent studies in Taiwan have reported that in areas where blackfoot disease is endemic, the standarized mortality ratio and the cumulative mortality rate were significantly higher for cancer of the bladder, kidney, skin, lung, liver, and colon (Chen et al., 1985). Other, less well known, manifestations of chronic arsenic poisoning have also been reported, such as systemic arterial disease resulting in myocardial infarction (Rosenberg, 1973; Moran et al., 1977), changes in electromyographic patterns (Hindmarsh et al., 1977) and moderate effects in the respiratory system (Borgoño et al., 1977). The mutagenic and teratogenic effects of arsenic exposure in human beings are still the subject of debate. Other studies have not found increases in the prevalence of cutaneous signs in places regarded as having drinking water with high arsenic concentrations, such as California (Goldsmith, 1972), Oregon (Morton et al., 1976), Alaska (Harrington, 1978), and Utah (Southwick et al., 1983) in the United States. However, in these studies, the number of people examined was small and the time of exposure was shorter than in the Taiwanese study, possibly explaining the absence of positive data. It is noteworthy that in the studies that have reported positive data, arsenic concentrations in drinking water were above 0.200 mg/L. It remains to be established if chronic exposure to concentrations near the limit (0.050 mg/L) set by the World Health Organization (WHO) increases the prevalence of neoplastic lesions.

On the other hand, above-average levels of exposure are associated with the smelting of copper and zinc, which often releases inorganic arsenic and cadmium into the air (Baker et al., 1977; Landrigan and Baker, 1981; Hartwell et al., 1983). Thus, workers and nearby residents are exposed to high levels, both through the air and as a result of atmospheric deposition in soil and water. Elevated levels of arsenic and other elements in soil are of particular concern for small children, who may swallow small amounts of soil while playing.

In Mexico, arsenic exposure is mainly by ingestion of drinking water naturally contaminated with arsenic or by inhalation and ingestion of contaminated soil. These are the subjects of this review.

2. ARSENIC EXPOSURE VIA DRINKING WATER: THE CASE OF THE REGION LAGUNERA

Chronic arsenic exposure via drinking water has been reported in six areas in Mexico. In Chihuahua, Puebla, Nuevo León, and Hidalgo there is no available information with which to assess the magnitude of this problem. In Morelos, the Ministry of Health has reported that the inhabitants of five villages in the county of Quitamula have been drinking water containing arsenic in concentrations ranging from 0.12 to 0.20 mg/L (Secretaría de Salubridad y Asistencia, 1991). In some parts of the Region Lagunera, located in the central part of northern Mexico, chronic arsenic poisoning is endemic (Chávez et al., 1964; Bracho, 1971; Sánchez de la Fuente et al., 1976), and acute outbreaks have affected both human beings (Ortiz-Mariote et al., 1963; Cantellano et al., 1964) and animals (Torres de Navarro, 1976). This region has a population of about two million people, of which 70% live in urban areas and the rest in rural areas. It is an important cotton-producing area, and other main economic activities are related to dairy products and industrial metals. The climate is very dry, with an average temperature of 25 °C in summer and 16 °C in winter; for five months a year, the temperature reaches more than 35 °C for at least 4 hr a day, a factor to remember when estimating the total ingested dose. Most of this region is reputed to have a substratum rich in arsenic, and it is known to produce high arsenic levels in well water. The use of organoarsenical pesticides before 1945 has been mentioned as an other possible source of contamination [Subsecretaría del Mejoramiento del Ambiente/Secretaría de Salubridad y Asistencia (SMA/SSA), 1977].

Albores et al. (1979) and Cebrián et al. (1983) reported on the prevalence of several signs and symptoms of chronic arsenic poisoning in two rural populations. The main objective was to estimate the effects on health and the risks associated with ingestion of arsenic-contaminated water over a long period. The arsenic concentration was 0.410 mg/L in the drinking water of the exposed population and 0.007 mg/L in the control population. In the control population, the prevalence of skin pigmentation changes was 2.2% (7/318). In the exposed population, 21.6% (64/296) of the sample showed at least one of the cutaneous signs of chronic arsenic poisoning, including 5.1% with Bowen's disease, the malignancy of which is still the subject of debate, and 1.4% with lesions clinically diagnosed as skin cancer. It was found that the proportion of individuals (per age group) affected with cutaneous lesions increased with age until the age of 50. The shortest time of exposure after which lesions were detected was 8 years for hypopigmentation, 12 years for hyperpigmentation and palmoplantar keratosis, 25 years for papular keratosis, and 38 years for ulcerative lesions. The authors did not find the high prevalence and severity of skin lesions in children that were reported in Antofagasta, since people under 20 years of age accounted for only 9.6% of the individuals with skin lesions, whereas Borgoño and Greiber (1971) and Zaldivar (1974) reported 78.8% in that age group. The relative risk of suffering a particular manifestation of poisoning ranged from 1.9 to 36 times higher in the exposed population. Apart from skin lesions, other symptoms were noted that were not specific (nausea, epigastric and colic abdominal pain, etc.); these were more prevalent in the exposed population and occurred more frequently in those individuals with the classic skin manifestations. The relative risk of suffering nonspecific symptoms ranged from 1.9 to 4.8%. Peripheral vascular alterations were also observed in the exposed population, including a 0.7% prevalence of blackfoot disease leading to amputation. A 4% prevalence of peripheral vascular alterations in several stages of progress was also found, but no alterations were found in children (Cebrián, 1987). Garcia-Salcedo et al. (1984) studied five villages with well-water arsenic concentrations ranging from 0.27 to 0.51 mg/L and reported a global prevalence of 0.75% (1.18% in men and 0.27% in women).

Del Razo et al. (1990) reported on the magnitude of the arsenic contamination problem in the Region Lagunera in terms of geographical extent and number of people affected. They also reported on the oxidation state of the arsenic compounds present in well water. One hundred and twenty-eight water wells were sampled in 11 counties in the states of Durango and Coahuila, which make up the Region Lagunera. The range of total arsenic concentrations was from 0.008 to 0.624 mg/L. Fifty percent of the samples (64) had arsenic levels greater than 0.05 mg/L, the current WHO drinking-water standard [World Health Organization (WHO), 1984]. These results indicate that a high proportion of water wells in the Region Lagunera are contaminated with arsenic and that those with higher arsenic concentrations are located in the northeastern part of the region, which is mostly rural. The data suggest that arsenic concentrations have been steadily increasing during the last 10 years. Conservative estimates suggest that 400,000 inhabitants of rural areas were exposed to arsenic via drinking water in concentrations higher than 0.05 mg/L. In this study, no evidence was found that water supplies in the main urban areas of the Region Lagunera, that is, the cities of Torreón in the state of Coahuila and Gómez Palacio and Lerdo in the state of Durango (which have a combined population of one million), are contaminated with arsenic.

In all samples, arsenic was present mainly in its inorganic form. In 93% of the samples, As(V) was the predominant species and As(III) was predominant in the rest. The range of As(V) concentrations was 0.004 to 0.604 mg/L, while that of As(III) was from trace amounts to 0.217 mg/L. The As(III) : As(V) ratio also had a wide range, a finding consistent with other studies. Organic arsenicals were generally present in very small amounts, since none of the samples had MMA (monomethylarsonate) concentrations greater than 0.003 mg/L. DMA (dimethylarsinate) concentrations were in the range from trace amounts to 0.020 mg/L.

The authors concluded that although the samples were from well water, they might not truly represent the oxidation state of arsenic within the aquifer or at the moment of ingestion, since samples could not be protected from changes occurring during the extraction and transportation processes. They also stated that if little information exists on the factors determining the oxidation state of arsenic in individual aquifers, much less is known on how to account for exposure to the different oxidation states of arsenic when assessing the health risks of arsenic exposure. From the toxicological point of view, As(III) is between 2.6 and 59 times more potent than As(V), depending on the test system used to evaluate toxicity. Unfortunately, no information is available on the differences in their respective ability to produce skin cancer, internal malignancies, or blackfoot disease, the main health problems associated with arsenic exposure via drinking water.

3. BIOCHEMICAL EFFECTS

The study of the human health effects of chronic arsenic ingestion via drinking water has mostly been focused on dermatological and peripheral vascular alterations. However, arsenic is well known for its reactivity with the sulfhydryl groups of proteins (Webb, 1966); thus, other metabolic effects are likely. On this assumption, we started a series of pilot studies aimed at identifying the biochemical markers of damage, which could help in the early diagnosis and prevention of diseases related to environmental contamination.

3.1. Urinary Excretion of Porphyrins

A pilot study was conducted by García-Vargas et al. (1991) to investigate whether the porphyrinuria produced by arsenic in rodents was present in humans chronically exposed to arsenic via drinking water. Twenty-one individuals were selected from Santa Ana, Coahuila, a town with drinking-water arsenic concentrations of 0.39 mg/L. The nonexposed group (19 individuals) was chosen from Luján, Durango, where the concentration was 0.012 mg/L. These concentrations have been relatively constant since 1980. The individuals selected had lived in their respective towns for at least 15 years. The parameters studied were the urinary concentrations of uroporphyrin, coproporphyrin, and total arsenic.

There were no significant differences in the urinary excretion of total porphyrins between exposed individuals and those in the control group. A significant decrease in coproporphyrin excretion was observed in the exposed group. The median uroporphyrin concentration in urine was 3.36 μg/mg creatinine in the control group, as compared to 6.81 in the exposed group; however, this change was not significant ($0.1 > P > 0.05$). As a consequence of these effects, a significant decrease in the coproporphyrin:uroporphyrin (COPRO:URO) ratio was observed in the exposed group. A significantly higher proportion of individuals in the exposed group had a COPRO:URO ratio below 1 (76%, or 16/21), as compared to the control group (32%, or 6/19). Furthermore, 43% of the exposed group had COPRO:URO ratios below 0.5. The median value of total arsenic in the urine of exposed individuals was 1.35 mg/L, as compared to 0.06 mg/L in the control group. No linear relationships were found between porphyrin-related variables and arsenic excretion in the exposed individuals.

Although chronic exposure to these arsenic levels did not result in porphyria, there were indications of an effect on the heme synthesis pathway. Attempting to explain their findings, the authors agreed with Doss (1979) that the COPRO:

URO ratio is an important element in the diagnosis of chronic hepatic porphyria and that hepatic porphyrin metabolism disturbances unfold progressively, beginning with inversion of the COPRO:URO ratio and progressing to chronic hepatic porphyria. If this were so, a progression of biochemical alterations in the affected individuals could be expected as exposure to arsenic continued. However, the possibility of an adaptive response, which would preclude progression of biochemical lesions beyond a certain point, was not ruled out by the authors, since an adaptive response to the effects of arsenic on heme metabolism, similar to that observed in rodents subchronically exposed (Cebrián et al., 1988), could be present in humans. The authors listed possible explanations for the absence of demonstrable correlations between porphyrin excretion, the COPRO:URO ratio, and total arsenic concentration in urine, including (a) the small sample size of the study; (b) the wide interindividual differences in porphyrin excretion; (c) changes in arsenic contaminated water intake, rapidly reflected in urinary arsenic output but not in the intracellular concentrations or heme metabolism effects; and (d) interindividual differences in arsenic metabolism, which could contribute to the adaptive mechanisms discussed earlier.

A study investigating the urinary excretion pattern of arsenic metabolites in humans chronically exposed to arsenic in drinking water (0.40 mg/L) reported that DMA was the main metabolite excreted by the control and exposed groups. However, the proportion of MMA excreted in urine was significantly higher in the exposed group and was accompanied by a small but significant decrease in the proportion of DMA, which resulted in significant differences in the MMA:DMA ratio. The authors concluded that individuals chronically exposed to high concentrations of inorganic arsenic via drinking water appear to have a decreased ability to further methylate MMA, and that further studies are needed to assess the value of MMA/DMA and inorganic arsenic/MMA as indicators of arsenic metabolism alterations; these, in turn, could be used to identify individuals at high risk of suffering toxic effects (Del Razo et al., 1992).

3.2. Genetic Markers

Rosales (1987) reported that chronic exposure to 0.5 mg As/L in drinking water did not result in an increased frequency of sister-chromatid exchanges (SCEs). The frequency of chromosomal aberrations was not significantly different in exposed children compared to control children, but it was significantly higher in youngsters and doubtful results were obtained in adults. Gómez-Arroyo et al. (1988) reported on the induction of SCEs in *Vicia faba* by arsenic-contaminated drinking water samples from several towns in the Region Lagunera. Arsenic concentrations were from 0.11 to 0.695 ppm. In all cases, SCE frequencies were significantly different from control values. Concentrations of sodium arsenite and arsenate (0.2–1.0 ppm) were also tested and, except for 0.2 ppm of arsenate, produced significant SCE frequencies.

In the search for biological markers with which to detect genetic damage, a pilot study on a chronically exposed population was conducted by Ostrosky-

Wegman et al. (1991). The exposed group consisted of 11 individuals from Santa Ana, Coahuila, where the drinking water contained 0.390 mg As/L, 98% in the pentavalent and the rest in the trivalent form. The 11 control individuals were chosen from Nuevo León, Coahuila, where arsenic levels ranged from 0.019 to 0.026 mg/L. Mean arsenic levels in urine samples from the control group were 0.12 μg/mL, as compared to 1.57 μg/mL in the exposed subjects. In both, 70 to 80% of the arsenic in urine was in its organic form (MMA and DMA). No significant differences on the induction of SCEs were found in this study.

In contrast, induction of SCEs was reported in 13 individuals chronically exposed to arsenic in water in Taiwan (Wen et al., 1981). In this study, no significant differences were found in the frequency of chromosomal aberrations between the two groups studied. However, the frequency of complex chromosomal aberrations (dicentrics, rings, and translocations) found in the exposed population was 0.73%, while the frequency in the control group and the laboratory controls was 0.16% and 0.30%, respectively. However, these differences were not significant. On average, the frequency of lymphocytes resistant to thioguanine [HGPRT (hypoxantine guaninephosphoribosyltransferase) locus assay] in the exposed subjects was twice as high as in the control individuals. However, this increase was not significant. These results agreed with those from in vitro studies, which had also failed to show the mutagenicity of arsenic [Environmental Protection Agency (EPA), 1984]. Results for genetic damage induced by urine samples in the *Bacillus subtilis rec* assay were negative for all samples. The main finding of this study was related to lymphocyte proliferation kinetics, which was analyzed at 48 and 72 hr of culture. A higher percentage of first divisions was found in cultures of the exposed subjects, who also displayed a low proportion of third divisions. This difference was significant at 72 hr. The average generation time was approximately 19 hr in laboratory controls and in the control group, while in the exposed group the time was approximately 28 hr. These results are in agreement with earlier reports showing that sodium arsenite and arsenate inhibit the in vitro proliferation of human and bovine lymphocytes. Therefore, the authors suggested that arsenic impairs the cellular immune response. Interestingly, Walder et al. (1971) reported that immunosuppressed patients developed skin lesions and skin cancers similar to those observed in individuals chronically exposed to arsenic.

The authors concluded that arsenic has been related to some malignant diseases in humans, but it seems unlikely that the mechanism is the direct induction of damage to DNA. Arsenic could either act co-carcinogenically with a variety of agents or it could interfere with the immunological ability of individuals.

4. EXPOSURE TO OTHER CONTAMINANTS

4.1. Pesticides

Mexico is one of the most important producers of DDT [1, 1, 1-trichloro-2, 2-bis(*p*-chlorophenyl) ethane], toxaphene, and BHC (benzene hexachloride) in

Latin America. DDT and other pesticides have been widely used in the Region Lagunera since 1948. Analyses of persistent pesticide residues is foods and feeds have shown that the levels were the highest in Mexico, exceeding the practical limits recommended by FAO/WHO (Albert and Saval, 1977; Albert and Reyes, 1978). Therefore, Albert et al. (1980) conducted a study to determine organochlorine pesticide residues in adipose tissue taken from autopsies in Torreón, Mexico City, and Puebla. The main difference was the significantly high concentration of DDE [1,1-dichloro-2,2-bis (chlorophenyl) ethylene] in samples from the Region Lagunera. The earlier use of large quantities of DDT in this region was also reflected in the values obtained for the ratio of DDE (as DDT) to the total equivalent DDT. The values reported in the literature range from 0.43 to 0.70, the low values belonging to countries where DDT was used intensively at the time of the study. In this study, the value was 0.95. The mean concentration of DDE (18.36 μg/g, lipid basis) in the samples from the Region Lagunera was among the highest previously reported for a general population. Acute intoxications with organophosphorous pesticides, both occupational and accidental, are endemic in this area (Reyes-Najera and Sánchez de la Fuente, 1975).

4.2. Fluoride

Del Razo et al. (1993) reported on the concentrations of and geographical relationships between fluoride and arsenic in well water from the Region Lagunera. One hundred and twenty-nine water wells were sampled in 11 counties in the states of Durango and Coahuila, which make up the region. Fluoride concentrations in well water ranged from less than 0.5 to 3.7 mg/L. Twenty-five samples (19.4%) had fluoride levels above 1.5 mg/L, the current maximum limit for drinking water in Mexico. The "optimum" fluoride concentration calculated for this area was 0.77 mg/L, and higher concentrations were found in about 44% (52/119) of the samples. Thus, a considerable proportion of water wells in this region were contaminated with fluoride, and those with higher concentrations were located in the northeastern part of the region, which is mostly rural. However, no information is available on the prevalence of diseases related to fluorosis. The range of total arsenic concentrations was 0.008 to 0.624 mg/L, and 50% (64) had arsenic levels above 0.050 mg/L, the current WHO drinking-water standard. [The implications of this finding were discussed in an earlier paper by Del Razo et al. (1990).] A linear regression analysis of total arsenic and fluoride concentrations showed a highly positive correlation ($r = 0.774$), whereas a lower correlation ($r = 0.380$) was found with trivalent arsenic. These data are consistent with the geographical distribution of the concentrations. The highest concentrations of arsenic and fluoride in well water were located in the rural, northeastern part of the region, whereas the lowest concentrations were found in the southwestern part and in the cities of Torreón in Coahuila and Gómez Palacio and Lerdo in Durango. In consequence, people exposed to high arsenic concentrations were also exposed to fluoride at levels above the drinking-water standard. The authors mentioned the possibility of toxicologic interactions between fluorides and arsenic, since both elements affect enzyme activity in the glycolytic, tricarboxylic, and heme metabolism pathways. The authors concluded their paper with a question: does fluoride play a role in the prevalence of signs and symptoms that, so far, have been attributed solely to arsenic?

5. CURRENT DRINKING WATER SUPPLIES IN THE REGION LAGUNERA

Before 1960, cities and villages obtained their water from local wells. However, as wells supplying villages in the north of the region were known to be contaminated with arsenic, water distribution systems had to be built and five new wells were drilled in noncontaminated areas. In the mid-1970s, these wells, in turn, were reported to be contaminated. In June 1988, a new source for the water distribution system serving the rural part of the Region Lagunera was inaugurated. The source for the system were 17 water wells drilled in the margins of the Rio Nazas. According to information given by the Mexican Ministry of Agriculture, it provided potable water to 104 villages (280, 300 inhabitants) in the counties of Francisco I. Madero (70,690), Finisterre (19,637), and San Pedro (150,700) in the state of Coahuila, and in the country of Tlahualilo (39,273) in the state of Durango (Excelsior, 1988), apparently solving the problem of arsenic contamination for 70% of the villages. The range of arsenic concentrations in those 17 wells was from 0.010 to 0.080 mg/L, and only one well had arsenic levels higher than the WHO limit. The new wells started operations gradually, and the old arsenic-contaminated wells continued in operation in order to provide sufficient water to the villages. Thus, during one year the villages received a mixture of waters containing variable arsenic concentrations, and their inhabitants drank water with arsenic concentrations higher than 0.05 mg/L but lower than before. By the end of 1989, arsenic levels in samples taken from storage tanks along the distribution system were below 0.020 mg/L.

6. ARSENIC EXPOSURE VIA SOIL INGESTION AND INHALATION: THE CASE OF SAN LUIS POTOSÍ

In San Luis Potosí City, located in the central part of Mexico, a smelting complex consisting of a copper smelter and a zinc electrolytic refinery is located within an urban zone. The metropolitan zone has 600,000 inhabitants and its main economic activities are related to metallurgy and the smelting industry. The annual capacity of the complex is about 90,000 tons of copper, 16,800 tons of lead by-products, and 8500 tons of arsenic trioxide. The zinc refinery and its cadmium smelting plant have an annual capacity of 90,000 tons of zinc, 600 tons of cadmium, and 140,000 tons of sulfuric acid (IMMSA, 1990). Thus, Diaz-Barriga et al. (1993) conducted a study to assess environmental contamination by arsenic and cadmium and its possible contribution to an increased body burden of these elements in children.

The exposed population was in Morales, an urban area within 1.5 km of the smelter complex; the urban control population was in Graciano, located 7 km from the complex and against prevailing winds; Mexquitic was the rural control, a small town located against prevailing winds 25 km away. Arsenic and cadmium were determined in the environment, as well as in the urine and hair of children. Seventy-five children from Morales, 35 from Graciano, and 25 from Mexquitic were studied.

6.1. Environmental Monitoring

The average concentration of arsenic in the air for a six-month period in Morales (0.46 $\mu g/m^3$) was twice the value found in Graciano and 10 times that found in Mexquitic. Arsenic concentrations in soil samples from Morales were extremely high (117–1396 ppm) compared to those found in Graciano or Mexquitic. Arsenic concentrations in dust collected from window sills in homes in Morales were also extremely high (514–2625 ppm) compared to those found in Graciano or Mexquitic. Ambient-air cadmium concentrations in Morales were not significantly different from those found in Graciano. Cadmium concentrations in the soil from Morales (17–28 ppm) were well above those found in Graciano or Mexquitic. Cadmium concentrations in dust collected from window sills in Morales were extremely high (28–214 ppm) compared to those found in Graciano or Mexquitic. Mean arsenic and cadmium levels in soil in Morales were higher in the area within 600 m of the smelter complex (1503 ± 31 ppm and 28.0 ± 0.3 ppm, respectively) than in the area between 600 and 1200 m of the complex (832 ± 27 ppm and 16.9 ± 0.3 ppm, respectively). None of the drinking-water samples had arsenic concentrations above the limit set by WHO. A low and high arsenic exposure scenario of 1.1 $\mu g/kg/day$ and 19.8 $\mu g/kg/day$, respectively, was developed for children living in Morales. Both estimates were above the reference dose of 1 $\mu g/kg/day$ calculated by the EPA (1988). Estimates of cadmium exposure were below the reference dose for this element. Soil ingestion was the major route of total ingestion.

6.2. Biological Monitoring

Mean arsenic levels in urine samples from Morales (232.3 $\mu g/L$) were twice as high as those from the control populations. Furthermore, the proportion of children having arsenic concentrations in the urine higher than 100 $\mu g/L$, considered to be the upper end of the normal environmental exposure range [Agency for Toxic Substances and Disease Registry (ATSDR), 1988], was 62/75 (82.6%), as compared to 6/33 (18%) in Graciano and 3/25 (12%) in Mexquitic. Average arsenic levels in hair samples from Morales (9.9 $\mu g/g$) were 12 and 21 times as high as those from the urban and control populations, respectively. All children had hair arsenic concentrations higher than 1 ppm, considered the upper end of the environmental exposure range (ATSDR, 1988), as compared to 11/28 (39%) in Graciano and 3/24 (12%) in Mexquitic. A direct relationship was found between arsenic concentrations in urine and those in hair samples from Morales. Cadmium levels in urine samples from Morales (1.04 μg/g creatinine) were not significantly different from those found in the control groups. Although mean cadmium levels in hair samples from Morales (0.91 μg/g) were twice as high as those from the control populations, they were within the range reported as normal. These data indicate that children living in Morales have absorbed considerable quantities of arsenic, reflecting the high level of environmental exposure, and suggest that if exposure continues, these children could suffer adverse health effects, since arsenic concentrations around 200 μg/g creatinine in urine (the median value in Morales) have been related to increases in the standard mortality ratio for respiratory cancer in smelter workers (Enterline et al., 1987). Similarly, individuals having concentrations in urine ranging from 392 to 2270 μg As/g creatinine after exposure to arsenic via drinking water have been reported to suffer from skin alterations and disturbances in heme metabolism (García-Vargas et al., 1991). At the time of this study, children in Morales spent most of their time playing on unpaved streets, increasing the risk of exposure. After the study was finished, the smelter company implemented dust-control methods and measures to reduce soil pollution. The authors concluded that further studies were needed to identify pathologies that might already be present in the exposed population, since the literature suggests that the cardiovascular and nervous systems are early targets for arsenic toxicity.

7. CONCLUSIONS

This paper reviews arsenic exposure in Mexico, where exposure is mainly by ingestion of drinking water naturally contaminated with arsenic or by inhalation and ingestion of contaminated soil around a smelter complex. Chronic arsenic exposure via drinking water has been reported in five areas of Mexico, namely Morelos, Chihuahua, Puebla, Nuevo León and Hidalgo, where no information is available to assess the magnitude of the problem. However, in some parts of the Region Lagunera, chronic arsenic poisoning has been known since 1958. A high proportion (50%) of water wells were contaminated with arsenic, of which those with higher concentrations were located in the rural northeastern part of the region. Most of the arsenic was in its inorganic form and pentavalent arsenic was the predominant species in 93% of the samples. However, variable percentages (20–50%) of trivalent arsenic were found in 36% of the samples. Conservative estimates suggest that 400,000 inhabitants of rural areas were exposed to arsenic via drinking water in concentrations higher than 0.05 mg As/L. The prevalence of the signs and symptoms of chronic arsenic poisoning was compared in two rural populations, respectively having 0.410 mg As/L and 0.005 mg As/L in the drinking water. In the control population, the prevalence of skin pigmentation changes was 2.2% (7/318). In the exposed population, 21.6% (64/296) showed at least one of the cutaneous signs of chronic arsenic poisoning, including 5.1% with Bowen's disease and 1.4% with lesions clinically diagnosed as skin cancer.

In contrast to animals, chronic human exposure to arsenic levels of 0.40 mg/L did not appear to result in increased urinary porphyrin excretion. However, there were indications of an altered urinary porphyrin excretion pattern, reflecting an effect on the heme synthesis pathway. In another study, the proportion of MMA excreted in urine by individuals chronically exposed to high concentrations of arsenic in drinking water was significantly higher compared to a control group. This was accompanied by a small but significant reduction in the proportion of DMA excreted, suggesting that these individuals had a decreased ability to further methylate MMA. Biological markers of genetic damage have been studied; most parameters gave negative or doubtful results except for a decrease in lymphocyte proliferation kinetics, suggesting impairments in the cellular immune response in exposed individuals.

Exposure to other contaminants has been documented in this region, and high levels of organochlorine pesticide residues in adipose tissue indicated that the inhabitants had been heavily exposed to DDT and BHC. High levels of fluoride in well water, correlating well with those of arsenic, have also been reported. In consequence, people exposed to high arsenic concentrations are also exposed to fluoride levels above the drinking-water standard. In June 1988, a new source for the water distribution system serving the rural part of the Region Lagunera was inaugurated, solving the arsenic contamination problem for 70% of the villages affected.

Regarding exposure to arsenic and cadmium via inhalation and soil ingestion in a smelter community (Morales, San Luis Potosí), it was concluded that (1) high arsenic concentrations in the urine and hair indicated that children in Morales have absorbed considerable quantities of these elements, reflecting the high level of environmental exposure, and (2) children living in Morales are at high risk of suffering cancer and other adverse health effects if exposure continues.

ACKNOWLEDGMENTS

This work was supported by the U.S. Environmental Protection Agency (1D3344NAFX). We thank Ms. Rosalinda Flores M. for secretarial assistance.

REFERENCES

Agency for Toxic Substances and Disease Registry (ATSDR) (1988). *Toxicological Profile for Arsenic.* U.S. Public Health Service, Atlanta, GA.

Albert, L., and Reyes, R. (1978). Plaguicidas organoclorados. II. Contaminacion de algunos quesos Mexicanos por plaguicidas organoclorados. *Rev. Soc. Quim. Mex.* **22**, 65–72.

Albert, L., and Saval, S. (1977). Residuos de plaguicidas organoclorados en alimentos balanceados. *Congr. Naci. Ing. Bioquim., 2nd,* Mexico.

Albert, L., Cebrián, M. E., Méndez, F., and Portales, A. (1980). Organochlorine pesticide residues in human adipose tissue in México: Results of a preliminary study in three Mexican cities. *Arch. Environ. Health* **35**, 262–269.

Albores, A., Cebrián, M. E., Tellez, I., and Valdez, B. (1979). Estudio comparativo de hidroarsenicismo crónico en dos comunidades rurales de la Región Lagunera de México. *Bol. of. Sanit. Panam.* **86**, 196–205.

Astolfi, E. (1971). Estudio de arsenicismo en agua de consumo. *Prensa Méd. Argent.* **58**, 1342–1343.

Baker, E. L., Hayes, C. G., Landrigan, P. J., Handke, J. L., Leger, R. T., Housworth, W. J., and Harrington, J. M. (1977). A nation wide survey of heavy metal absorption in children living near primary copper, lead and zinc smelters. *Am. J. Epidemiol.* **106**, 261–273.

Borgoño, J. M., and Greiber, R. (1971). Estudio epidemiológico del arsenicismo en la ciudad de Antofagasta. *Rev. Med. Chile* **99**, 702–707.

Borgoño, J. M., Vincent, P., Venturino, H., and Infante, A. (1977). Arsenic in the drinking water of the city of Antofagasta: Epidemiological and clinical study before and after the installation of a treatment plant. *Environ. Health Perspect.* **19**, 103–105.

Bracho, A. R. (1971). Arsenicismo crónico en la Comarca Lagunera de Coahuila. Tesis Recepcional, Facultad de Medicina, Universidad Autónoma de Coahuila, México.

Cantellano, A. L., Viniegra, G., Eslava, G.R., and Alvarez, A. J. (1964). El arsenicismo en la Comarca Lagunera. *Salud Publ. Mex.* **6**, 375–385.

Cebrián, M. E. (1987). Some potential problems in assessing the effects of chronic arsenic exposure in north Mexico. *Prepr. Pap. Natl. Meet., Div. Environ. Chem. Am. Chem. Soc., 194th Meet.* **27**, 114–116.

Cebrián, M. E., Albores, A., Aguilar, M., and Blakely, E. (1983). Chronic arsenic poisoning in the north of Mexico. *Hum. Toxicol.* **2**, 121–133.

Cebrián, M. E., Albores, A., Connelly, J. C., and Bridges, J. W. (1988). Assessment of arsenic effects on cytosolic heme status using tryptophan pyrrolase as an index. *J. Biochem. Toxicol.* **3**, 77–86.

Chávez, A., Pérez, C., Tovar, E., and Gramilla, M. (1964). Estudios en una comunidad con arsenicismo crónico endémico. *Salud. Publ. Mex.* **6**(3), 421–433.

Chen, C. J., Chuang, Y. C., Lin, T. M., and Wu, H. Y. (1985). Malignant neoplasms among residents of a Blackfoot disease-endemic area in Taiwan: High-arsenic artesian well water and cancers. *Cancer Res.* **45**, 5895–5899.

Del Razo, L. M., Arellano, M. A., and Cebrián, M. E. (1990). The oxidation states of arsenic in well water from a chronic arsenicism area of northern Mexico. *Environ. Pollut.* **64**, 143–153.

Del Razo, L. M., Corona, J. C., Garcia-Vargas, G. G., Albores, A., and Cebrián, M. E. (1993). Fluoride levels in well-water from a chronic arsenicism area of Northern Mexico. *Environ. Pollut.* **80**, 91–94.

Del Razo, L. M., Hernández, J. L., García-Vargas, G. G., Ostrosky-Wegman, P., Cortinas de Nava, C., and Cebrián, M. E. (1992). The urinary excretion of inorganic arsenic and its metabolites in a human population chronically exposed to arsenic via drinking water. *4th Annu. Meet., Int. Soc. Environ. Epidemiol./Int. Soc. Exposure Anal.* Cuernavaca, Mor. México.

Diaz-Barriga, F., Santos, M. A., Mejía, J. J., Batres, L., Yañez, L., Carrizales, L., Vera, E., Del Razo, L. M., and Cebrián, M. E. (1993). Arsenic and cadmium absorption in children living near a smelter complex in San Luis Potosi, México. *Environ. Res.* **62**, 242–250.

Doss, M. (1979). Chronic hepatic porphyrias in humans (endogenic factors). In J. J. T. W. A. Strik and J. H. Koeman (Eds.), *Chemical Porphyria in Man.* Elsevier/North-Holland Biomedical Press, Amsterdam, p. 11.

Enterline, P. E., Henderson, V. L., and Marsh, G. M. (1987). Exposure to arsenic and respiratory cancer. A reanalysis. *Am. J. Epidemiol.* **125**, 929–938.

Environmental Protection Agency (EPA) (1984). *Health Assessment Document for Inorganic Arsenic*, Final Report, EPA 600/8-83-021F. USEPA, Research Triangle Park, NC.

Environmental Protection Agency (EPA) (1988). *Risk Assessment Forum. Special Report on Ingested Inorganic Arsenic: Skin Cancer; Nutritional Essentiality*, EPA/625/3-87/013. USEPA, Research Triangle Park, NC.

Excelsior (1988). Sect. A, p. 44, México.

Feinglass, E. J. (1973). Arsenic intoxication from well water in the U.S. *N. Eng. J. Med.* **228**, 828-830.

García-Salcedo, J. J., Portales, A., Blakely, E., and Diaz R. (1984). Estudio transversal de una cohorte de pacientes con vasculopatia por intoxicación crónica arsenical en poblados de los municipios de Francisco I. Madero y San Pedro, Coah. México. *Rev. Fac. Med. (Torreón)* **1**, 12-16.

García-Vargas, G. G., García-Rangel, A., Aguilar-Romo, M., García-Salcedo, J., del Razo, L.M., Ostrosky-Wegman, P., Cortinas de Nava, C., and Cebrián, M. E. (1991). A pilot study on the urinary excretion of porphyrins in human populations chronically exposed to arsenic in México. *Hum. Exp. Toxicol.* **10**, 189-193.

Goldsmith, J. R. (1972). Evaluation of health implications of arsenic in well water. *Water Res.* **6**, 1113-1136.

Gómez-Arroyo, S., Hernández-Garcia, A., and R. Villalobos-Pietrini (1988). Induction of sister chromatid exchanges in *Vicia faba* by arsenic contaminated drinking water. *Mutat. Res.* **208**, 219-224.

Grantham, S. A., and Jones, J. P. (1977). Arsenic contamination of water wells in Nova Scotia. J. *Am. Water Works Assoc.* **69**(12), 653-657.

Harrington, J. M. (1978). A survey of a population exposed to high concentrations of arsenic in well water in Fairbanks, Alaska. *Am. J. Epidemiol.* **108**(5), 377-385.

Hartwell, T. D., Handy, R. W., Harris, B. S., Williams, S. R., and Gehibach, S. H. (1983). Heavy metal exposure in populations living around zinc and copper smelters. *Arch. Environ. Health* **38**, 284-295.

Hindmarsh, J. T., McLetchie, O. R., Heffernan, L. P. M., Hayne, O. A., Ellenberger, H. A., McCurdy, R. F., and Thiebaux, H. J. (1977). Electromyographic abnormalities in chronic environmental arsenicalism. *J. Anal. Toxicol.* **1**, 270-276.

IMMSA (1990). *Industrial Minera México*, Boletín Informativo.

Landrigan, P. J., and Baker, E. L. (1981). Exposure of children to heavy metals from smelters: Epidemiology and toxic consequences. *Environ. Res.* **25**, 204-224.

Moran, S., Maturana, G., Rosenberg, H., Casanegra, P., and Dubernet, I. (1977). Occlusions coronariennes liées á une intoxication arsenicale chronique. *Arch. Mal. Coeur Vaiss.* **70**, 1115-1120.

Morton, W., Starr, G., Pohl, D., Stoner, J., Wagner, S., and Weswig, P. (1976). Skin cancer and water arsenic in Lane County, Oregon. *Cancer (Philadelphia)* **37**, 2523-2532.

Ortiz-Mariote, C., Olvera, R., and Verduzco, E. (1963). Intoxicación colectiva por arsénico en Torreón, Coahuila, México. II. Valoración final. *Bol. Epidemiol.* (*Mexico City*) **27**(4), 221-252.

Ostrosky-Wegman, P., Gonsebatt, M. E., Montero R., Vega, L., Barba, H., Espinoza, J., Palao, A., Cortinas, C., García-Vargas, G., del L. M., Razo, and Cebrián, M. E. (1991). Lymphocyte proliferation kinetics and genotoxic findings in a pilot study on individuals chronically exposed to arsenic in México. *Mutat. Res.* **250**, 477-482.

Reyes-Najera, R., and Sánchez de la Fuente, E. (1975). Intoxicación por plaguicidas en la Comarca Lagunera durante el ciclo agrícola de 1974. *Salud Publ. Mex.* **17**, 687-698.

Rosales, M. G. (1987). Determinación del daño cromosómico en individuos con hidroarsenicismo endémico. Tesis Recepcional, Facultad de Medicina, Universidad Autónoma de Coahuila, Torreón Coah, México.

Rosenberg, H. G. (1973). Systemic arterial disease with myocardial infarction. *Circulation* **47**, 270-275.

Sánchez de la Fuente, E. et al. (1976). Arsenicismo crónico en la zona rural de la Comarca Lagunera. *34th Reun. Anu. Asoc. Front. Méx.-Estadounidense Salubridad.*

Secretaria de Salubridady Asistencia (1991). Boletín Informativo. *Secretaría de Salud, México.*

Southwick, J. W., Western, A. E., Beck, M. M., Whiteley, T., Isaacs, R., Petajan, J., and Hansen, C. D. (1983). An epidemiological study of arsenic in drinking water in Millar County, Utah. In W. H. Lederer and R. J. Fensterheim (Eds.) *Arsenic, Industrial, Biomedical and Environmental Perspectives.* Van Nostrand-Reinhold, New York, p. 210.

Subsecretaría del Mejoramiento del Ambiente/Secretaría de Salubridad y Asistencia (SMA/SSA) (1977). *Minuta de la 1a. Reunión del Grupo Interinstitucional sobre el Problema del Arsenicismo en la Comarca Lagunera.* SMA/SSA, México.

Torres de Navarro, E. (1976). Intoxicación arsenical en el ganado vacuno. *Salud Publ. Mex.* **18**, 1037–1044.

Tseng, W. P., Chu, H. M., How, S. W., Fong, J. J., Lin, C. S., and Yeh, S. (1968). Prevalence of skin cancer in an endemic area of chronic arsenicism in Taiwan. *J. Natl. Cancer Inst. (U.S.)* **40**, 453–463.

Walder, B. K., Robertson, M. R., and Jeremy, J. (1971). Skin cancer and immunosuppression. *Lancet* **2**, 1282–1283.

Webb, J. L. (1966). *Enzyme and Metabolic Inhibitors.* Academic Press, New York, Vol. 3, p. 595.

Wen, W.-N., Lieu, T.-L., Chang, H.-J., Wuu, S. W., Yau, M.-L., and Jan, K. Y. (1981). Baseline and sodium arsenite-induced sister chromatid exchanges in cultured lymphocytes from patients with blackfoot disease and healthy persons. *Hum. Genet.* **59**, 201–203.

WHO (World Health Organization). (1984). *Guidelines for Drinking Water Quality*, Vol. 2: Health Criteria and Other Supporting Information. Geneva, Switzerland.

Zaldivar, R. (1974). Arsenic contamination of drinking water and foodstuffs causing endemic chronic poisoning. *Beitr. Pathol.* **151**, 384–400.

6

HUMAN CARCINOGENICITY AND ATHEROGENICITY INDUCED BY CHRONIC EXPOSURE TO INORGANIC ARSENIC

Chien-Jen Chen*,† and Li-Ju Lin*

**Institute of Public Health, College of Public Health, National Taiwan University 10018; †Institute of Biomedical Sciences, Academia Sinica 11529, Taipei, Taiwan*

Arsenic in the Environment, Part II: Human Health and Ecosystem Effects,
Edited by Jerome O. Nriagu.
ISBN 0-471-30436-0

1. INTRODUCTION

Arsenic is a ubiquitous element present in various compounds throughout the earth's crust. It is widely distributed in the environment, and all humans are exposed to low levels of arsenic. For most people, food constitutes the largest source of arsenic intake, with smaller amounts coming from drinking water and air. Some edible fish and shellfish contain elevated levels of arsenic, but this is predominantly in an organic form that has low toxicity. Above-average levels of exposure are usually observed among people who live in areas where drinking water has an elevated level of inorganic arsenic because of natural mineral deposits or contaminations from human activities, among workers and nearby residents of copper and other metal smelters, among persons who manufacture or use arsenic-containing pesticides, and among patients treated with inorganic arsenic (such as Fowler's solution, which contains 1% potassium arsenite) for diseases such as asthma and psoriasis [World Health Organization (WHO), 1981; U.S. Public Health Service, 1989].

Arsenic enters the human body through ingestion, inhalation, or skin absorption. Most ingested and inhaled arsenic is well absorbed through the gastrointestinal tract and lung into the bloodstream. It is distributed in a large number of organs including the lungs, liver, kidneys, and skin (Hunter et al., 1942). Most arsenic absorbed into the body is converted by the liver to less toxic methylated forms that are efficiently excreted in the urine. The rate of decrease of arsenic in the skin appears to be especially low compared with the rate for other organs (WHO, 1981; U.S. Public Health Service, 1989).

Inorganic arsenic has been recognized as a human poison since ancient times. Acute and subacute effects of inorganic arsenic may involve many organ systems, including the respiratory, gastrointestinal, cardiovascular, nervous, and hematopoietic systems. An ingested dose of 50 to 300 mg of arsenic oxide has been reported to be fatal in human beings (U.S. Public Health Service, 1989). Long-term exposure to inorganic arsenic has toxic effects on a large number of organs (WHO, 1981). Inorganic arsenic has both carcinogenic and atherogenic effects on humans. Perhaps the single most characteristic systemic effect of oral exposure to inorganic arsenic is a pattern of skin abnormalities including hyperpigmentation, hyperkeratosis, and cancers. Inhalation exposure to inorganic arsenic is well documented to induce lung cancer among smelter workers. In 1979, a working group of the International Agency for Research on Cancer concluded that there was sufficient evidence that inorganic arsenic compounds were skin and lung carcinogens in humans, but that data for other sites were inadequate for evaluation [International Agency for Research on Cancer (IARC), 1980]. Current epidemiological studies have shown a significant dose-response relationship between chronic exposure to inorganic arsenic in drinking water and mortality from cancers of the skin, lung, liver, bladder, kidney, and prostate (C.J. Chen et al., 1988a, 1992a; Chen and Wang, 1990).

Arsenic is involved in the development of several cancers in humans without showing any organotropism. However, limited evidence shows the carcinogenicity of arsenic to experimental animals. Sodium arsenate and sodium arsenite induce morphological transformation of cultured cells, but they are inactive or too weak to induce gene mutations at specific genetic loci. Arsenic causes chromosomal aberration and sister-chromatid exchanges, and its ability to induce gene amplification was recently documented. Arsenic also potentiates cytotoxicity and mutagenicity of several chemicals. However, the exact mechanism of arsenic-induced carcinogenicity remains to be elucidated. Integrated multidisciplinary efforts are urgently needed (IARC, 1982, 1987).

Chronic exposure to inorganic arsenic also has toxic effects on the cardiovascular system. Increased risk of both coronary heart diseases and peripheral vascular diseases has been reported in some areas of the world where heavy exposure due to ingestion of inorganic arsenic has occurred. Systematic atherosclerosis associated with chronic arsenicism has been well documented in autopsy studies of patients affected by blackfoot disease, a unique peripheral vascular disease observed in an endemic area where drinking water from artesian wells has a high arsenic concentration (Yeh and How, 1963).

In this chapter, the human carcinogenic and atherogenic effects associated with inorganic arsenic exposure mainly through inhalation and ingestion will be reviewed. Possible mechanisms related to the multiplicity of arsenic carcinogenesis and atherosclerosis will also be discussed.

2. CARCINOGENICITY

2.1. Arsenic-Induced Skin Lesions

Arsenic induces a wide range of skin lesions including hyperpigmentation, hyperkeratosis, and various cancers. Significant dose-response relationships have been well documented between level of ingested inorganic arsenic and prevalence of various skin lesions. Both hyperpigmentation and hyperkeratosis can be used as biological markers to indicate exposure to arsenic. Arsenic-induced skin cancers, including Bowen's disease, basal cell carcinoma, and squamous cell carcinoma, have been observed in patients treated with Fowler's solution, in vintners, in workers in pesticide production, and in residents of areas where the drinking water has an elevated level of arsenic.

2.1.1. *Hyperpigmentation and Hyperkeratosis*

A number of skin lesions have been attributed to chronic exposure to inorganic arsenic compounds. Hyperpigmentation of the skin often associated with depigmentation is a pathologic hallmark of chronic arsenic exposure and may occur anywhere on the body. It is not considered to be a malignant neoplasm or a precancerous lesion [U.S. Environmental Protection Agency (USEPA), 1988]. Hyperkeratosis is a characteristic lesion that occurs most frequently on the palms and soles but may also occur at other sites. It usually appears as a small corn-like elevation, 0.4 to 1.0 cm in diameter. In the majority of cases, arsenic keratoses are morphologically benign, showing very little cellular atypia (Yeh, 1973). Although there has been controversy about the distinguishability between arsenical keratosis and Bowen's disease (Yeh et al., 1968; Hugo and Conway, 1967), the consensus is that some arsenical keratoses may develop into invasive squamous cell carcinoma (Shannon and Strayer, 1987). Both hyperpigmentation and hyperkeratosis are indicators of arsenic exposure, and can be used as biological markers. These lesions have been reported where levels of arsenic in drinking water were elevated in regions of Argentina (Arguello et al., 1938; Biagini et al., 1974), Chile (Borgoño et al., 1977; Zaldivar, 1974), China (Huang et al., 1985), Japan (Yoshikawa et al., 1960), Mexico (Albores et al., 1979; Alvarado et al., 1964; Cebrian et al., 1983; Chavez et al., 1964), and Taiwan (Tseng et al., 1968). They were also found in patients treated with arsenic-containing Fowler's solution (Fierz, 1965) and in workers exposed to airborne arsenic in a pharmaceutical plant (Watrous and McCaughey, 1945), in a sheep-dip factory (Perry et al., 1948), and at an insecticide manufacturer (Hamada and Horiguchi, 1976). Teenagers exposed to arsenic-tainted milk in infancy have also been affected with hyperpigmentation and hyperkeratosis (Yamashita et al., 1972).

2.1.2. *Skin Cancers*

Hutchinson (1888) first discussed the possibility that medication with inorganic arsenic was an etiological factor for skin cancer. Several types of neoplastic changes of the skin, including Bowen's disease and basal and squamous cell carcinomas, have been associated with arsenic exposure. These arsenic-induced cancer lesions possess no unique histological features (Deng and How, 1977). The arsenic cancers that occur on every part of the body are different from skin lesions induced by other carcinogens, such as UV radiation and polyaromatic hydrocarbons, which are limited to the areas of exposure. In other words, arsenic-induced skin cancers can be distinguished from those resulting from exposure to other skin carcinogens by the distribution of lesions on the body. Although arsenical keratoses may develop into squamous cell carcinoma, they are not always precursors to malignant lesions, and some cancerous lesions arise *de novo*.

There is clear evidence that chronic oral exposure to inorganic arsenic through contaminated drinking water or medication increases the risk of skin cancer (WHO, 1981; USEPA, 1988). Ingestion has usually taken place over several decades, with daily doses of several milligrams of arsenic. The largest study of arsenic-induced skin cancer was carried out in Taiwan (Tseng et al., 1968). A total of 40,421 inhabitants were surveyed in this study, and a dose-response relationship between the arsenic content in well water and the prevalence of skin cancer was observed. Assuming a daily intake of 2 liters of water, it has been estimated that a total ingested dose of about 20 g of inorganic arsenic over a lifetime would correspond to a skin prevalence of roughly 6% (WHO, 1981).

In a study of skin cancer prevalence in patients treated with Fowler's solution (Fierz, 1965), there was an increased risk of skin cancer as the total ingested dose of Fowler's solution increased. In patients who had ingested between 200 and 800 mL of Fowler's solution (i.e., between 1.5 and 6.0 g of arsenic), the prevalence of skin cancer ranged from 5 to 10%. While patients treated with Fowler's solution at a dose equivalent to 7.6 g of arsenic had a skin cancer prevalence of 20%, the corresponding prevalence among residents in the area of endemic chronic arsenicism in Taiwan was less than 3% (WHO, 1981). This discrepancy could be due to factors such as differences in exposure regimen and medium, the valence states of arsenic, the presence of other chemicals, genetic composition, and cultural and socioeconomic conditions. Possible biases in subject selection and response of patients treated with arsenical medication to follow-up should also be taken into consideration.

Based on the data from Taiwan and a generalized multistage model, it has been estimated that the skin cancer risk ranges from 3×10^{-5} to 7×10^{-5} for a person weighing 70 kg who consumes 2 L of water contaminated with 1 μg/L arsenic per day. In other words, the estimate of maximum risk due to 1 μg/kg/day of arsenic intake ranges from 1×10^{-3} to 2×10^{-3} (USEPA, 1988; Brown et al., 1989). The model developed for the Taiwan data was used to predict prevalence of skin cancer for residents in two rural Mexican towns, one with arsenic-contaminated drinking water (Cebrián et al., 1983). The prediction of skin cancer risk was consistent with observations.

A dose-response relationship between arsenic concentration in drinking water and skin cancer mortality has also been documented in the area of endemic blackfoot disease in Taiwan (Chen et al., 1985, 1988a; Wu et al., 1989). A significant ecological correlation was also observed in a recent study of the association between arsenic level in well water and age-adjusted mortality from skin cancer in 314 townships throughout Taiwan island. The increase in age-adjusted mortality per 100,000 person-years for every 0.1 ppm increase in arsenic level in well water was 0.9 and 1.0, respectively, for men and women (Chen and Wang, 1990). A recent survey has shown that residents in villages of hyperendemic blackfoot disease still had an extraordinarily high prevalence of skin cancer after more than

15 years free of ingestion of high-arsenic artesian well water (C. J. Chen et al., 1992a). In addition to the cumulative effects of exposure to ingested inorganic arsenic, abnormal liver function was also found to play a significant role in the development of skin cancer (S. Y. Chen, 1992).

Inhaled inorganic arsenic has also been documented to induce skin cancer among workers producing sheep-dip powder from sodium arsenite. The statistically significant standardized mortality ratio was as high as 25 (Hill and Faning, 1948). Skin cancer has also been observed among vineyard workers (Luchtrath, 1983; Thiers et al., 1967).

2.2. Respiratory Cancers

The association between exposure to inorganic arsenic and risk of respiratory cancers is well documented. In 1979, the working group of the International Agency for Research on Cancer concluded that there was sufficient evidence that inorganic arsenic compounds were carcinogens to the human lung (IARC, 1980). An increased risk of lung cancer has been associated with long-term exposures to inorganic arsenic through inhalation and ingestion. It was recently documented in an ecological correlation study that exposure to inorganic arsenic in drinking water is related to mortality from nasal-cavity cancer (Chen and Wang, 1990).

2.2.1. Lung Cancer

An excess of deaths due to lung cancer has been observed among workers exposed through inhalation to inorganic arsenic in the production and use of pesticides (Hill and Faning, 1948; Roth, 1958; Ott et al., 1974; Mabuchi et al., 1979), in gold mining (Osburn, 1957, 1969), and in the smelting of nonferrous metals, especially copper (Lee and Fraumeni, 1969; Milham and Strong, 1974; Tokudome and Kuratsune, 1976; Pinto et al., 1977, 1978; Rencher et al., 1977; Pershagen et al., 1977; Axelson et al., 1978; Enterline and Marsh, 1980, 1982; Lubin et al., 1981; Higgins et al., 1982; Lee-Feldstein, 1983, 1986; Brown and Chu, 1983; Enterline et al., 1987). There was also an increased risk of lung cancer for individuals living within several kilometers of inorganic arsenic-emitting industries (Blot and Fraumeni, 1975; Matanoski et al., 1981; Cordier et al., 1983; Brown et al., 1984; Pershagen, 1985). Most of these occupational and environmental exposures to inorganic arsenic through inhalation involved other chemicals. The effects observed were limited to some degree by confounding factors such as smoking and exposure to other chemicals. However, the weight of evidence that inhaled inorganic arsenic is a risk factor for lung cancer is convincing.

The increased lung cancer risk among workers engaged in the production of insecticides containing inorganic arsenic compounds was primarily observed in three studies. A significant excess proportion (31.8%) of deaths attributable to cancers of respiratory organs was observed among factory workers producing sheep-dip powder from sodium arsenite, compared to 15.9% among workers in other occupational groups in the environs of the plant (i.e., agricultural workers, general labors, artisans, and shopworkers) (Hill and Faning, 1948). A dose-response relationship between daily 8-hr time-weighted average airborne arsenic concentration and mortality due to respiratory malignancies was observed among workers producing arsenic-containing insecticides (Ott et al., 1974). Workers employed in a pesticide factory producing many arsenic compounds were reported to have a significant excess mortality from cancers of the trachea, bronchus, and lung, with a standardized mortality ratio of 1.7 (Mabuchi et al., 1979).

The dose-response relationship between inhaled inorganic arsenic and lung cancer risk in copper smelter workers was observed mainly in two large cohorts in Anaconda, Montana (Lee and Fraumeni, 1969; Lubin et al., 1981; Higgins et al., 1982; Lee-Feldstein, 1983, 1986; Brown and Chu, 1983), and in Tacoma, Washington (Pinto et al., 1977, 1978; Enterline and Marsh, 1980, 1982; Enterline et al., 1987). The Anaconda cohort was originally studied by Lee and Fraumeni (1969). The cohort, comprising 8047 white male smelter workers who had been employed for a year or more before 1957, was followed up for an additional 14 years by Lee-Feldstein (1986) to examine the mortality experience from 1938 to 1977. A sample of 1800 workers from the same cohort was followed up by Higgins et al. (1982) using different exposure classifications and different methods of analysis. Based on the Anaconda data reported by Lee-Feldstein, the effect of arsenic exposure has been evaluated using the multistage theory of carcinogenesis (Brown and Chu, 1983). It has been suggested that arsenic appears to exert a definite effect at a late stage of the carcinogenic process, although an additional effect at the initial stage cannot be ruled out.

The Tacoma study, involving 2802 men employed at the smelter for a year or more during 1940 to 1964, was conducted to analyze the mortality experience of smelter workers followed up to 1976 (Enterline and Marsh, 1982). Data of this cohort has recently been reanalyzed using urinary excretion as the exposure marker, and a higher lung cancer risk than indicated previously in other studies was observed (Enterline et al., 1987). Analysis by multistage modeling incorporating a time-dependent exposure pattern has also indicated a late-stage effect of arsenic. But an additional early-stage effect still cannot be ruled out (Mazumdar et al., 1989).

Assuming a person who weighs 70 kg and inhales 20 m^3 of air per day with 100% absorption, the crude lung cancer risk due to 1 μg/kg/day arsenic intake from air was estimated at 0.5, 1.0, 1.7 and 2.4%, respectively, for the data of Brown and Chu (1983), Lee-Feldstein (1986), Higgins et al. (1982), and Enterline and Marsh (1982).

Several epidemiologic studies have reported an increased risk of lung cancer among people living near industries that produce arsenic-containing pesticides (Matanoski et al., 1981) or near copper, lead, or zinc smelters and refineries (Blot and Fraumeni, 1975; Cordier et al., 1983; Brown et al., 1984; Pershagen, 1985).

Significant associations between ingested inorganic arsenic and lung cancer risk have been observed in patients treated with arsenic-containing medicine (Sommers and McManus, 1953; Calnan, 1954; Braun, 1958; Robson and Jelliffe,

1963; Fierz, 1965; Goldman, 1973), in Moselle vintners exposed to arsenic pesticides (Roth, 1957; Luchtrath, 1983), and in persons exposed to inorganic arsenic from well water (Biagini et al., 1978; Chen et al., 1985, 1986, 1988a, 1992a; Wu et al., 1989; Chen and Wang, 1990; Tsuda et al., 1989, 1990). Most observations of increased lung cancer risk associated with ingested inorganic arsenic from medications were case reports that may have been subject to selection bias. In an autopsy study of lung cancer among Moselle vintners and in comparison postmortem series, a statistically significant relative risk estimate of 20.9 was observed for vintners (Luchtrath, 1983). Two Japanese studies have shown a significantly elevated risk of lung cancer for arsenic-poisoned subjects, with a standardized mortality ratio of 9.7 (Tsuda et al., 1989) and 6.5 (Tsuda et al., 1990), respectively.

Excess mortality from lung cancer has been observed among residents in the area of endemic blackfoot disease in Taiwan, with a significant standardized mortality ratio of 5.3 for men and 6.5 for women (Chen et al., 1985). A significant dose-response relationship between duration of high-arsenic artesian well water ingestion and mortality from lung cancer has also been documented in the area of endemic blackfoot disease (Chen et al., 1986). The age-adjusted lung cancer mortality rate per 100,000 person-years was 35.1, 64.7, and 87.9 for male residents living in villages where the arsenic concentration in drinking water was < 0.30, 0.30–0.59, and ≥ 0.60 ppm, respectively. The corresponding figures for female residents were 26.5, 40.9, and 83.8, respectively (Chen et al., 1988a). Based on lung cancer mortality by age and arsenic level in drinking water, the lifetime risk of developing lung cancer due to an intake of 10 μg/kg/day was 0.012 for males and 0.013 for females (C. J. Chen et al., 1992a). In a large-scale ecological correlation study of the association between arsenic concentration in well water and age-adjusted lung cancer mortality in 314 townships throughout Taiwan, a significant increase in lung cancer mortality of 5.3 per 100,000 person-years was observed in both men and women for every 0.1 ppm increase in well-water arsenic level (Chen and Wang, 1990).

2.2.2. *Nasal-Cavity Cancer*

Perforation of the nasal septum has been documented among smelter workers exposed to high levels of arsenic (WHO, 1981). A significant association between arsenic concentration in drinking water and age-adjusted mortality from nasal-cavity cancer has recently been reported. For every 0.1 ppm increase in well-water arsenic level, there was an increase of 0.7 and 0.4 per 100,000 person-years in mortality from nasal-cavity cancer for males and females, respectively (Chen and Wang, 1990).

2.3. Liver Cancers

Both hepatic angiosarcoma and hepatocellular carcinoma have been associated with long-term exposure to ingested and inhaled inorganic arsenic. Most studies on arsenic-related hepatic angiosarcoma were case reports. Exposure was from contaminated wine, drinking water, Fowler's solution, copper smelting, and arsenic-containing pesticides. Hepatocellular carcinoma has been associated with long-term arsenic exposure through contaminated wine, drinking water, and copper smelting.

2.3.1. *Hepatic Angiosarcoma*

Several cases of hepatic angiosarcoma have been reported among Moselle vintners (Roth, 1957). The workers had been exposed to arsenic-containing insecticides for several years and had typical arsenic-induced skin lesions. There have also been several case reports that patients treated with Fowler's solution were affected by hepatic angiosarcoma (Rosset, 1958; Regelson et al., 1968; Lander et al., 1975; Popper et al., 1978; Falk et al., 1981a; Roat et al., 1982; Kasper et al., 1984). Drinking high-arsenic water has also been related to the development of hepatic angiosarcoma (Rennke et al., 1971). A 20-month-old child of an arsenic-exposed worker was reported to be affected with hepatic angiosarcoma. The case lived near an arsenic-emitting copper smelter (Falk et al., 1981b). Workers exposed to arsenical pesticides were also found to develop hepatic angiosarcoma (IARC, 1987).

Because hepatic angiosarcoma is a very rare disease, its association with exposure to inorganic arsenic does not seem likely to be a matter of chance. However, the possibility of selection bias and uncertainty regarding exposure dose make it difficult to estimate the risk of hepatic angiosarcoma induced by inorganic arsenic.

2.3.2. *Hepatocellular Carcinoma*

Long-term exposure to inorganic arsenic has been documented to induce hepatocellular carcinoma among Moselle vintners (Falk et al., 1981a; Luchtrath, 1983). Because these cases may have been heavy alcohol drinkers, their hepatocellular carcinoma may be alcohol-related. A significantly increased mortality from hepatocellular carcinoma has also been reported among copper smelter workers in Japan, with a standardized mortality ratio of 3.4 (Tokudome and Kuratsune, 1976).

In a series of studies in Taiwan, a significant dose-response relationship between ingested inorganic arsenic and mortality from liver cancer, mainly hepatocellular carcinoma, was observed. A significantly increased mortality from liver cancer has been observed in the area of endemic blackfoot disease in Taiwan, with a standardized mortality ratio of 1.7 in men and 2.3 in women (Chen et al., 1985). There was a significant dose-response relationship between duration of high-arsenic artesian well-water ingestion and risk of liver cancer (Chen et al., 1986). The age-adjusted liver cancer mortality rate per 100,000 person-years was 32.6, 42.7, and 68.8 for male residents living in villages where the arsenic concentration in drinking water was < 0.30, 0.30–0.59, and $\geqslant 0.60$ ppm, respectively. The corresponding figures for female residents were 14.2, 18.8, and 31.8, respectively (Chen et al., 1988a). Based on liver cancer mortality by age and drinking-water arsenic level, the lifetime risk of developing liver cancer due to an intake of 10 μg/kg/day was 0.0043 for males and 0.0036 for females (C. J. Chen et al., 1992a). A significant association between arsenic concentration in well water and age-adjusted liver cancer mortality in 314 townships throughout Taiwan has been observed in a large-scale ecological correlation study. An increase in liver cancer mortality of 6.8 and 2.0 per 100,000 person-years, respectively, for males and females was observed for every 0.1 ppm increase in well-water arsenic level (Chen and Wang, 1990).

2.4. Genitourinary Cancers

Both inhaled and ingested inorganic arsenic have been documented to induce genitourinary cancers. Increased risk of bladder cancer was observed among patients treated with Fowler's solution and among Moselle vintners, workers in nonferrous-metal smelters, refineries, and mines, and residents of areas where drinking water had an elevated level of arsenic. A significant association has been documented between risk of kidney cancer and level of inorganic arsenic exposure through drinking water, arsenic-containing medicine, and copper smelting. Recent studies have also found increased mortality from prostate cancer among residents in the area of endemic chronic arsenicism in Taiwan.

2.4.1. Bladder Cancer

Bladder cancer has been documented among patients treated with Fowler's solution in several case reports (Sommers and McManus, 1953; von Roemeling et al., 1979; Nagy et al., 1980; Robertson and Low-Beer, 1983). In a cohort study of 478 patients treated with Fowler's solution for various lengths of time ranging from 2 weeks to 12 years, significantly increased mortality from bladder cancer was observed, with a standardized mortality ratio of 3.1 (Cuzick et al., 1992). An increased but not statistically significant risk of bladder cancer was also reported among Moselle vintners, with a standardized mortality ratio of 2.8 (Luchtrath, 1983).

Increased mortality from bladder cancer has been documented among residents in the area of endemic blackfoot disease, with a standardized mortality ratio of 11.0 for men and 20.1 for women (Chen et al., 1985). The bladder cancer risk was found to increase with the duration of high-arsenic artesian well-water ingestion in a dose-response relationship (Chen et al., 1986). A significant dose-response relationship between arsenic concentration in drinking water and mortality from bladder cancer has also been observed. The age-adjusted bladder cancer mortality rate per 100,000 person-years was 15.7, 37.8, and 89.1 for male residents living in villages where the arsenic concentration in drinking water was < 0.30, 0.30–0.59, and $\geqslant 0.60$ ppm, respectively. The corresponding figures for female residents were 16.7, 35.1, and 91.5, respectively (Chen et al., 1988a). Based on bladder cancer mortality by age and arsenic level in drinking water, the lifetime risk of developing bladder cancer due to an intake of 10 μg/kg/day was 0.012 for males and 0.017 for females (C. J. Chen et al., 1992a). In a large-scale ecological correlation study including 314 townships throughout

Taiwan, an increase in bladder cancer mortality of 3.9 and 4.2 per 100,000 person-years, respectively, for males and females was observed for every 0.1 ppm increase in well-water arsenic level (Chen and Wang, 1990).

Inhaled inorganic arsenic has also been found to be associated with an increased risk of bladder cancer. Copper smelter workers in the United States were reported to have an increased risk of bladder cancer, with a standardized mortality ratio around 2.0 (Welch et al., 1982; Lee-Feldstein, 1983). A standardized mortality ratio as high as 7.7 has been observed among workers at a copper refinery in Japan (Tsuda et al., 1990).

2.4.2. *Kidney Cancer*

Excess mortality from kidney cancer, with a standardized mortality ratio of 2.0, was observed among workers employed at a copper smelter before respirators were implemented in the factory (Enterline and Marsh, 1982). A patient treated with an arsenical medicine was also reported to be affected by kidney cancer (Nurse, 1978).

Residents in the area of endemic chronic arsenicism have been reported to have a significantly increased risk of kidney cancer, with a standardized mortality ratio of 7.7 for men and 4.2 for women (Chen et al., 1985). A significant dose-response relationship between ingested inorganic arsenic and kidney cancer mortality has also been observed. The age-adjusted kidney cancer mortality rate per 100,000 person-years was 5.4, 13.1, and 21.6 for male residents living in villages where the arsenic concentration in drinking water was < 0.30, 0.30–0.59, and $\geqslant 0.60$ ppm, respectively. The corresponding figures for female residents were 3.6, 12.5, and 33.3, respectively (Chen et al., 1988a). Based on kidney cancer mortality by age and drinking-water arsenic level, the lifetime risk of developing kidney cancer due to an intake of 10 μg/kg/day was 0.0042 for males and 0.0048 for females (C. J. Chen et al., 1992a). A significant increase in bladder cancer mortality of 1.1 and 1.7 per 100,000 person-years, respectively, for men and women was observed for every 0.1 ppm increase in well-water arsenic level in a large-scale ecological correlation study including 314 townships throughout Taiwan (Chen and Wang, 1990).

2.4.3. *Prostate Cancer*

There was also a significant association between ingested inorganic arsenic and prostate cancer mortality among residents in the areas of endemic blackfoot disease in Taiwan. The age-adjusted prostate cancer mortality rate per 100,000 person-years was 0.5, 5.8, and 8.4 for male residents living in villages where the arsenic concentration in drinking water was < 0.30, 0.30–0.59, and $\geqslant 0.60$ ppm, respectively (Chen et al., 1988a). In a large-scale ecological correlation study including 314 townships throughout Taiwan, an increase in prostate cancer mortality of 0.5 per 100,000 person-years was observed for every 0.1 ppm increase in well-water arsenic level (Chen and Wang, 1990).

2.5. Other Internal Cancers

Patients treated with Fowler's solution have been documented to have an increased risk of internal cancers at all sites, with a standardized mortality ratio of 3.3 (Moller et al., 1975). Chronic exposure to inorganic arsenic has also been documented to be significantly associated with internal cancers other than cancers of the skin, respiratory system, liver, and genitourinary system.

2.5.1. Gastrointestinal Cancers

Significant excess mortality from cancers of the digestive tract has been observed among copper smelter workers in Anaconda, with a standardized mortality ratio of 1.3 (Lee-Feldstein, 1983); only a slight excess in mortality from digestive tract cancer was observed among smelter workers in Tacoma (Enterline and Marsh, 1982). A significantly increased mortality from stomach cancer has been reported among copper smelter workers in Sweden with a standardized mortality ratio of 1.7 (Wall, 1980), and among Moselle vintners with a standardized mortality ratio of 2.4 (Luchtrath, 1983). Colon cancer mortality has also been documented to be significantly associated with chronic exposures to inorganic arsenic among copper smelter workers in Tacoma with a statistically significant standardized mortality ratio of 2.1 for those who employed before respirators were implemented in the smelter (Enterline and Marsh, 1982), and among smelter workers in Japan with a standardized mortality ratio of 5.1 (Tokudome and Kuratsune, 1976).

There was no increase in stomach cancer mortality among residents in the area of endemic blackfoot disease in Taiwan, but significantly increased colon cancer mortality, with a standardized mortality ratio of 1.6 for men and 1.7 for women, was observed in the area (Chen et al., 1985). The age-adjusted colon cancer mortality rate per 100,000 person-years was 7.9, 8.3, and 12.5 for male residents aged 20 years or more who lived in villages where the arsenic concentration in drinking water was < 0.30, 0.30–0.59, and $\geqslant 0.60$ ppm, respectively. The corresponding figures for female residents aged 20 years or more were 9.1, 8.2, and 17.2, respectively (Wu et al., 1989).

2.5.2. Hematolymphatic Malignancies

Workers exposed to arsenic in a pesticide manufacturing plant were reported to have significant excess mortality from malignant neoplasms (other than leukemia) of the lymphatic and hematopoietic tissues, with a standardized mortality ratio of 3.9 (Ott et al., 1974). Some malignant neoplasms of the lymphatic and hematopoietic tissues was also observed among copper smelter workers in Tacoma (Enterline and Marsh, 1982). Patients affected with leukemia and myeloma had been reported as more likely to be exposed to arsenic than controls (Axelson et al., 1978). There was no increase in leukemia mortality among residents in the area of endemic blackfoot disease in Taiwan (Chen et al., 1985).

2.5.3. *Malignant Neoplasms of the Brain and Nervous System*

Significantly increased mortality from malignant neoplasms of the brain and nervous system was observed among copper smelter workers in Sweden, with a standardized mortality ratio of 1.9 (Wall, 1980).

3. ATHEROGENICITY

3.1. Peripheral Vascular Diseases

Chronic exposure to inorganic arsenic has been documented to induce the development of blackfoot disease, a unique peripheral vascular disorder identified in the area of endemic chronic arsenicism in Taiwan (Wu et al., 1961). Clinically, the disease starts with numbness or coldness of one or more extremities and intermittent claudication, which progress to black discoloration, ulceration, and gangrene. In the end stages of the disease, spontaneous or surgical amputation of the distal parts of affected extremities is common (Tseng et al., 1961). Extensive pathological study showed that 30% of blackfoot disease patients had histological lesions compatible with thromboangiitis obliterans, and 70% showed changes of arteriosclerosis obliterans (Yeh and How, 1963). Marked generalized atherosclerosis was observed in all autopsied cases of blackfoot disease, and the fundamental vascular changes of the disease represent an unduly developed severe arteriosclerosis.

Blackfoot disease has long been associated with the high-arsenic drinking water derived from artesian wells in the area (K. P. Chen and Wu, 1962; Chi and Blackwell, 1968). There was a dose-response relationship between arsenic concentration in drinking water and the prevalence of blackfoot disease (Tseng, 1977). In villages where the arsenic concentration in drinking water was < 0.30, 0.30–0.59, and $\geqslant 0.60$ ppm, blackfoot disease prevalence was 0.5, 1.3, and 1.4%, respecitvely, for residents aged 20 to 39 years; 1.1, 3.2, and 4.7%, respectively, for residents aged 40 to 59 years; and 2.0, 3.2, and 6.1%, respectively, for residents over age 60. A significantly increased prevalence of arsenic-related skin lesions, including hyperpigmentation, hyperkeratosis, and cancers, has been observed among blackfoot disease patients. The coexistence of arsenic skin lesions and blackfoot disease further supports the role of inorganic arsenic in the induction of blackfoot disease.

Based on the dose-response relationship between level of inorganic arsenic in drinking water and prevalence of blackfoot disease, it has been estimated that a total ingested dose of about 20 g of inorganic arsenic corresponds to a blackfoot disease prevalence of about 3% (WHO, 1981).

In a recent study of multiple risk factors and related malignant neoplasms of blackfoot disease, undernourishment was found to be associated to a significant degree with the development of blackfoot disease, in addition to consumption of high-arsenic artesian well water and a familial history of blackfoot disease (Chen et al., 1988b). Blackfoot disease patients have also been reported to have significantly increased mortality from cardiovascular diseases and cancers of the bladder, skin, lung, liver, and colon.

Fluorescent substances were postulated as the etiological factor in blackfoot disease (Lu et al., 1975). However, evidence for an ergot-like action of fluorescent substances was considered inadequate (WHO, 1981) and was subsequently disregarded by investigators. Humic substances were then proposed instead of ergotamine as the determinant of blackfoot disease (Lu, 1990). When one of the substances was injected peritoneally into mice for 3 to 5 weeks, half of the experimental animals developed ulceration, necrosis, and gangrene in the extremities, resembling the outer appearance of blackfoot disease. However, it was pointed out that the pathological changes in these mice were completely different from those observed in blackfoot disease patients (Chen, 1990). The lesions developed in mice resulted from the occlusion of peripheral arteries induced by acute thrombosis via an action on blood coagulation, rather than chronic atherosclerosis characterized by severe atheroma, calcification, bone formation, hyalinization, and increased elastic fibers in the intimal coat, as observed in blackfoot disease. Injection of pure sodium hydroxide (pH 12.0) solution into animals was found to have the same effect observed in animal experiments with humic substances (W. Y. Chen and Lien, 1963). Although symptomatically similar, these animal models do not demonstrate any pathological lesions comparable to those of blackfoot disease.

Humic substances resulting from the decomposition of organic matter, particularly dead plants, are widespread contaminants of water supplies and are definitely not confined to the area of endemic blackfoot disease. No correlation between humic substances and peripheral vascular disorder has ever been documented. On the contrary, peripheral vascular disease resembling blackfoot disease has been reported among arsenic-exposed vintners and the residents of areas where the arsenic concentration in drinking water was elevated.

Peripheral vascular lesions have been described in vintners chronically exposed to arsenic through contaminated wine (Butzengeiger, 1940; Roth, 1957; Grobe, 1976). Distinct peripheral vascular lesions were observed in more than 60% of the vintners, but in only 1 to 2% of an unexposed control group (Grobe, 1976). Peripheral vascular disease comparable to blackfoot disease has also been observed in people exposed to high-arsenic drinking water in Chile (Borgoño et al., 1977; Borgoño and Greiber, 1972) and Mexico (Salcedo et al., 1984; Cebrián, 1987). Altered blood vessel function and Raynaud's phenomenon have also been observed in copper smelter workers (Lagerkvist et al., 1986).

3.2. Coronary Heart Disease

Significantly increased mortality from cardiovascular disease has been observed among copper smelter workers in Anaconda, with a standardized mortality ratio 1.2 (Lee and Fraumeni, 1969). The standardized mortality ratio remained significant in a follow-up study of this smelter cohort, showing a dose-response relationship between level of arsenic exposure and mortality from ischemic heart disease (Welch et al., 1982).

Significant excess mortality from cardiovascular disease has also been found among copper smelter workers in Sweden, showing an overall standardized mortality ratio of 2.0 and a dose-response relationship to arsenic exposure (Axelson et al., 1978). Arsenic-exposed chimney sweeps also had significantly increased mortality from ischemic heart disease, with a standardized mortality ratio of 1.35 in Sweden (Gustavsson et al., 1987) and 2.2 in Denmark (Hansen, 1983). These chimney sweeps were exposed to soot, which contains polycyclic aromatic hydrocarbons and inorganic carcinogens, including arsenical compounds.

Chronic exposure to inorganic arsenic from drinking water has been well documented to be associated with the development of cardiovascular disease. Arsenic-related myocardial infarction and arterial thickening have been reported in several autopsy studies in Antofagasta, Chile (Rosenberg, 1974; Zaldivar, 1974, 1980; Moran et al., 1977). An autopsy of a Japanese worker chronically exposed to arsenic revealed lesions compatible with ischemic heart disease (Tanimoto et al., 1990). Significantly increased mortality from ischemic heart disease has also been observed among certified arsenic-poisoned patients in Toroku, Japan, showing a standardized mortality ratio fo 2.1 (Tsuda et al., 1990).

Significant excess mortality from cardiovascular disease has been reported among residents in the area of endemic blackfoot disease in Taiwan (Wu et al., 1989). The age-adjusted mortality rates per 100,000 person-years for male residents aged 20 years or more were 125.9, 154.0, and 259.5 in villages where the arsenic levels in drinking water were < 0.30, 0.30–0.59, and $\geqslant 0.60$ ppm, respectively. The corresponding figures for female residents were 91.1, 153.1, and 144.7, respectively. Blackfoot disease patients have also been found to have significantly increased mortality from cardiovascular disease, with a standardized mortality ratio of 2.1 (Chen et al., 1988b).

3.3. Cardiovascular Disease Risk Factors

Recent studies have documented that the risk factors of cardiovascular diseases, including hypertension and diabetes mellitus, are significantly associated with chronic exposure to inorganic arsenic from drinking water. An increased prevalence of hypertension has been observed among residents in the area of endemic blackfoot disease in Taiwan (C. J. Chen et al., 1992b). The prevalence odds ratios of hypertension were 1.0, 3.7, and 5.5 for residents who had a cumulative arsenic exposure of < 0.1, 0.1–15.0, and > 15.0 ppm-years respectively, after adjusting for the effects of age, sex, and other associated factors of hypertension. Increased prevalence of arterial hypertension has been observed among patients affected with chronic arsenicism compared with nonaffected controls in Antofagasta, Chile (Zaldivar, 1980).

Residents in the area of endemic blackfoot disease also had a significantly elevated prevalence of diabetes mellitus (C. J. Chen et al., 1992c). After being adjusted for the effects of age, sex, body mass index, and physical activity at work, the prevalence odds ratios of diabetes mellitus were 1.0, 4.0, and 6.3 for residents with a cumulative arsenic exposure of < 0.1, 0.1–15.0, and > 15.0 ppm-years, respectively.

4. MECHANISMS

4.1. Animal Experiments and In Vitro Studies

Various arsenic compounds have been tested for their roles in initiating and promoting carcinogenicity by perinatal treatment of mice, by intratracheal instillation in hamsters and rats, by skin application to mice and rabbits, by implantation into the stomach of rats, and by oral administration to rats, mice, and dogs. Only limited evidence shows carcinogenic effects of arsenic on experimental animals (IARC, 1987). Species differences in the metabolism of arsenic may be the reason for the discrepancy in arsenic-induced carcinogenicity between humans and experimental animals. There has never been a study on the atherogenic effect of inorganic arsenic on animals (WHO, 1981).

Inorganic arsenic has been tested for its genotoxicity in a variety of test systems ranging from bacteria to peripheral lymphocytes of exposed human beings. The Reproductive Effects Assessment Group of the U.S. Environmental Protection Agency has reported the following conclusions on the effects of inorganic arsenic: (1) Arsenic is either inactive or too weak to induce gene mutations in vitro; (2) arsenic is clastogenic and induces sister-chromatid exchanges in a variety of cell types; (3) arsenic does not seem to induce chromosome aberrations in vivo in experimental animals; (4) human beings exposed to arsenic have higher frequencies of sister-chromatid exchanges and chromosomal aberrations in peripheral lymphocytes; and (5) arsenic may affect DNA by inhibition of the repair process or by occasionally substituting for phosphorus in the DNA backbone (Jacobson-Kram and Montalbano, 1985). Arsenic also induces morphological transformation of cultured cells (IARC, 1987). Arsenic has recently been shown to induce gene amplification in mouse cells in culture (Lee et al., 1988). Further in vitro studies are needed to establish and characterize human cell lines from various stages of arsenic-induced neoplastic changes and to investigate cytogenetic anomalies, biochemical and cytological changes, and alterations of oncogenes and tumor suppressor genes in such cell lines; to examine the effects of arsenic on DNA replication and repair; and to compare arsenic-related metabolism, resistant mechanism, and oxidative stress induction in human and animal cells.

4.2. Multiplicity of Carcinogenesis

Arsenic has been found to be a human carcinogen in various organs. The multiplicity of inorganic-induced carcinogenicity that does not show any organotropism deserves further investigation. There is considerable evidence indicating the multistage nature of carcinogenesis (Weinstein, 1988). Multiple changes in oncogenes and tumor suppressor genes have been documented to be involved in the development of common cancers such as lung and colorectal cancers (Weston et al., 1989; Volgelstein et al., 1989). Such gene changes, especially the loss of tumor suppressor genes at the late stages resulting from increased chromosomal aberrations and sister-chromatid exchange frequencies, could also be involved in arsenic-induced cancers of the skin and other internal cancers. It is important to determine whether the changes occur and are consistent in different arsenic-induced cancers as well as to examine whether the changes are similar to those induced by other carcinogens.

Arsenic could also exert its human carcinogenicity without having any genotoxic effects. It might induce human cancers by inducing cell proliferation through its inhibition of thiol-dependent enzyme systems. From the viewpoint of multifactorial etiology, there might exist other factors that interact with arsenic in the induction of human cancers. These might play a role in genotoxicity, mitogenicity, or both. Further investigations of multiple risk factors of arsenic-induced cancers are recommended.

4.3. Systematic Atherosclerosis

Atherosclerosis has been well documented to be a multistage progression from fatty streaks to fibrous plaques to raised lesions. In an updated review of the development of atherosclerosis, Ross (1986) noted that there are two principal hypotheses: (1) the "response-to-injury" hypothesis, which suggests that an endothelial injury leads to intimal proliferation and a subsequent atherosclerotic lesion, and (2) the "monoclonal expansion" hypothesis, which suggests that each cell in an atherosclerotic lesion results from the proliferation of a single smooth-muscle cell. These two distinct hypotheses are not mutually exclusive and might just emphasize two major events in the multistage process of atherosclerosis. The monoclonal proliferation of smooth-muscle cells in the atherosclerotic process is similar to a neoplastic process.

The dual effect of arsenic on atherosclerosis and carcinogenesis might result from a common mechanism involving somatic mutations. In addition to arsenic, ionizing radiation, vinyl chloride monomer, tobacco smoke, and various industrial combustion effluents containing polycyclic aromatic hydrocarbons also cause both cancers and atherosclerotic diseases (Hansen, 1990). Tobacco smoke and combustion effluents of fuel contain many potentially hazardous agents, and it is difficult to identify a common etiological factor for these two categories of disease. In other words, arsenic, vinyl chloride monomer, and ionizing radiation are unusual in their potential to induce both carcinogenesis and atherosclerosis. Whether arsenic induces these two pathogenic processes through a common mechanism or through two distinct pathways deserves further evaluation.

Because arsenic is also associated with the development of hypertension and diabetes mellitus, it is important to determine whether arsenic induces atherosclerosis through an indirect effect on these cardiovascular disease risk factors or through a direct effect on the atherosclerotic process.

REFERENCES

Albores, A., Cebrián, M. E., Tellez, I., and Valdez, B. (1979). Comparative study of chronic hydroarsenicism in two rural communities in the lagoon region of Mexico. *Bol. Of. Sanit. Panam.* **86**, 196–203.

Alvarado, L. C., Viniegran, G., Garcia, R. E., and Acevedo, J. A. (1964). Arsenicism in the lake region. An epidemiologic study of arsenicism in the colonies of Miguel-Aleman and Eduardo Guerra of Toreeon, Coahvila (Mexico). *Salud Publ. Mex.* **6**(3), 375–385.

Arguello, A., Cenget, D., and Tello, E. (1938). Regional endemic cancer and arsenical intoxication in Cordoba. *Argent. Rev. Dermatosyphilol.* 22, Pt. 4.

Axelson, O., Dahlgren, E., Jansson, C. D., and Rehnlund, S. O. (1978). Arsenic exposure and mortality: A case reference study from a Swedish copper smelter. *Br. J. Ind. Med.* **35**, 8–15.

Biagini, R. E., Quiroga, G. C., and Elias, V. (1974). Chronic hydroarsenism in Urutau. *Archi. Argent. Dermatol.* **24**(1), 8–11.

Biagini, R. E., Rivero, M., Salvador, M., and Cordoba, S. (1978). Chronic arsenism and lung cancer. *Archi. Argent. Dermatol.* **48**, 151–158.

Blot, W. J., and Fraumeni, J. R., Jr. (1975). Arsenical air pollution and lung cancer. *Lancet* **2**, pp. 142–144.

Borgoño, J. M., and Greiber, R. (1972). Epidemiological study of arsenism in the city of Antofagasta. *Trace Subst. Environ. Health,* **5**, 13–24.

Borgoño, J. M., Vincent, P., Venturino, H., and Infante, A. (1977). Arsenic in the drinking water of the city of Antofagasta: Epidemiological and clinical study before and after the installation of the treatment plant. *Environ. Health Perspect,* **19**, 103–105.

Braun, W. (1958). Carcinoma of the skin and the internal organs caused by arsenic: Delayed occupational lesions due to arsenic. *Ger. Med. Mon.* **3**, 321–324.

Brown, C. C., and Chu, K. C. (1983). Implication of the multistage theory of carcinogenesis applied to occupational arsenic exposure. *JNCI, J. Natl. Cancer Inst.* **70**, 455–463.

Brown, K. G., Boyle, K. E., Chen, C. W., and Gibb, H. J. (1989). A dose-response analysis of skin cancer from inorganic arsenic in drinking water. *Risk Anal.* **9**, 519–528.

Brown, L. M., Pottern, L. M., and Blot, W. J. (1984). Lung cancer in relation to environmental pollutants emitted from industrial sources. *Environ. Res.* **34**, 250–261.

Butzengeiger, K. H. (1940). Peripheral circulation disorders in arsenism. *Klin. Wochenschr* **19**, 523–527.

Calnan, C. D. (1954). Arsenical keratoses and epitheliomas with bronchial carcinoma . *Proc. R. Soc. Med.* **47**, 405–406.

Cebrián, M. E. (1987). Some potential problems in assessing the effects of chronic arsenic poisoning in North Mexico. *Prep. Pap. Natl. Meet., Div. Environ. Chem., Am. Chem. Soc., 194th Meet.*, Abstr., pp. 114–116.

Cebrián, M. E., Albores, A., Aquilar, M., and Blakely, E. (1983). Chronic arsenic poisoning in the north of Mexico. *Hum. Toxicol.* **2**, 121–133.

Chavez, A., Perez Hidalgo, C., Tovar, E., and Garmilla, M. (1964). Studies in a commuity with chronic endemic arsenic poisoning. *Salud. Publ. Mex.* **6**(3), 435–442.

Chen, C. J. (1990). Blackfoot disease (Letter). *Lancet* **2**, 442.

Chen, C. J., and Wang, C. J. (1990). Ecological correlation between arsenic level in well water and age-adjusted mortality from malignant neoplasms. *Cancer Res.* **50**, 5470–5474.

Chen, C. J., Chuang, Y. C., Lin, T. M., and Wu, H. Y. (1985). Malignant neoplasms among residents of a blackfoot disease-endemic area in Taiwan: High-arsenic artesian well water and cancers. *Cancer Res.* **45**, 5895–5899.

Chen, C. J., Chuang, Y. C., You, S. L., Lin, T. M., and Wu, H. Y. (1986). A retrospective study on malignant neoplasms of bladder, lung, and liver in blackfoot disease endemic area in Taiwan. *Br. J. Cancer* **53**, 399–405.

Chen, C. J., Kuo, T. L., and Wu, M. M. (1988a). Arsenic and cancers. *Lancet* **1**, 414–415.

Chen, C. J., Wu, M. M., Lee, S. S., Wang, J. D., Cheng, S. H., and Wu, H. Y. (1988b). Atherogenicity and carcinogenicity of high-arsenic artesian well water: Multiple risk factors and related malignant neoplasms of blackfoot disease. *Arteriosclerosis*, **8**, 452–460.

Chen, C. J., Chen, C. W., Wu, M. M., and Kuo, T. L. (1992a). Cancer potential in liver, lung, bladder, and kidney due to ingested inorganic arsenic in drinking water. *Br. J. Cancer* **66**, 888–892.

Chen, C. J., Hus, M. P., Chen, S. Y., and Wu, M. M. (1992b). Increased prevalence of hypertension among residents in hyperendemic villages of chronic arsenicism in Taiwan. *Abstr. 4th Annu. Meet. Int. Soc. Environ. Epidemiol.*, Cuernavaca, Mexico.

Chen, C. J., Lai, M. S., Hsu, M. P., Chen, S. Y., Wu, M. M., Lu, S. N., Wu, T. J., and Tai, T. Y. (1992c). Dose-response relationship between ingested inorganic arsenic and diabetes mellitus. *Abstr. 4th Annu. Meet. Int. Soc. Environ. Epidemiol.*, Cuernavaca, Mexico.

Chen, K. P., and Wu, H. Y. (1962). Epidemiologic studies on blackfoot disease: II. A study of source of drinking water in relation to the disease. *J. Formosan Med. Assoc.* **61**, 611–617.

Chen, S. Y., Yu., H.S ., Chen, K. H., Kuo, T. L., and Chen, C. J. (1992). Early arsenic exposure and skin cancer— a community-based case-control study in an endemic area of blackfoot disease. *J. Natl. Public Health Assoc. (ROC)* **12**, 1–9.

Chen, W. Y., and Lien, W. P. (1963). Experimental studies on the drinking water blackfoot endemic area I: Studies on the change of limbs of experimental rats receiving repeated injections of drinking water. *J. Formosan Med. Assoc.* **62**, 794–805.

Chi, I. C., and Blackwell, R. Q. (1968). A controlled retrospective study of blackfoot disease, an endemic peripheral gangrene disease in Taiwan. *Am. J. Epidemiol.* **88**, 7–24.

Cordier, S., Theriault, G., and Iturra, H. (1983). Mortality patterns in a population living near a copper smelter. *Environ. Res.* **31**, 311–322.

Cuzick, J., Sasieni, P., and Evans, S. (1992). Ingested arsenic, Keratosis, and bladder cancer. *Am. J. Epidemiol*, **136**, 417–421.

Deng, J. S., and How, S. W. (1977). Comparison of the fine structure of basal cell carcinomas in ordinary patients and patients with chronic arsenicalism. *J. Formosan Med. Assoc.* **76**, 685–692.

Enterline, P. E., and Marsh, G. M. (1980). Mortality studies of smelter workers. *Am. J. Ind. Med.* **1**, 251–259.

Enterline, P. E., and Marsh, G. M. (1982). Mortality among workers exposed to arsenic and other substances in a copper smelter. *Am. J. Epidemiol.* **116**, 895–910.

Enterline, P. E., Henderson, V. L., and Marsh, G. M. (1987). Exposure to arsenic and respiratory cancer. A reanalysis. *Am. J. Epidemiol.* **125** (6), 929–938.

Falk, H., Caldwell, C. G., and Ishak, K. G. (1981a). Arsenic-related hepatic angiosarcoma. *Am J. Ind. Med.* **2**, 43–50.

Falk, H., Herbert, J. T., and Edmonds, L. (1981b). Review of four cases of childhood hepatic angiosarcoma—elevated environmental arsenic exposure in one case. *Cancer* (*Philedelphia*) **47**, 383–391.

Fierz, U. (1965). Catamnestic investigations of the side effects of therapy of skin diseases with inorganic arsenic. *Dermatologica* **131**, 41–58.

Goldman, A. L. (1973). Lung cancer in Bowen's disease. *Am. Rev. Respir. Dis.* **108**, 1205–1207.

Grobe, J. W. (1976). Periphere Druchblutungsstorungen und Akrocyanose bei arsengeschadigten Moselwinzern. (Peripheral circulatory disorders and acrocyanosis with arsenic.) *Berufs-Dermatosen* **24**, 78–84.

Gustavsson, P., Gustavsson, A., and Hogstedt, C. (1987). Excess mortality among Swedish chimney sweeps. *Br. J. Ind. Med.* **44**, 738–743.

Hamada, T., and Horiguchi, S. (1976). Occupational chronic arsenical poisoning. On the cutaneous manifestations. *Jpn. J. Ind. Health* **18**, 103–105.

Hansen, E. S. (1983). Mortality from cancer and ischemic heart disease in Danish chimney sweeps: A five-year follow-up. *Am. J. Epidemiol.* **117**, 160–164.

Hansen, E. S. (1990). Shared risk factors for cancer and atherosclerosis: A review of the epidemiological evidence. Mutat. Res. **239**, 163–179.

Higgins, I., Welch, K., and Burchfiel, C. (1982). *Mortality of Anaconda Smelter Workers in Relation to Arsenic and Other Exposures.* Department of Epidemiology, University of Michigan, Ann Arbor.

Hill, A. B., and Faning, E. L. (1948). Studies on the incidence of cancer in a factory handling inorganic compounds of arsenic. I. Mortality experience in the factory. *Br. J. Ind. Med.* **5**, 1–6.

Huang, Y. Z., Qian, X. C., and Wang, G. Q. (1985). Endemic chronic arsenism in Xinjiang. *Chin. Med. J.* **98**, 219–222.

Hugo, N. E., and Conway, H. (1967). Bowen's disease: Its malignant potential and relationship to systemic cancer. *Plast. Reconstr. Surg.* **39**, 190–194.

Hunter, F. T., Kip A. F., and Irvine, W. (1942). Radioactive tracer studies on arsenic injected as potassium arsenite. *J. Pharmacol. Exp. Ther.* **76**, 207.

Hutchinson, J. (1888). On some examples of arsenic-keratosis of the skin and of arsenic-cancer. *Trans. Pathol. Soc. London.* **39**, 352–363.

International Agency for Research on Cancer. (IARC) (1980). *Arsenic and Its Compounds,* Vol. 23. IARC, Lyon.

International Agency for Research on Cancer (IARC) (1982). *Evaluation of the Carcinogenic Risk of Chemical to Humans, Supplement 4.* IARC, Lyon, pp. 55–51.

International Agency for Research on Cancer (IARC) (1987). *Evaluation of Carcinogenic Risks to Humans, supplement 7.* IARC, Lyon, pp. 100–106.

Jacobson-Kram, D., and Montalbano, D. (1985). The Reproductive Effects Assessment Group's report on the mutagenicity of inorganic arsenic. *Environ. Mutagen,* **7**, 787–804.

Kasper, M. L., Schoenfield, L., Strom, R. L., and Theologides, A. (1984). Hepatic angiosarcoma and bronchiolalveolar carcinoma induced by Fowler's solution. *JAMA, J. Am. Med. Assoc.* **252**, 3407–3408.

Lagerkvist, B., Linderholm, H., and Nordberg, G.F. (1986) . Vasospastic tendency and Raynaud's phenomenon in smelter workers exposed to arsenic. *Environ. Res.* **39**, 463–474.

Lander, J. J., Stanley, R. J., Sumner, H. W., Dee, C., Boswell, D. C., and Arch, R. D. (1975). Angiosarcoma of the liver associated with Fowler's solution (potassium arsenite). *Gastroenterology* **68**, 1582–1586.

Lee, A. M., and Fraumeni, J. F., Jr. (1969). Arsenic and respiratory cancer in man: An occupational study. *J. Natl. Cancer Inst. (U.S.)* **42**, 1045–1052.

Lee, T. C., Thanka, N., Lamb, W., Gilmer, T. M., and Barrett, J. C. (1988). Induction of gene amplification by arsenic. *Science,* **241**, 79–81.

Lee-Feldstein, A. (1983). Arsenic and respiratiory cancer in man: Follow-up of an occupational study. In W. Lederer and R. Fensterheim (Eds), *Arsenic: Industrial, Biomedical, and Environmental Perspectives.* Van Nostrand-Reinhold, New York, pp. 245–254.

Lee-Feldstein, A. (1986). Cumulative exposure to arsenic and its relationship to respiratory cancer among copper smelter employees. *J. Occup. Med.* **28**, 296–302.

Lu, F. J. (1990). Blackfoot disease: Arsenic or humic acid? (Letter.) *Lancet* **2**, 115–116.

Lu, F. J., Yang, C. K., and Lin, K. H. (1975). Physico-chemical characteristics of drinking water in blackfoot endemic areas in Chia-I and Tainan Hsiens. *J. Formosan Med. Assoc.* **79**, 21–22.

Lubin, J. H., Pottern, L. M., Blot, W. J., Tokudome, S., Stone, B. J., and Fraumeni, J. F., Jr. (1981).

Respiratory cancer among copper smelter workers: Recent mortality statistics. *J. Occup. Med.* **23**, 779–784.

Luchtrath, H. (1983). The consequences of chronic arsenic poisoning among Moselle wine growers: pathoanatomical investigations of post-mortem examinations performed between 1960 and 1977. *J. Cancer Res. Clin. Oncol.* **105**, 173–182.

Mabuchi, K., Lilienfeld, A., and Snell, L. (1979). Lung cancer among pesticide workers exposed to inorganic arsenicals. *Arch. Environ. Health.* **34**, 312–320.

Matanoski, G., Landau, E., and Tonascia, J. (1981). *Lung Cancer Mortality in Proximity to a Pesticide Plant.* American Public Health Association, Environmental Protection Agency Office of Toxic Substances, Washington, DC.

Mazumdar, S., Redmond, C. K., Enterline, P. E., Marsh, G. M., Costantino, J. P., Zhou, S. Y. J., and Patwardhan, R. N. (1989). Multistage modeling of lung cancer mortality among arsenic-exposed copper-smelter workers. *Risk Anal.* **9**, 551–563.

Milham, S., Jr., and Strong, T. (1974). Human arsenic exposure in relation to a copper smelter. *Environ. Res.* **7**, 176–182.

Moller, R., Nielsen, A., and Reymann, F. (1975). Multiple basal cell carcinoma and internal malignant tumors. *Arch. Dermatol.* **111**, 584–585.

Moran, S., Maturana, G., Rosenberg, H., Casanegra, P., and Dubernet, J. (1977). Occlusions coronariennes liées à une intoxication arsenical chronique. *Arch. Mal. Coeurvaiss.* **70**(10), 1115–1120.

Nagy, G., Nemeth, A., Bodor, F., and Ficsor, E. (1980). Cases of bladder cancer caused by chronic arsenic poisoning. *Orv. Hetil.* **121**, 1009–1011.

Nurse, D.S. (1978). Hazards of inorganic arsenic. *Med. J. Aust.* **1**, 102.

Osburn, H.S. (1957). Cancer of the lung in Gwanda. *Cent. Afr. J. Med.* **3**, 215–223.

Osburn, H. S. (1969). Lung cancer in a mining district in Rhodesia. *S. Afr. Med. J.* **43**, 1307–1312.

Ott, M. G., Holder, B. B., and Gordon, H. I. (1974). Respiratory cancer and occupational exposure to arsenicals. *Arch. Environ. Health.* **29**, 250–255.

Perry, K., Bowler, R. G., Buckell, H. M., Druett, H. A., and Schilling, RSF. (1948). Studies in the incidence of cancer in a factory handling inorganic compounds of arsenic. II: Clinical and environmental investigations. *Br. J. Ind. Med.* **5**, 6–15.

Pershagen, G. (1985). Lung cancer mortality among men living near an arsenic-emitting smelter. *Am. J. Epidemiol.* **122**, 684–696.

Pershagen, G., Elinder, C. G., and Bolander, A. M. (1977). Mortality in a region surrounding an arsenic emiting plant. *Environ. Health Perspect.* **19**, 133–137.

Pinto, S. S., Enterline, P. E., Henderson, V., and Varner, M. O. (1977). Mortality experience in relation to a measured arsenic trioxide exposure. *Environ. Health Perspect.* **19**, 127–130.

Pinto, S. S., Henderson, V., and Enterlin, P. E. (1978). Mortoality experience of arsenic-exposed workers. *Arch. Environ. Health.* **33**, 325–331.

Popper, H., Thomas, L. B., Telles, N. C., Falk, H., and Selikoff, I. J. (1978). Development of hepatic angiosarcoma in man induced by vinyl chloride, thorotrast, and arsenic. *Am. J. Pathol.* **92**, 349–376.

Regelson, W., Kim, U., Ospina, J., and Holland, J. F. (1968). Hemangioendothelial sarcoma of liver from chronic arsenic intoxication by Fowler's solution. *Cancer (Philadelphia)* **21**, 514–522.

Rencher, A. C., Carter, M. W., and McKee, D. W. (1977). A retrospective epidemiological study of mortality at a large western copper smelter. *J. Occup. Med.* **19**, 754–758.

Rennke, H., Part, G. A., Etcheverry, R. B., Katz, R. U., and Donoso, S. (1971). Malignant hemangioendothelioma of the liver and chronic arsenicism. *Rev. Med. Chile.* **99**(9), 664–668.

Roat, J. W., Wald, A., Mendelow, H., and Pataki, K. I. (1982). Hepatic angiosarcoma associated with short-term arsenic ingestion. *Am. J. Med.* **73**, 933–936.

Robertson, D. A., and Low-Beer, T. S. (1983). Long term consequences of arsenical treatment for multiple sclerosis. *Br. Med. J.* **286**, 605–606.

Robson, A. O., and Jelliffe, A. M. (1963). Medicinal arsenic poisoning and lung cancer. *Br. Med. J.* **2**, 204–209.

Rosenberg, H. G. (1974). Systemic arterial disease and chronic arsenicism in infants. *Arch. Pathol.* **97**, 360.

Ross, R. (1986). The pathogenesis of atherosclerosis: An update. *N. Engl. J. Med.* **314**, 488–500.

Rosset, M. (1958). Arsenical keratoses associated with carcinomas of the internal organs. *Can. Med. Assoc. J.* **78**, 416–419.

Roth, F. (1957). Concerning the delayed effects of chronic arsenic of the Moselle wine growers. *Dtsch. Med. Wochenschr.* **82**, 211–217.

Roth, F. (1958). The sequelae of chronic arsenic poisoning in Moselle vintners. *Ger. Med. Mon.* **2**, 172–175.

Salcedo, J. C., Portales, A., Landecho, E., and Diaz, R. (1984). Transverse study of a group of patients with vasculopathy from chronic arsenic poisoning in communities of the Francisco de Madero and San Pedro Districts, Coahuila, Mexico. *Rev. Fac. Med. (Torreon)* **12**, 16.

Shanhon, R. L., and Strayer, D. S. (1987). *Arsenic-Induced Skin Toxicity.* U.S. Environmental Protection Agency, Washington, DC.

Sommers, S. C., and McManus, R. G. (1953). Multiple arsenical cancers of the skin and internal organs. *Cancer (Philadelphia)* **6**, 347–359.

Tanimoto, A., Hamada, T., Kanesaki, H., Matsuno, K., and Koide, O. (1990). Multiple primary cancers in a case of chronic arsenic poisoning: An autopsy report. *J. UOEH* **12**, 89–99.

Thiers, H., Colomb, D., Moulin, C., and Colin, L. (1967). Le cancer cutane arsenical des viticulteurs du Beaujolais. *Ann. Dermatol. Syphiligr.* **94**, 133–158.

Tokudome, S., and Kuratsune, M. (1976). A cohort study on mortality from cancer and other causes among workers at a metal refinery. *Int. J. Cancer* **17**, 310–317.

Tseng, W. P. (1977). Effects and dose-response relationships of skin cancer and blackfoot disease with arsenic. *Environ. Health Perspect.* **19**, 109–119.

Tseng, W. P., Chen, W. Y., Sung, J. L., and Chen, J. S. (1961). A clinical study of blackfoot disease in Taiwan: An endemic peripheral vascular disease. *Mem. Coll. Med. Natl. Taiwan Univ.* **7**, 1–18.

Tseng, W. P., Chu, H. M., How, S. W., Fong, J. M., Lin, C. S., and Yeh, S. (1968). Prevalence of skin cancer in an endemic area of chronic arsenicism in Taiwan. *J. Natl. Cancer Inst. (U.S.)* **40**(3), 453–463.

Tsuda, T., Nagira, T., Yamamoto, M., Kurumatani, N., Hotta, N., Harada, M., and Aoyama, H. (1989). Malignant neoplasma among residents who drank well water contaminated by arsenic from a King's Yellow factory. *Sangyo Ika Daigaku Zasshi* **11**, 289–301.

Tsuda, T., Nagira, T., Yamamoto, M., and Kume, Y. (1990). An epidemiological study on cancer in certified arsenic poisoning patients in Toroku. *Ind. Health* **28**, 53–62.

U.S. Environmental Protection Agency (USEPA) (1988). *Risk Assessment Forum. Special Report on Ingested Inorganic Arsenic: Skin Cancer, Nutritional Essentiality.* USEPA, Washington, DC.

U.S. Public Health Service (1989). *Toxicological Profile for Arsenic.* USPHS, Washington, DC.

Vogelstein, B., Fearon, E. R., Kern, S. E., Hamilton, S. R., Preisinger, A. C., Nakamura, Y., and White, R. (1989). Allelotype of colorectal carcinoma. *Science* **244**, 207.

von Roemeling, R., Hartwich, G., and König, H. (1979). Occurrence of tumours at different locations after arsenic therapy. *Med. Welt* **30**, 1928–1929.

Wall, S. (1980). Survival and mortality pattern among Swedish smelter workers. *Int. J. Epidemiol.* **9**, 73–87.

Watrous, R. M., and McCaughey, M. B. (1945). Occupational exposure to arsenic in the manufacture of arsphenamine and related compounds. *Ind. Med.* **14**, 639–646.

Weinstein, I. B. (1988). The origins of human cancer: Molecular mechanisms of carcinogenesis and their implications for cancer prevention and treatment. *Cancer Res.* **48**, 4135.

Welch, K., Higgins, I., Oh, M., and Burchfiel, C. (1982). Arsenic exposure, smoking, and respiratory cancer in copper smelter workers. *Arch. Environ. Health* **37**, 325–335.

Weston, A., Willey, J. C., Modali, R., Sugimura, H., McDowell, E. M., Resau, J., Light, B., Haugen, A., Mann, D. L., Trump, B. F., and Harris, C. C. (1989). Differential DNA sequence deletions from chromosomes 3, 11, 13, and 17 in squamous-cell carcinoma, large-cell carcinoma, and adenocarcinoma of human lung. *Proc. Natl. Acad. Sci. U.S.A.* **86**, 5099–5103.

World Health Organization (WHO) (1981). *Environmental Health Criteria 18: Arsenic.* WHO, Geneva.

Wu, H. Y., Chen, K. P., Tseng, W. P., and Hsu, C. L. (1961). Epidemiologic studies on blackfoot disease: I. Prevalence and incidence of the disease by age, sex, year, occupation and geographical distribution. *Mem. Coll. Med. Natl. Taiwan Univ.* **7**, 33–50.

Wu, M. M., Kuo, T. L., and Hwang, Y. H., and Chen, C. J. (1989). Dose-response relation between arsenic concentration in well water and mortality from cancer and vascular disease. *Am. J. Epidemiol.* **130**, 1123–1131.

Yamashita, N., Doi, M., Nshio, M., Hojo, H., and Masato, T. (1972). Current state of Kyoto children poisoned by arsenic-tainted Morinaga dry milk. *Jpn. J. Hyg.* **27**(4), 364–399.

Yeh, S. (1973). Skin cancer in chronic arsenicism. *Hum. Pathol.* **4**(4), 469–485.

Yeh, S., and How, S. W. (1963). A pathological study on blackfoot disease in Taiwan. *Rep. Ins. Pathol. Natl. Taiwan Univ.* **14**, 25–73.

Yeh, S., How, S. W., and Lin, C. S. (1968). Arsenical cancer of skin—histologic study with special reference to Bowen's disease. *Cancer, (Philadelphia)* **21** (2), 312–339.

Yoshikawa, T., Utsumi, J., Okada, T., Moriuchi, M., Ozawa, K., and Kaneko, Y. (1960). Concerning the mass outbreak of chronic arsenic toxicosis in Niigata Prefecture. *Chiryo* **42**, 1739–1749.

Zaldivar, R. (1974). Arsenic contamination of drinking water and foodstuffs causing endemic chronic poisoning. *Beitr. Pathol.* **151**, 384–400.

Zaldivar, R. (1980). A morbid condition involving cardiovascular, bronchopulmonary, digestive and neural lesions in children and young adults after dietary arsenic exposure. *Zentralbl. Bakteriol. Abt. I, Orig. B* **170**, 44–56.

7

EFFECTS OF ARSENIC ON DNA SYNTHESIS IN HUMAN LYMPHOCYTES

Ziqiang Meng

Department of Environmental Science, Shanxi University, Taiyuan 030006 China

1. INTRODUCTION

Arsenic is a metalloid and is widely distributed in nature. Epidemiological studies have shown that arsenic exposure is indeed correlated with the increased incidence of skin, lung, and possibly liver cancers in human beings. However,

Arsenic in the Environment, Part II: Human Health and Ecosystem Effects,
Edited by Jerome O. Nriagu.
ISBN 0-471-30436-0

there is no reliable evidence that arsenic induces tumors in experimental animals (Leonard and Lauwerys, 1980; Leonard, 1984). In some experiments, arsenicals have not produced any tumors (Hueper and Payne, 1962), and these compounds have even been reported to minimize the induction of tumors (Boutwell, 1963; Milner, 1969). Therefore, arsenicals are considered to be comutagens and cocarcinogens. A possible mechanism of the comutagenic and cocarcinogenic properties of arsenic could be processes that affect DNA synthesis or that inhibit DNA repair, resulting in abnormal DNA. Trivalent and pentavalent arsenicals are known to inhibit DNA, RNA, and protein synthesis (Sibatani, 1959; Petres et al., 1977; Nakamuro and Sayato, 1981). Recently, arsenic was reported to stimulate the synthesis of eight proteins and to inhibit the synthesis of two proteins in rat kidney proximal tubule epithelial cells (Aoki et al. 1990). To gain a better understanding of the mechanisms by which arsenic affects target cell populations, we studied the effects of arsenic on DNA synthesis in human peripheral blood lymphocytes from adult males and females unexposed to arsenic. Our results show that arsenic at very low concentrations enhances DNA synthesis in human lymphocytes stimulated by phytohemagglutinin (PHA) and at high concentrations inhibits DNA synthesis.

2. MATERIALS AND METHODS

2.1. Test Substances

Sodium arsenite ($NaAsO_2$), sodium arsenate (Na_2HAsO_4), and arsenic trioxide (As_2O_3) were dissolved in distilled water and added directly to the culture medium. The final concentrations of sodium arsenite in the cultures ranged from 8×10^{-9} to 1×10^{-2} M, those of arsenic trioxide ranged from 8×10^{-9} to 1×10^{-5} M, and those of sodium arsenate ranged from 1×10^{-7} to 1×10^{-2} M.

2.2. Lymphocyte Preparation

For the study of DNA synthesis and mitogenesis of lymphocytes treated with arsenic for 72 hr, heparinized whole peripheral blood from healthy adult males and females unexposed to arsenic was used. For the study of DNA synthesis in lymphocytes treated with arsenic for 1 hr, peripheral blood lymphocytes were obtained from heparinized blood after separation on a Ficoll-Paque density gradient, as described by Friedmann and Rogers (1980). Cells harvested from the interphase were washed with phosphate-buffered saline (PBS) solution and then resuspended in culture medium RPMI 1640.

2.3. Mitogenesis of Lymphocytes

Heparinized whole blood (0.30 mL) was added to 4.70 mL RPMI 1640 medium supplemented with 20% heat-inactivated fetal-calf serum, penicillin (100 units/mL), streptomycin (100 μg/mL), and phytohemagglutinin P (PHA-P) (7.5 μg/mL) for 72 hr at 37 °C in a 5% CO_2 atmosphere. Arsenic compounds at the indicated concentrations were added at culture initiation. The lymphocyte mitotic indexes were measured after 72 hr of incubation to assess the effect of arsenic on the mitogenesis of PHA-stimulated human blood lymphocytes. For each test point, 2000 lymphocytes were analyzed, and the cell mitotic index was calculated as follows:

$$\text{Lymphocyte mitotic index} = (\text{No. of mitotic lymphocytes})/(\text{No. of total lymphocytes scored})$$

2.4. DNA Synthesis

2.4.1. Lymphocytes Treated with Arsenic for 72 Hours

Heparinized whole blood (0.30 mL) was added to 4.70 mL RPMI 1640 medium supplemented with 20% heat-inactivated fetal-calf serum, penicillin (100 units/mL), streptomycin (100 μg/mL), and PHA-P (7.5 μg/mL) for 72 hr at 37 °C in a 5% CO_2 atmosphere. Inorganic arsenic compounds at the indicated concentrations were added at culture initiation. Twenty hours before interrupting the cultures, 7.4×10^4 Bq of ^{3}H-thymidine (^{3}H-TdR) (specific activity: 6.66×10^{11} Bq/mM) in 100 μL saline was added to each tube. The tubes were then put in an icewater bath and 10 mL of distilled water was added to each tube. The cells were then collected onto glass-fiber filters (Friedmann and Rogers, 1980) and washed with distilled water, 5% trichloroacetic acid (TCA), and then absolute ethanol. The filters were dried overnight at room temperature or at 50 °C for 2 hr. Incorporated radioactivity was counted in a Packard liquid scintillation spectrometer using a scintillation cocktail of 4 g ppo and 0.4 g popop dissolved in 1 L toluene. The cultures were prepared in triplicate.

2.4.2. Lymphocytes Treated with Arsenic for 1 Hour

Lymphocytes separated at a density of 5×10^5 cells/mL were cultured in RPMI 1640 medium supplemented with 20% heat-inactivated fetal-calf serum, penicillin (100 units/mL), streptomycin (100 μg/mL), and PHA-P (7.5 μg/mL) for 48 hr at 37 °C. The cells cultured in all the tubes were harvested together and resuspended at a density of 2×10^6 cells/mL in RPMI 1640 medium to which penicillin (100 units/mL) and streptomycin (100 μg/mL), but not fetal-calf serum and PHA, had been added. One milliliter of the cell suspension was added to each tube. Inorganic arsenic compounds at the indicated concentrations were added to each tube. The cells were incubated for 1 hr at 37 °C and then washed twice with RPMI 1640 medium without arsenic. ^{3}H-TdR (3.70×10^5 Bq) (specific activity: 6.66×10^{11} Bq/mM) was added to each tube, and the cells were then incubated for 30 minutes at 37 °C. The cells were collected on glass-fiber filters and the filters were washed, dried, and counted. The cultures were prepared in triplicate.

3. RESULTS

The effects of exposure to trivalent arsenic (As_2O_3 and $NaAsO_2$) and pentavalent arsenic (Na_2HAsO_4) for 72 hr on the cell mitotic index and on the incorporation of ^{3}H-TdR into human lymphocytes stimulated by PHA in vitro are shown in Tables 1, 2, and 3. In all cases, there was an initial stimulation of ^{3}H-TdR incorporation into the lymphocytes by arsenic at very low concentrations, followed by an inhibition of ^{3}H-TdR incorporation into the cells as the concentration of arsenic was raised. The concentrations of As_2O_3, $NaAsO_2$, and Na_2HAsO_4 at which maximum stimulation of ^{3}H-TdR incorporation occurred were 0.1×10^{-6}, 0.5×10^{-6} to 1.0×10^{-6}, and 2×10^{-6} M, respectively. However, no difference between the cell mitotic indexes of lymphocytes unexposed and exposed to arsenic at very low concentrations was found, so arsenic at these concentrations did not enhance mitogenesis of the lymphocytes. Hence, the increase in ^{3}H-TdR incorporation into the lymphocytes was due to stimulation of DNA synthesis in the lymphocytes by arsenic at very low concentrations rather than to augmentation of mitogenesis of the cells.

Tables 1, 2, and 3 also show that individuals might vary in their susceptibility to arsenic stimulation of DNA synthesis in lymphocytes.

Tables 4, 5, and 6 summarize the effects of a 1 hr treatment with arsenic compounds on DNA synthesis in PHA-stimulated human lymphocytes. It was shown that As_2O_3, $NaAsO_2$, and Na_2HAsO_4 at very low concentrations all increased DNA synthesis in the lymphocytes, followed by a decrease in DNA synthesis as

Table 1 DNA Synthesis and Mitogenesis of PHA-Stimulated Human Lymphocytes Exposed to Sodium Arsenite for 72 Hr

Donor	$NaAsO_2$ ($\times 10^{-6}$ M)	Cell Mitotic Index (%)	Incorporation of ^{3}H-TdR (cpm) ($\bar{X} \pm$ SD)	Increase (%)	Decrease (%)
A	0	3.85 ± 0.11	29594 ± 946	—	—
	0.1	3.81 ± 0.13	36688 ± 1111**	23.97	
	0.5	3.86 ± 0.15	49655 ± 1830**	67.79	
	1.0	3.78 ± 0.12	56583 ± 1946**	91.20	
	2.0	3.40 ± 0.11	29754 ± 1002**	0.54	
	5.0	1.81 ± 0.18	5511 ± 340**		81.38
B	0	3.14 ± 0.13	5863 ± 92	—	—
	0.1	3.20 ± 0.11	6410 ± 572	9.33	
	0.5	3.18 ± 0.11	8023 ± 212**	36.84	
	1.0	3.17 ± 0.14	6753 ± 187**	15.17	
	2.0	2.90 ± 0.15	6067 ± 188	3.49	
	5.0	1.12 ± 0.19	756 ± 59**		87.11
C	0	4.55 ± 0.11	70439 ± 1171	—	—
	0.008	4.61 ± 0.14	72514 ± 1052	2.28	
	0.1	4.58 ± 0.11	84140 ± 1024**	15.25	
	0.5	4.59 ± 0.15	91316 ± 1185**	21.70	
	1.0	4.48 ± 0.17	83093 ± 1081**	13.75	
	2.0	2.43 ± 0.31	25774 ± 1201**		46.14

** $p < 0.01$ versus controls by t test.

Table 2 DNA Synthesis and Mitogenesis of PHA-Stimulated Human Lymphocytes Exposed to Arsenic Trioxide for 72 Hr

Donor	As_2O_3 ($\times 10^{-6}$ M)	Cell Mitotic Index (%)	Incorporation of ^{3}H-TdR (cpm) ($\bar{X} \pm SD$)	Increase (%)	Decrease (%)
D	0	3.55 ± 0.13	13244 ± 1060	—	—
	0.1	3.51 ± 0.12	18001 ± 1046**	35.92	
	0.5	3.56 ± 0.15	17514 ± 1012**	32.24	
	1.0	3.01 ± 0.19	7130 ± 117**		46.16
	2.0	2.78 ± 0.34	823 ± 22**		93.79
	5.0	0.00	159 ± 19**		98.80
E	0	3.78 ± 0.14	16794 ± 1025	—	—
	0.1	3.77 ± 0.11	18180 ± 1029	8.25	
	0.5	3.80 ± 0.17	16522 ± 1287		1.62
	1.0	2.21 ± 0.45	6623 ± 105**		60.56
	2.0	1.01 ± 0.63	623 ± 40**		96.29
	5.0	0.00	438 ± 39**		97.39
F	0	5.12 ± 0.16	68753 ± 989	—	—
	0.008	5.22 ± 0.38	68745 ± 1144	—	—
	0.1	5.14 ± 0.22	74677 ± 1152**	8.62	
	0.5	5.13 ± 0.21	70483 ± 1229	2.52	
	1.0	2.82 ± 0.87	30348 ± 1138**		55.86
	2.0	0.50 ± 0.20	8168 ± 146**		88.12
G	0	3.15 ± 0.11	1452 ± 132	—	—
	0.008	3.14 ± 0.13	1930 ± 123*	32.92	
	0.1	3.01 ± 0.11	2964 ± 132**	104.13	
	0.5	3.01 ± 0.17	2342 ± 145**	61.29	
	1.0	2.73 ± 0.58	1046 ± 124*		27.96
	2.0	0.12 ± 0.06	321 ± 56**		77.89

*$p < 0.05$. ** $p < 0.01$ versus controls by t test.

Table 3 DNA Synthesis and Mitogenesis of PHA-Stimulated Human Lymphocytes Exposed to Sodium Arsenate for 72 Hr

Donor	Na_2HAsO_4 ($\times 10^{-6}$ M)	Cell Mitotic Index (%)	Incorporation of ^{3}H-TdR (cpm) ($\bar{X} \pm SD$)	Increase (%)	Decrease (%)
H	0	4.24 ± 0.15	58266 ± 1261	—	—
	0.5	4.28 ± 0.14	59217 ± 1273	1.63	
	1.0	4.17 ± 0.15	67562 ± 1380**	15.95	
	2.0	4.20 ± 0.13	78582 ± 1377**	34.87	
	5.0	4.23 ± 0.16	67167 ± 1106**	15.28	
	10.0	3.94 ± 0.86	28626 ± 1393**		50.87
	100.0	1.12 ± 0.38	8980 ± 916**		84.59

** $p < 0.01$ versus control by t test.

Table 4 DNA Synthesis in PHA-Stimulated Human Lymphocytes Exposed to Sodium Arsenite for 1 Hr

Concentration of $NaAsO_2$ (M)	Incorporation of ^{3}H-TdR (cpm) ($\bar{X} \pm SD$)	Enhancement (%)	Inhibition (%)
0	10944 ± 2001	—	—
1×10^{-7}	11797 ± 998	7.79	
1×10^{-6}	15855 ± 1960**	44.86	
1×10^{-5}	13020 ± 985	18.97	
1×10^{-4}	5242 ± 200**		52.10
1×10^{-3}	4195 ± 195**		61.67
1×10^{-2}	6219 ± 198*		43.17

*$p < 0.05$. **$p < 0.01$ versus control by *t* test.

Table 5 DNA Synthesis in PHA-Stimulated Human Lymphocytes Exposed to Arsenic Trioxide for 1 Hr

Concentration of As_2O_3 (M)	Incorporation of ^{3}H-TdR (cpm) ($\bar{X} \pm SD$)	Enhancement (%)	Inhibition (%)
0	27647 ± 1131	—	—
0.5×10^{-7}	30912 ± 1084[b]	11.80	
0.1×10^{-6}	31995 ± 1506[b]	15.73	
0.5×10^{-6}	33439 ± 1145[a]	20.95	
1.0×10^{-6}	26722 ± 1264		3.35
2.0×10^{-6}	25561 ± 639[b]		7.55
1.0×10^{-5}	14428 ± 723[a]		44.81

[b] $p < 0.05$. [a] $p < 0.01$ versus control by *t* test.

Table 6 DNA Synthesis in PHA-Stimulated Human Lymphocytes Exposed to Sodium Arsenite for 1 Hr

Concentration of Na_2HAsO_4 (M)	Incorporation of ^{3}H-TdR (cpm) ($\bar{X} \pm SD$)	Enhancement (%)	Inhibition (%)
0	12946 ± 1143	—	—
1×10^{-7}	14107 ± 1146	8.97	
1×10^{-6}	18849 ± 1085**	45.60	
1×10^{-5}	21103 ± 1056**	63.00	
1×10^{-4}	12747 ± 1259		1.54
1×10^{-3}	6760 ± 177**		47.79
1×10^{-2}	5444 ± 145**		57.95

**$p < 0.01$ versus control by *t* test.

the concentration of these chemicals was raised. The concentrations of As_2O_3, $NaAsO_2$, and Na_2HAsO_4 at which maximum stimulation of DNA synthesis occurred were 0.5×10^{-6}, 1×10^{-6}, and 1×10^{-5} M, respectively. The concentrations of As_2O_3, $NaAsO_2$, and Na_2HAsO_4 at which about 50% inhibition of DNA synthesis in the controls was found were 1×10^{-5}, 1×10^{-4}, and 1×10^{-3} M, respectively. The rate of DNA synthesis in the lymphocytes changed after the cells were exposed to the arsenicals for only 1 hr. This implies that both trivalent and pentavalent arsenic compounds can be rapidly uptaken into human lymphocytes and immediately stimulate or inhibit DNA synthesis in the cells.

4. DISCUSSION

In this study, it was shown that the effects of inorganic arsenic compounds on DNA synthesis in PHA-stimulated human lymphocytes in vitro are biphasic: the chemicals at very low concentrations stimulated DNA synthesis, whereas at higher concentrations they inhibited DNA synthesis. Further experiments are required to interpret these results, although Nordenson and Beckman (1991) have indicated that the genotoxic effect of arsenic on cultured human lymphocytes may be mediated by free oxygen radicals. Our study emphasized the importance of studying the toxicological or biological effects of arsenic at very low concentrations in order to obtain as accurate and comprehensive a picture as possible.

For the general population, the daily inhalation and intake of arsenic from ambient air and foodstuffs does not exceed 10 μg, because most food contains little arsenic (< 0.25 mg/kg) (Leonard, 1984). However, arsenic can be accumulated in certain (aquatic) food chains and is a cumulative poison in the human body. In our study, the results show that arsenic at very low concentrations is able to enhance DNA synthesis in human lymphocytes. It is therefore possible that among some individuals in the general population, arsenic can accumulate in some tissues and cells in sufficient concentrations to affect DNA synthesis or other cellular biochemical processes. For this reason, studies of the biological effects of arsenic at very low concentrations are very important.

McCabe and coworkers (1983) reported that sodium arsenite and sodium arsenate augment mitogenesis of human and bovine lymphocytes stimulated by PHA in vitro through an increase in ^{3}H-TdR incorporation into the lymphocytes by arsenic at very low concentrations. However, in our study, no difference between the cell mitotic indexes of human lymphocytes unexposed and exposed to the arsenic compounds at very low concentrations for 72 hr was found, although in this case the chemicals did enhance incorporation of ^{3}H-TdR into the cells. Our results suggest that the 72 hr exposure to arsenic at very low concentrations enhanced DNA synthesis in PHA-stimulated lymphocytes rather than mitogenesis of the cells. We also found that incorporation of ^{3}H-TdR into PHA-stimulated human lymphocytes is significantly enhanced immediately after exposure to the arsenicals at very low concentrations for only 1 hr. These results provided further evidence that arsenic at very low concentrations significantly enhances DNA synthesis in PHA-stimulated human lymphocytes, but not mitogenesis of the cells.

The results of our study also indicate large interindividual variations in the effects of arsenic on DNA synthesis in human lymphocytes, particularly its stimulating effect on DNA synthesis at very low concentrations. This may be related to differences in metabolic activity and sensitivity to arsenic.

In general, trivalent arsenic is more lethal and clastogenic than pentavalent (Jacobson-Kram and Montalbano, 1985; Aposhian, 1989). Our results also show that both the stimulating and inhibiting effects of trivalent arsenic on DNA synthesis are far greater than those of pentavalent arsenic. The weakness of the biological effects of pentavalent arsenic might be due to the great quantity of pentavalent phosphorus compounds (e.g., phosphate) present in cells. These compounds are strong competitors against pentavalent arsenic compounds by inhibiting their contact with enzymes and other biological active molecules. The inhibitory effects of pentavalent phosphorus against pentavalent arsenic might be due to the fact that it interacts more effectively with enzymes and proteins with an SH group, though to a less extent than trivalent arsenic.

The biphasic effects of arsenic on DNA synthesis imply that the toxicological or biological role of arsenic in human health is very complex. Under certain conditions, arsenic at very low concentrations might be considered to enhance replications of DNA fragment(s) of mutated gene(s) induced by chemicals or other factors to result in mutagenesis or carcinogenesis of the cells. In this case, arsenic as comutagen or cocarcinogen might enhance cancer development through its stimulation of DNA synthesis. Many studies have indicated that arsenic as a comutagen inhibits DNA repair (Jung et al., 1969; Rossman et al., 1977; Okui and Fujiwara, 1986; Li and Rossman, 1989b; Snyder and Lachmann, 1989) and potentiates mutagenicity of methyl methanesulfonate (MMC) (Lee et al., 1986b), *N*-methyl-*N*-nitrosourea (MNN) (Li and Rossman, 1989a), *cis*-diamminedichloroplatinum(11) (Lee et al., 1986a), and UV radiation (Rossman, 1981; Lee et al., 1985). However, some studies have reported that in certain conditions, arsenic reduces the mutagenicity of MMC (Lee et al., 1986b) and UV radiation (Rossman et al., 1975, 1977; Nunoshiba and Nishioka, 1987). In some experiments, arsenic fed to experimental animals was not carcinogenic (Hueper and Payne, 1962), and even minimized induction of cancers (Boutwell, 1963; Milner, 1969). These observations and the present study raise the question of whether arsenic at very low concentrations might enhance, under certain conditions, replication of gene(s) related to DNA repair and lead to enhancement of DNA repair. Arsenic as an inhibitor of mutagenesis or carcinogenesis might reduce the incidence of cancers induced by DNA-damaging agents. At very low concentrations, arsenic would therefore have a dual effect on mutagenesis as both comutagen and inhibitor. Naturally, this remains to be confirmed by further studies.

REFERENCES

Aoki, Y., Lipsky, M. M., and Fowler, B. A. (1990). Alteration in protein synthesis in primary cultures of rat kidney proximal tubule epithelial cells by exposure to gallium, indium, and arsenite. *Toxicol. Appl. Pharmacol.* **106**, 462–468.

Aposhian, H. V. (1989). Biochemical toxicology of arsenic. *Rev. Biochem. Toxicol.* **1**, 265–299.

Boutwell, R. K. (1963). A carcinogenicity evaluation of potassium arsenite and arsenilic acid. *J. Agric. Food Chem.* **11**, 381–385.

Friedmann, P. S., and Rogers, S. (1980). Photochemotherapy of psoriasis: DNA damage in blood lymphocytes. *J. Invest. Dermatol.* **74**, 440–443.

Hueper, W. C., and Payne, W. W. (1962). Experimental studies in metal carcinogenesis: Chromium, nickel, iron and arsenic. *Arch. Environ. Health* **5**, 445–462.

Jacobson-Kram, D., and Montalbano, D. (1985). The reproductive effects assessment group's report on the mutagenicity of inorganic arsenic. *Environ. Mutagen.* **7**, 787–804.

Jung, E. G., Trachsel, B., and Immich, H. (1969). Arsenic as an inhibitor of the enzymes concerned in cellular recovery (dark repair). *Ger. Med. Mon.* **14**, 614–616.

Lee, T. C., Huang, R. Y., and Jan, K. Y. (1985). Sodium arsenite enhances the cytotoxicity, clastogenicity and 6-thioguanine-resistant mutagenicity of ultraviolet light in Chinese hamster ovary cells. *Mutat. Res.* **148**, 83–89.

Lee, T. C., Lee, K. C., Tzeng, Y. J., Huang, R. Y., and Jan, K. Y. (1986a). Sodium arsenite potentiates the clastogenicity and mutagenicity of DNA crosslinking agents. *Environ. Mutagen.* **8**, 119–128.

Lee, T. C., Wang-Wuu, S., Huang, R. Y., Lee, K. C. C., and Jan, K. Y. (1986b). Differential effects of pre- and posttreatment of sodium arsenite on the genotoxicity of methyl methanesulfonate in Chinese hamster ovary cells. *Cancer Res.* **46**, 1854–1857.

Leonard, A. (1984). Recent advances in arsenic mutagenesis and carcinogenesis. *Toxicol. Environ. Chem.* **7**, 241–250.

Leonard, A., and Lauwerys, R. R. (1980). Carcinogenicity, teratogenicity and mutagenecity of arsenic. *Mutat. Res.* **75**, 49–62.

Li, J. H., and Rossman, T. G. (1989a). Mechanism of comutagenesis of sodium arsenite with *N*-methyl-*N*-nitrosourea. *Biol. Trace Elem. Res.* **21**, 373–381.

Li, J. H., and Rossman, T. G. (1989b). Inhibition of DNA ligase activity by arsenite: A possible mechanism of its comutagenesis. *Mol. Toxicol.* **2**, 1–9.

McCabe, M., Maguire, D., and Nowak, M. (1983). The effects of arsenic compounds on human and bovine lymphocyte mitogenesis in vitro. *Environ. Res.* **31**, 323–331.

Milner, J. E. (1969). The effects of ingested arsenic on methyl cholanthrene induced skin tumours in mice. *Arch. Environ. Health* **18**, 7–11.

Nakamuro, K., and Sayato Y. (1981). Comparative studies of chromosomal aberration induced by trivalent and pentavalent arsenic. *Mutat. Res.* **88**, 73–81.

Nordenson, I., and Beckman, L. (1991). Is the genotoxic effect of arsenic mediated by oxygen free radicals? *Hum. Hered.* **41**, 71–73.

Nunoshiba, T., and Nishioka, H. (1987). Sodium arsenite inhibits spontaneous and induced mutations in *Escherichia coli. Mutat. Res.* **184**, 99–105.

Okui, T., and Fujiwara, Y. (1986). Inhibition of human excision DNA repair by inorganic arsenic and the co-mutagenic effect in V79 Chinese hamster cells. *Mutat. Res.* **172**, 69–76.

Petres, J., Baron, D., and Hagedorn, M. (1977). Effects of arsenic cell metabolism and cell proliferation: Cytogenetic and biochemical studies. *Environ. Health Perspect.* **19**, 223–227.

Rossman, T. G. (1981). Enhancement of UV-mutagenesis by low concentrations of arsenite in *Escherichia coli. Mutat. Res.* **91**, 207–211.

Rossman, T. G., Meyn, M. S., Troll, W. (1975). Effects of sodium arsenite on the survival of UV-irradiated *Escherichia coli:* Inhibition of a recA-dependent function. *Mutat. Res.* **30**, 157–162.

Rossman, T. G., Meyn, M. C., and Troll, W. (1977). Effects of arsenite on DNA repair in *Escherichia coli. Environ. Health Perspect.* **19**, 229–233.

Sibatani, A. (1959). In vitro incorporation of ^{32}P into nucleic acids of lymphocytic cells: Effects of x-irradiation and some other agents. *Exp. Cell Res.* **17**, 131–138.

Snyder, R. D., and Lachmann, P. J. (1989). Thiol involvement in the inhibition of DNA repair by metals in mammalian cells. *J. Mol. Toxicol.* **2**, 117–128.

8

INDUCTION OF LUNG-SPECIFIC DNA DAMAGE BY METHYLARSENICS VIA THE PRODUCTION OF FREE RADIALS

Shoji Okada

Department of Radiobiochemistry, University of Shizuoka School of Pharmaceutical Sciences, 52-1 Yada, Shizuoka-shi, 422, Japan

Kenzo Yamanaka

Department of Biochemical Toxicology, Nihon University College of Pharmacy, 7-7-1 Narashinodai, Funabashi-shi, Chiba 274, Japan

Arsenic in the Environment, Part II: Human Health and Ecosystem Effects,
Edited by Jerome O. Nriagu.
ISBN 0-471-30436-0

---

1. INTRODUCTION

Although epidemiological investigations have shown that inorganic arsenics (i.e., arsenite and arsenate) are carcinogenic in human subjects, particularly in the skin and lungs, experimental studies on the carcinogenicity of inorganic arsenics have not had positive results [International Agency Research on Cancer (IARC), 1980, 1987; World Health Organization (WHO), 1981]. On the other hand, mutagenicity tests of inorganic arsenics using bacterial tester strains have provided positive, if not potent, and negative results (Nishioka, 1975; Rossman, 1981b). Rossman (1981a) and Rossman et al. (1977, 1986), and Li and Rossman (1991) proposed, on the basis of such discrepancies, that the inorganic arsenics are not direct mutagens but mutation-modifying agents. They observed two effects on bacterial genes depending on the concentration—a comutagenic effect appearing at low concentrations and an antimutagenic effect at high concentrations. The latter may be the result of inhibition of the error-prone SOS repair system. In fact, we have found that inorganic arsenics inhibit the error-prone repair of ultraviolet (UV)-induced DNA damage by retarding DNA replication and thus prolonging excision repair in the bacterial tester strains, possibly leading to a reduction in UV-induced mutation (Okada et al., 1983). However, no conclusive experimental proof has so far been provided for the induction of mutagenesis and carcinogenesis.

On the other hand, as shown in Scheme 1, inorganic arsenics are known to be metabolically reduced and then methylated to methanearsonic acid and subsequently to dimethylarsinic acid (DMAA) in mammals and microorganisms (McBride and Wolfe, 1971; Woolson, 1977; Crecelius, 1977; Tam et al., 1978,

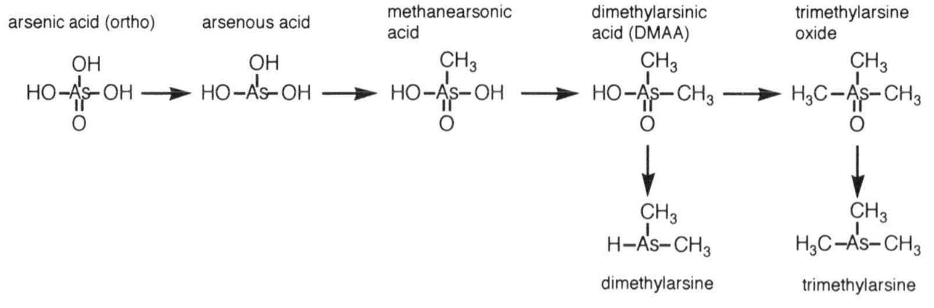

arsenic acid (ortho): $HO{-}As(OH)(=O){-}OH$ → arsenous acid: $HO{-}As(OH){-}OH$ → methanearsonic acid: $HO{-}As(CH_3)(=O){-}OH$ → dimethylarsinic acid (DMAA): $HO{-}As(CH_3)(=O){-}CH_3$ → trimethylarsine oxide: $H_3C{-}As(CH_3)(=O){-}CH_3$

dimethylarsinic acid (DMAA) ↓ dimethylarsine: $H{-}As(CH_3){-}CH_3$

trimethylarsine oxide ↓ trimethylarsine: $H_3C{-}As(CH_3){-}CH_3$

Scheme 1. Proposed metabolic pathway of arsenics. [Reprinted by permission from Yamanaka et al. (1989a).]

1979; Buchet et al., 1981; Marafante and Vahter, 1984; Buchet and Lauwerys, 1985; Vahter and Marafante, 1987), and even to trimethylated arsenics in some aerobic microorganisms (Cullen et al., 1977) and hamsters (Yamauchi and Yamamura, 1984). In mammals, the reduction and methylation of arsenics occur mainly in the blood and liver (Marafante and Vahter, 1984; Marafante et al., 1985; Buchet and Lauwerys, 1985; Vahter and Marafante, 1987). Methylated arsenics, metabolically produced from inorganic arsenics, have a higher rate of excretion and a lower affinity for tissues than original inorganic forms (Buchet et al., 1981; Vahter and Marafante, 1983; Vahter et al., 1984), suggesting that the methylation is a detoxification process. However, the genotoxic effect of these methylated arsenics has not been sufficiently investigated.

These facts suggest the importance of studying the genotoxic effect of the methylated metabolites of inorganic arsenics. We have, therefore, focused our attention on DMAA, which is the main metabolite of inorganic arsenics in mammals. It had already been demonstrated that dimethylarsine, produced in the further metabolism of DMAA, is potently mutagenic for *E. coli* tester strains under aerobic conditions (Yamanaka et al., 1989a), and that oral administration of DMAA induces lung-specific DNA damage in rats and mice (Yamanaka et al., 1989b, c). We then found that the DNA damage induced by dimethylated arsenics was due to the free radical species (the dimethylarsenic peroxyl radical and the active oxygens), which are produced by the reaction of molecular oxygen and dimethylarsine, a further metabolite of DMAA (Yamanaka et al., 1989b, c, 1990). These observations might explain the high risk of lung cancer induced by inorganic arsenics. In this article, we review these experimental studies on the mechanism of gene damage due to diemthylated arsenics, the metabolites of inorganic arsenic.

2. LUNG-SPECIFIC DNA DAMAGE INDUCED BY DMAA ADMINISTRATION IN MICE AND RATS

Male rats (Wistar strain, about 110 g) and male mice (ICR strain, about 25 g) were orally administered a sodium salt of DMAA at doses of 1950 mg/kg body weight and 1500 mg/kg body weight, respectively. The alkaline elution assay was employed to measure DNA strand breaks as a DNA damage indicator, according to the method of Kohn et al. (1981). In the alkaline elution assay, DNA contents are determined by fluorescence analysis using 3, 5-diaminobenzoic acid, which reacts with the deoxyribose moieties of DNA (Kohn et al., 1981).

Figures 1 and 2 show the alkaline elution profiles of DNA in the lungs of DMAA-dosed mice and rats, respectively. DNA single-strand breaks in the lung appeared 12 hr after administration in both mice and rats. The breaks then recovered partially at 15 hr and fully at 24 hr in mice. In rats, full recovery was observed even at 18 hr. These results indicate that the breaks are readily repairable. On the other hand, in the liver, kidney, and spleen, which also accumulate arsenics (though less than the lung), no DNA single-strand breaks

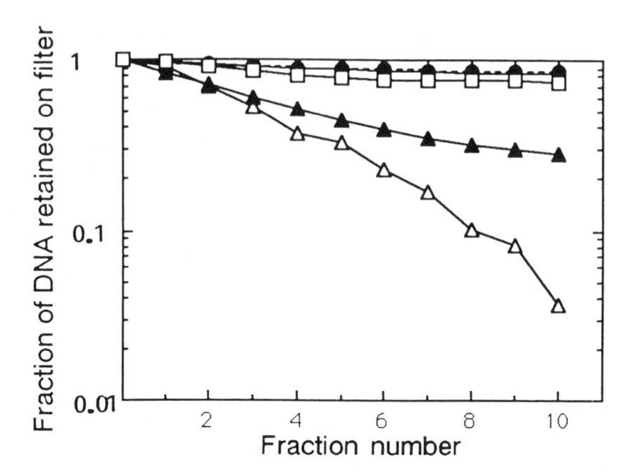

Figure 1. Single-strand breaks of pulmonary DNA induced by DMAA administration in mice. Mice were administered a sodium salt of DMAA (1500 mg/kg, p.o.). The pulmonary DNA was eluted at pH 12.1 using the alkaline elution assay. Key: Control (○ with dashed line), 9 hr (●), 12 hr (△), 15 hr (▲), 24 hr (□) after administration of DMAA. Note that control symbols (○) are at almost the same positions as the 9-hr symbols (●). [Reprinted by permission from Yamanaka et al. (1989c).]

appeared within 24 hr after administration in rats and mice. Thus, the lung seems to be a target organ for DNA damage induced by DMAA administration.

Some forms of base damage in DNA (e.g., alkylation at certain positions) can be recognized by specific DNA glycosylase, which catalyzes excision of the modified bases, leaving apurinic or apyrimidinic (AP) sites in the DNA (Hanawalt et al., 1979). It is known that AP sites are alkali-labile (Kohn et al., 1981). Therefore, alkali-labile sites of pulmonary DNA in mice after the administration of DMAA were determined by the alkaline elution assay at pH 12.6. The alkali-labile sites were hardly detected at 12 hr when the strand breaks of DNA were maximal, as seen in Figure 1. However, the formation of AP sites in DNA prior to the induction of DNA single-strand breaks in DMAA-treated human cultured cells (unpublished data by the authors) has been confirmed. The administration of DMAA thus modifies DNA bases prior to DNA strand scissions in the lungs of mice and rats.

3. INCREASE OF HETEROCHROMATIN IN VENULAR ENDOTHELIUM OF THE LUNG IN MICE AFTER ADMINISTRATION OF DMAA

To further confirm lung-specific DNA damage induced by oral administration of DMAA, morphological and pathological studies were performed (Nakano et al., 1992). In mice, a light microscopic survey showed no morphological change within 24 hr after the administration of DMAA in the liver, spleen, kidney, testes, bone marrow, and even in the lung, where DNA strand scissions were induced. However, under a transmission electron microscope, a marked condensation of

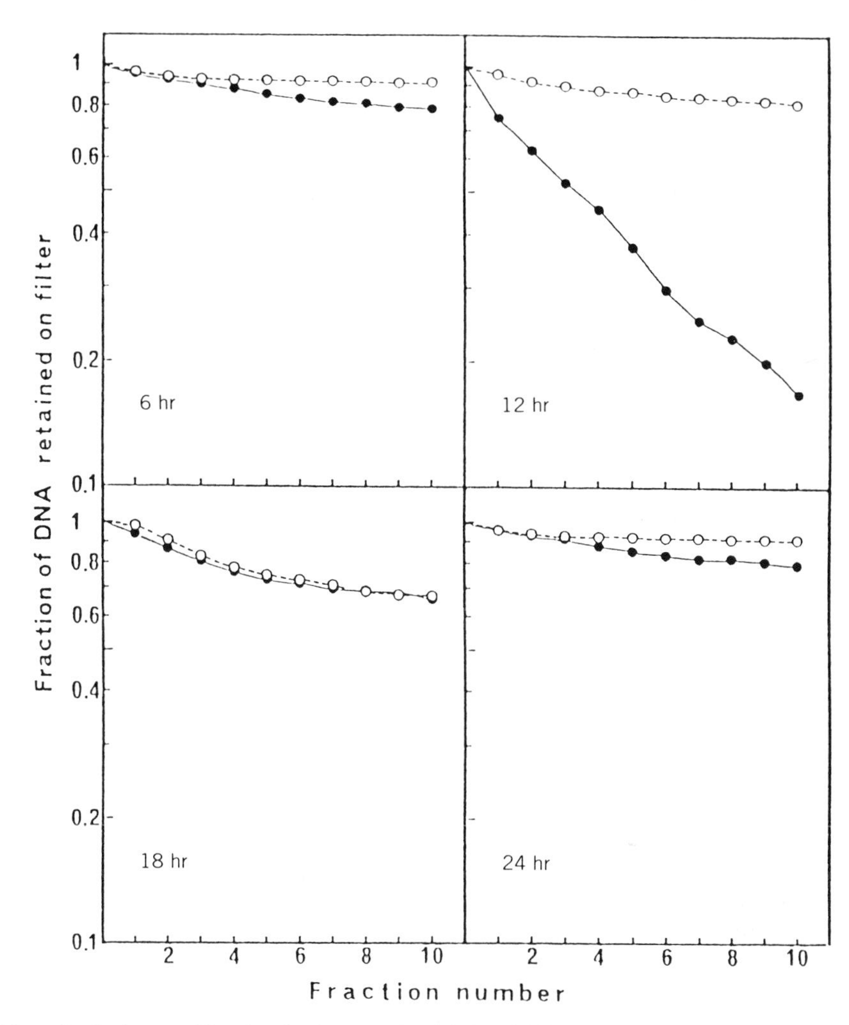

Figure 2. Single-strand breaks of pulmonary DNA induced by DMAA administration in rats. Rats were administered a sodium salt of DMAA (1950 mg/kg, p.o.). The pulmonary DNA was eluted at pH 12.1 using the alkaline elution assay. Key: Control (○), DMAA-dosed (●).

chromatin (increase of heterochromatin) was observed in the endothelial cells of alveolar wall capillaries in mice 12 to 48 hr after the administration of DMAA (Fig. 3), but not in the sinus endothelium of the liver. These results indicate that the morphological changes induced by DMAA administration in mice were also lung-specific, as were the DNA strand scissions.

The morphological changes in chromatin suggest that, besides DNA strand scissions, DNA-protein crosslinks were also induced. In fact, it is known that nickel exposure induces both DNA single-strand breaks and DNA-protein crosslinks in mammalian cultured cells, together with morphological changes in

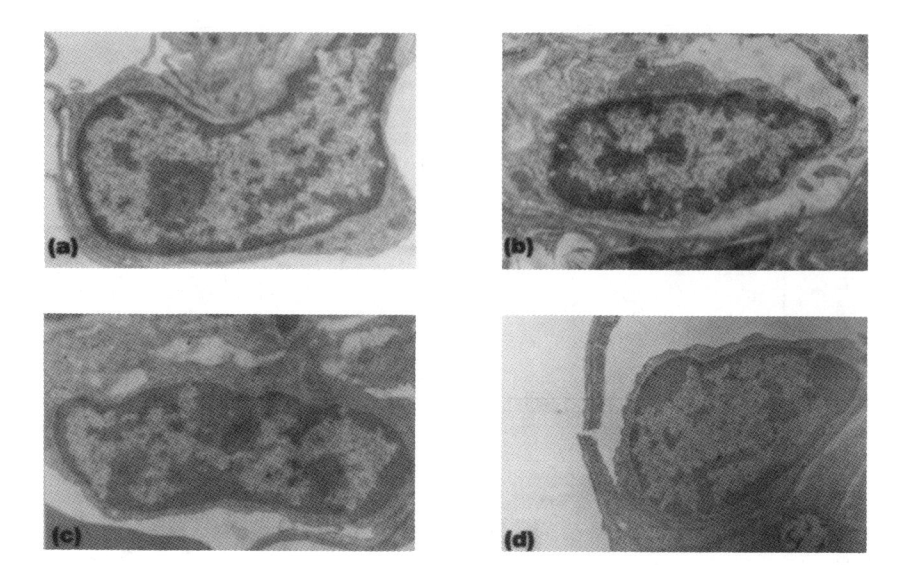

Figure 3. Transmission electron micrographs of alveolar capillary endothelial cells of mice. Mice were administered a sodium salt of DMAA (1500 mg/kg, p.o.). (a) control; (b) 12 hr; (c) 24 hr; (d) 48 hr after administration of DMAA. Percent area of heterochromatin to nucleus: Control, 31.1 ± 9.7; 12 hr, 46.1 ± 6.0**; 24 hr, 47.1 ± 8.0**; 48 hr, 35.8 ± 7.0*. Each value represents mean $\pm$ S.D; ($n = 21-48$); Significant difference (*$p < 0.05$, **$p < 0.001$) from control. [Plates a and b are reprinted by permission from Nakano et al. (1992).].

chromosomes (Ciccarelli et al., 1981; Robinson and Costa, 1982; Pramila and Costa, 1985; Costa, 1989). Actually, in vivo and in vitro experiments have shown that exposure to DMAA induces DNA-protein crosslinks as well as DNA strand scissions (Yamanaka et al., 1993).

In regard to endothelial tumors such as endothelial hyperplasia of the lymph nodes, spleen, and liver and hemangioendothelial sarcoma, many cases have been clinically observed after chronic ingestion of arsenic (Roth, 1957; Regelson et al., 1968). Furthermore, cell morphological evaluation suggests that an increase in heterochromatin contents may reflect early-phase gene damage. The early change in heterochromatin induced in the endothelium by DMAA administration might be related to the development of sarcomas such as hemangioendothelial sarcoma at a later stage. Whether or not such early changes relate to carcinogenesis is a crucial point to be determined in the future.

4. MECHANISMS OF LUNG-SPECIFIC DNA STRAND BREAKS INDUCED BY DMAA ADMINISTRATION IN MICE AND RATS

We undertook to clarify whether pulmonary DNA damage was induced by DMAA itself or by its further metabolites. As is well known, DMAA is metabolically reduced to dimethylarsine and, in some microorganisms, to trimethylarsine

(McBride and Wolfe, 1971; Woolson, 1977; Cullen et al., 1977). These arsines are volatile substances with a strong garlicky odor. We previously confirmed that dimethylarsine, a further metabolite of DMAA, was potently mutagenic for *E. coli* tester strains under aerobic conditions, while trimethylarsine was not mutagenic (Yamanaka et al., 1989a). Furthermore, DMAA-treated animals smelled garlicky for several hours. These facts suggest that some arsine—possibly dimethylarsine, which might induce lung-specific gene damage—was excreted into air expired after DMAA administration. To confirm this, the air expired by mice for 24 hr after DMAA administration was trapped in 5% hydrogen peroxide, which oxidizes dimethylarsine and trimethylarsine to DMAA and trimethylarsine oxide, respectively. Arsenic corresponding to 0.0032% of the total administered dose was recovered. Thin-layer chromatography indicated that the arsenic thus recovered in the 5% hydrogen peroxide was DMAA and not trimethylarsine oxide. Therefore, a major volatile metabolite of DMAA in the expired air of mice is very likely to be dimethylarsine.

An in vitro experiment was carried out to determine whether DMAA itself or dimethylarsine was responsible for the DNA strand breaks. DNA prepared from cell nuclei of mouse lung was exposed to DMAA or gaseous dimethylarsine, which was generated by the reduction of DMAA with sodium borohydride, and then subjected to the alkaline elution assay. Remarkably, dimethylarsine induced DNA single-strand breaks, while DMAA did not. This implies that DNA single-strand breaks induced by DMAA administration are not caused by DMAA itself but by dimethylarsine, the main metabolite of DMAA.

Dimethylarsine is a potent reducing agent (Dehn and Wilcox, 1906). In vitro experiments using the nitroblue tetrazolium method had demonstrated that the superoxide anion radical (O_2^-) is produced by one-electron reduction of molecular oxygen, which coexists with dimethylarsine (unpublished data by the authors). The effects of superoxide dismutase (SOD) and catalase on the induction of DNA single-strand breaks were therefore investigated using this in vitro system. In the presence of these enzymes, the breaks were remarkably suppressed. This suggests that the active oxygens (the superoxide anion radical, etc.) produced by the reaction of molecular oxygen and dimethylarsine are responsible for the DNA single-strand breaks in lung after the administration of DMAA (Yamanaka et al., 1989c).

5. PARTICIPATION OF ACTIVE OXYGEN SPECIES IN THE INDUCTION OF DNA DAMAGE BY DMAA ADMINISTRATION IN MICE

To further confirm the participation of the active oxygens in the induction of pulmonary DNA damage after oral administration of DMAA in mice, we investigated the occurrence of several protective responses to active oxygens in the pulmonary cells in in vivo experiments.

The results obtained from a series of experiments are summarized in Figure 4 (Yamanaka et al., 1991). Of the pulmonary enzymes that directly catalyze the

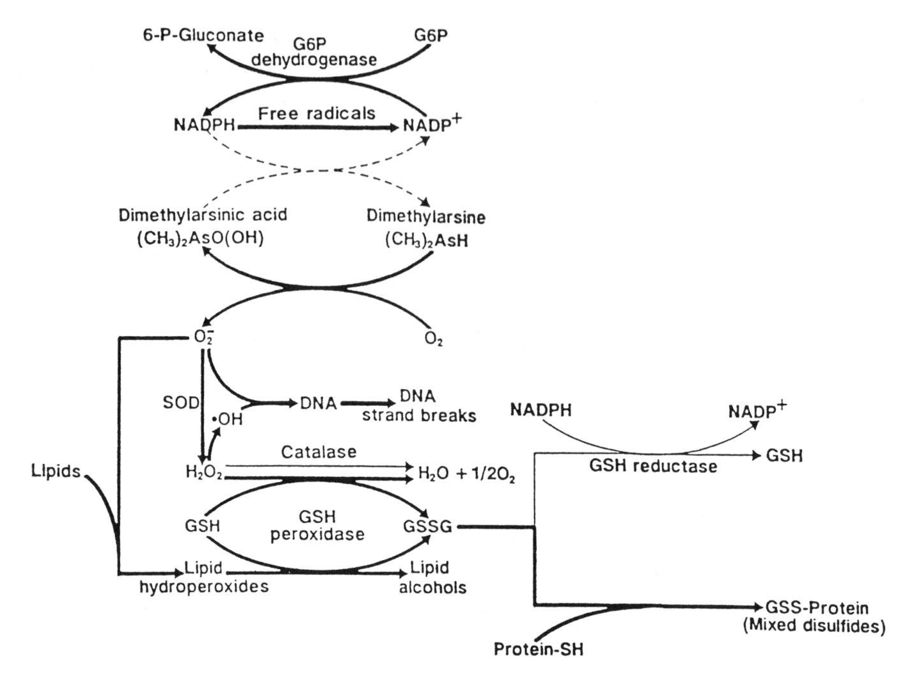

Figure 4. Proposed mechanism of the lung's cellular protective response against DNA damage induced by DMAA administration in mice. The pathways determined to be activated are indicated by boldface arrows, those not activated are indicated by lightface arrows, and those assumed to be activated are indicated by broken arrows. [Reprinted by permission from Yamanaka et al. (1991).]

protective reactions against the active oxygen species, the activities of Mn-SOD and glutathione peroxidase (GSH-Px) were elevated by the administration of DMAA. These elevations may indicate that DMAA administration produces the superoxide anion radical and hydrogen peroxide in the lung, which might cause DNA strand breaks. The superoxide anion radical is probably produced through one-electron reduction of molecular oxygen by dimethylarsine. The activity of SOD in the mitochondria (Mn-SOD) was elevated much more than in the cytoplasm (Cu, Zn-SOD), suggesting that the superoxide anion radical is produced mainly in mitochondria-rich Clara and alveolar type-II cells (Kistler et al., 1967). On the other hand, as shown in Figure 4, the increase in GSH-Px activity may contribute to the treatment of lipid hydroperoxides produced by the attack of active oxygens on lipids, or to the treatment of hydrogen peroxide formed from the superoxide anion radical.

Glutathione reductase (GSH-R) and glucose-6-phosphate dehydrogenase (G6P-DH) are known to be the enzymes that protect against the active oxygen species in the lung and that supply reduced glutathione (GSH) and NADPH, respectively. In general, free radicals including the active oxygens are apt to bring about an increase in G6P-DH activity and a decrease in NADPH level. We actually found such changes in the lung of mice after DMAA administration. It is likely that the elevation of G6P-DH activity replenishes NADPH diminished in the lung. Moreover, the marked decrease in NADPH level after DMAA administration suggests that NADPH acts as a proton donor in the metabolic reduction of DMAA to dimethylarsine. The redox cycle shown in Figure 4 might be formed to bring about this reduction. Although GSH levels markedly decreased in the lung after the administration of DMAA in a manner similar to the known cases of free radical-induced damage, GSH-R was not activated. Furthermore, the mixed disulfides significantly increased in the lung after DMAA administration. These results indicate that the oxidized glutathione (GSSG) produced in activated glutathione oxidation–reduction systems is employed in the formation of mixed disulfides, particularly with SH-protein, rather than reconverted into GSH by GSH-R.

These cellular events may lend support to the possibility that DNA-damaging free radical species such as the active oxygen species are produced in pulmonary cells after DMAA administration in mice.

6. POSSIBLE PRODUCTION OF THE DIMETHYLARSENIC PEROXYL RADICAL AND ITS ROLE IN DNA DAMAGE

To further elucidate the mechanisms of lung-specific DNA strand scissions in murine after oral administration of DMAA, substances other than the active oxygen species that might ultimately cause DNA damage were pursued in in vitro experiments using dimethylarsine, a further metabolite of DMAA. An incidental finding gave us a cue. In the alkaline elution assay of DNA, two methods for DNA analysis led to remarkably different results. In the fluorescence analysis using 3, 5-diaminobenzoic acid (DABA), which reacts with the deoxyribose moieties of DNA, the single-strand DNA breaks caused by exposure to dimethylarsine were markedly diminished by the addition of SOD and catalase, as mentioned previously (Fig. 5a). In contrast, when the assay was carried out by counting the ^{3}H-radioactivity of DNA in ^{3}H-thymidine-labeled L-132 cells [an established human embryonic cell line of alveolar type-II epithelial cells (Ichikawa and Yokoyama, 1977], the strand scissions were diminished only slightly by the addition of SOD and catalase (Fig. 5b). This result indicates that the deoxyribose moieties of DNA were attacked by free radicals formed by the reaction of molecular oxygen and dimethylarsine, and the formation of fluorescent quinaldine from DABA and deoxyribose moieties consequently decreased. In fact, we found that dimethylarsine readily reacted with the deoxyribose moieties of DNA and produced base-propanal derivatives that could not form fluorescent derivatives with DABA (unpublished data by the authors). It can therefore be proposed that the radiotracer measurement is more relevant in the determination of DNA damage than the fluorescence measurement currently utilized.

The results obtained by the radioactivity assay (i.e., SOD and catalase showed only a small effect on DNA strand scission caused by dimethylarsine exposure)

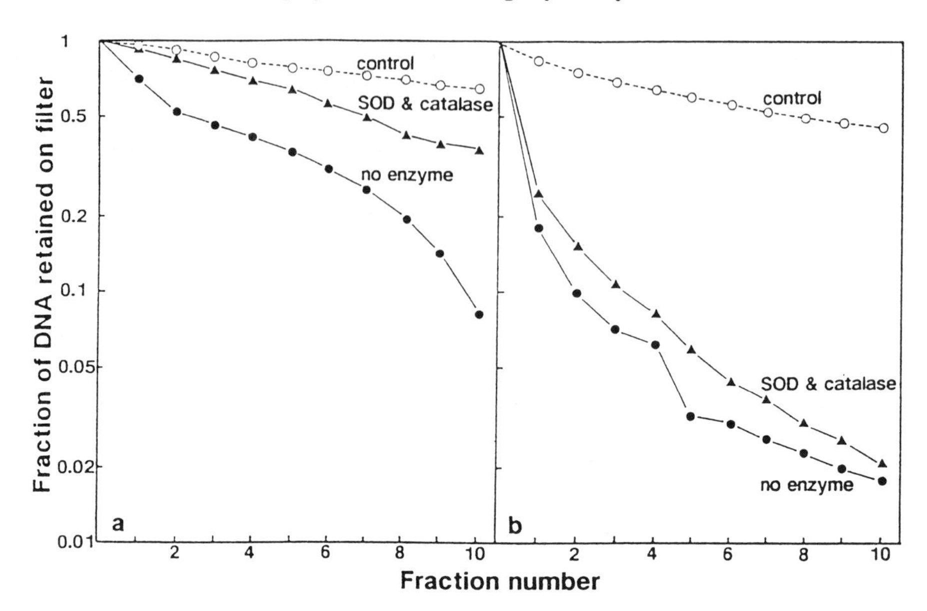

Figure 5. Single-strand breaks of DNA detected by the fluorescence assay (a) and the radioactivity assay (b) after in vitro exposure to dimethylarsine. DNA of L-132 cells was exposed to dimethylarsine generated from 0.2 mmol DMAA at 20 °C for 1 hr. After exposure of dimethylarsine, DNA was eluted at pH 12.1 using the alkaline elution assay. The determination of nonlabeled DNA was carried out by the fluorescence method using 3,5-diaminobenzoic acid. The radioactivity of ^{3}H-labeled DNA was measured in a liquid scintillation counter. [Reprinted by permission from Yamanaka et al. (1990).]

suggested that some radical species other than the active oxygens participate in DNA damage. To determine the radical species, electron spin resonance (ESR) analysis using the spin trapping method with 5, 5′-dimethyl-1-pyrroline *N*-oxide (DMPO) was performed. The ESR spectrum was determined about 3 min after the begining of the reaction of molecular oxygen and dimethylarsine in acetonitrile. The coupling constants of the DMPO spin-adduct ($a_N = 13.7\ G$ and $a_H = 11.2\ G$) were not in agreement with those reported for the DMPO spin-adducts of O_2^- and $\cdot OH$ (Kim et al., 1989; Harbour et al., 1974), but they were similar to those of the alkyl peroxyl radicals (Kim et al., 1989). We assumed that the most probable free radical detected here was the dimethylarsenic peroxyl radical (Yamanaka et al., 1990). We also tried to detect the superoxide anion radical by using the cytochrome c method. Immediately after exposure of dimethylarsine to cytochrome c solution under air, absorbance at 550 nm indicated that the reduction of cytochrome c increased gradually for approximately 60 minutes. In the presence of SOD, this gradual increase in absorbance was not observed. This suggests that the superoxide anion radical was produced relatively slowly by the reaction of molecular oxygen and dimethylarsine.

As shown in Scheme 2, dimethylarsine, which is metastable, gives an electron to molecular oxygen to form $(CH_3)_2As\cdot$ and O_2^- radicals rather slowly. The

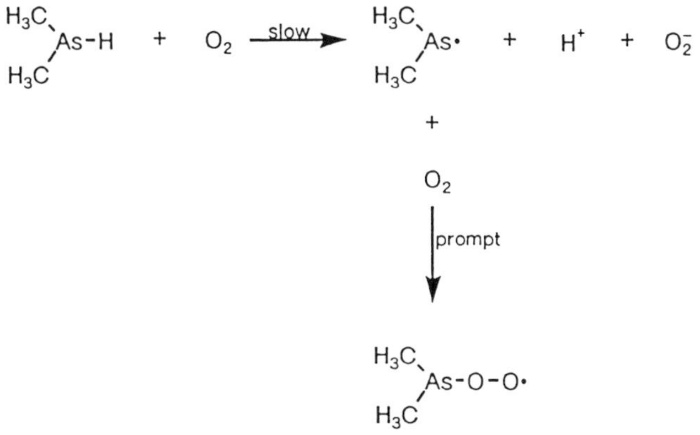

$$\begin{matrix} H_3C \\ \quad As\text{-}H \\ H_3C \end{matrix} + O_2 \xrightarrow{\text{slow}} \begin{matrix} H_3C \\ \quad As\cdot \\ H_3C \end{matrix} + H^+ + O_2^-$$

$$+$$

$$O_2$$

$$\downarrow \text{prompt}$$

$$\begin{matrix} H_3C \\ \quad As\text{-}O\text{-}O\cdot \\ H_3C \end{matrix}$$

Scheme 2. Proposed reaction pathway of dimethylarsine and molecular oxygen. [Reprinted by permission from Yamanaka et al. (1990).]

former radical promptly reacts with molecular oxygen to produce the dimethylarsenic peroxyl radical [$(CH_3)_2AsOO\cdot$], which is fairly stable, even in cells. This radical may attack the deoxyribose moieties of DNA to cause strand scissions in a manner similar to $\cdot OOH$, since the chemical properties of the two radicals seem to be analogous. We believe that this dimethylarsenic peroxyl radical, rather than the active oxygen species, may play a major role in DNA strand breaks, assumingly through the formation of DNA-adducts (Tezuka et al., 1993).

7. CONCLUSIONS

Although epidemiological studies have established the human carcinogenicity of inorganic arsenics, a number of experimental studies have failed to confirm it (IARC, 1980, 1987; WHO, 1981). This failure might be attributable to a lack of attention to the metabolism of inorganic arsenics. It is known that inorganic arsenics are readily metabolized to methyl arsenics in mammals and that DMAA is the main metabolite of inorganic arsenics (Crecelius, 1977; Tam et al., 1978, 1979; Buchet et al., 1981; Marafante and Vahter, 1984; Buchet and Lauwerys, 1985; Vahter and Marafante, 1987). The methylation pathway of inorganic arsenics has generally been considered to be a metabolic detoxication process. In fact, the value of acute LD_{50} of arsenic trioxide is about 1/100 that of DMAA (Fairchild et al., 1977). The high acute toxicity of inorganic arsenics makes it difficult to prove their carcinogenicity using experimental animals.

DMAA has a higher affinity for cell nuclei than inorganic arsenics (Vahter et al., 1984), suggesting that DMAA may inflict injury on the nucleus more readily than inorganic arsenics. We therefore focused our attention on DMAA and its further metabolites. We found that oral administration of DMAA induced gene damage in the lung such as DNA strand scissions, DNA-protein crosslinks, and increased heterochromatin. It is proposed that the induction of gene damage was due to the free radical species, particularly the dimethylarsenic peroxyl radical, which is produced as a metabolic by-product of DMAA in the presence of oxygen. These findings suggest that the methylation pathway, thought to be a detoxication process in view of acute toxicity, is a rather toxic process from the standpoint of genotoxicity.

Recent work on chemical carcinogenesis has pointed out the dominant role of active oxygen, generated in the interaction between carcinogens and cell constituents, in DNA damage. Of the active oxygen species, the hydroxyl radical is thought to be the primary cause of DNA strand breaks (Mello Filho et al., 1984). In the presence of iron or copper ions, the hydroxyl radical is likely to be formed by the Haber-Weiss reaction in cell nuclei from hydrogen peroxide, which is easily transported into the nuclei across the cell membrane and nuclear envelope. Besides the hydroxyl radical, interest has recently focused on the superoxide anion radical. In vitro experiments have shown that neither O_2^- nor hydrogen peroxide undergoes any chemical reaction with DNA when the strand breaks (Brawn and Fridovich, 1981; Lesko et al., 1980; Rowley and Halliwell, 1983) as shown by the chemical changes in the deoxyribose and base moieties of DNA (Aruoma et al., 1989a, b). It is reported that intracellular DNA receives strand scissions by O_2^-, possibly leading to carcinogenesis (Birnboim and Kanabus-Kaminska, 1985; Birmboim, 1986). In the case of arsenics, however, we propose that the dimethylarsenic peroxyl radical, rather than active oxygen, is the primary substance that ultimately causes DNA damage. This might explain the high risk of lung cancer by inorganic arsenics and, furthermore, provide useful information as to the mechanism of genotoxicity and carcinogenicity of several metals.

ACKNOWLEDGMENTS

We are grateful to Dr. Mikio Hoshino of the Institute of Physical and Chemical Research for ESR analysis and to Dr. Masayuki Nakano of Chiba University for electron microscopic elevation. We are also grateful to Drs. Akira Hasegawa and Ryoji Sawamura of Nihon University and Dr. Yutaka Kawazoe of Nagoya City University for helpful discussions.

REFERENCES

Aruoma, O. I., Halliwell, B., and Dizdaroglu, M. (1989a). Iron ion-dependent modification of bases in DNA by the superoxide radical-generating system hypoxanthine/xanthine oxidase. *J. Biol. Chem.* **264**, 13024–13028.

Aruoma, O.I., Halliwell, B., Gajewski, E., and Dizdaroglu, M. (1989b). Damage to the bases in DNA induced by hydrogen peroxide and ferric ion chelates. *J. Biol. Chem.* **264**, 20509–20512.

Birnboim, H. C. (1986). DNA strand breaks in human leukocytes induced by superoxide anion, hydrogen peroxide and tumor promoters are repaired slowly compared to breaks induced by ionizing radiation. *Carcinogenesis (London)* **7**, 1511–1517.

Birnboim, H. C., and Kanabus-Kaminska, M. (1985). The production of DNA strand breaks in human leukocytes by superoxide anion may involve a metabolic process. *Proc. Natl. Acad. Sci. USA* **82**, 6820–6824.

Brawn, M. K., and Fridovich, I. (1981). DNA strand scission by enzymically generated oxygen radicals. *Arch. Biochem. Biophys.* **206**, 414–419.

Buchet, J. P., and Lauwerys, R. (1985). Study of inorganic arsenic methylation by rat liver in vitro: Relevance for the interpretation of observations in man. *Arch. Toxicol.* **57**, 125–129.

Buchet, J. P., Lauwerys, R., and Roels, H. (1981). Comparison of the urinary excretion of arsenic metabolites after a single oral dose of sodium arsenite, monomethylarsonate, or dimethylarsinate in man. *Int. Arch. Occup. Environ. Health* **48**, 71–79.

Ciccareli, R. B., Hampton, T. H., and Jennette, K. W. (1981). Nickel carbonate induces DNA-protein links and DNA-strand breaks in rat kidney. *Cancer Lett.* **12**, 349–354.

Costa, M. (1989). Perspectives of the mechanism of nickel carcinogenesis gained from models of in vitro carcinogenesis. *Environ. Health Perspect.* **81**, 73–76.

Crecelius, E. A. (1977). Changes in the chemical speciation of arsenic following ingestion by man. *Environ. Health Perspect.* **19**, 147–150.

Cullen, W. R., Froese, C. L., Lui, A., McBride, B. C., Patmore, D. J., and Reimer, M. (1977). The aerobic methylation of arsenic by microorganisms in the presence of L-methionine-methyl-d_3. *J. Organomet. Chem.* **139**, 61–69.

Dehn, W. M., and Wilcox, B. B. (1906). Secondary arsines. *Am. Chem. J.* **35**, 1–54.

Fairchild, E. J., Lewis, R. J., and Tatken, R. L. (1977). *Registry of Toxic Effects of Chemical Substances.* USDHEW NIOSH, Cincinnati, OH.

Hanawalt, P. C., Cooper, P. K., Ganesan, A. K., and Smith, C. A. (1979). DNA repair in bacteria and mammalian cells. *Annu. Rev. Biochem.* **48**, 783.

Harbour, J. P., Chow, V., and Bolton, J. R. (1974). An electron spin resonance study of the spin adducts of OH and HO_2 radicals with nitrones in the ultraviolet photolysis of aqueous hydrogen peroxide solutions. *Can. J. Chem.* **52**, 3549–3553.

Ichikawa, I., and Yokoyama, E. (1977). Studies on the mechanism of lung surfactant secretion. Dipalmitoyl-lecithin secretion of the cultured L-132 cells. *J. Jpn. Med. Soc. Biol. Interface* **8**, 19–26.

International Agency for Research on Cancer (IARC) (1980). Some metals and metallic compounds. *IARC Monogr. Eval. Carcinog. Risk Chem. Man* **23**, 39–141.

International Agency for Research on Cancer (IARC) (1987). Overall evaluations of carcinogenicity: An updating of IARC monographs, Vols. 1 to 42. *IARC Monogr. Eval. Carcinog. Risk Man, Suppl.* **7**, 100–106.

Kim, Y. H., Lim, S. C., Hoshino, M., Ohtsuka, Y., and Ohishi, T. (1989). Spin trapping studies of peroxyl radicals. Detection of the reactive intermediates for oxidation generated from O_2^- and sulfonyl, sulfinyl, and phosphoryl chlorides. *Chem. Lett.* pp. 167–170.

Kistler, G. S., Caldwell, P. R. B., and Weibel, E. R. (1967). Development of fine structural damage to alveolar and capillary lining cells in oxygen-poisoned rat lungs. *J. Cell Biol.* **32**, 605–628.

Kohn, K. W., Ewig, R. A. G., Erickson, L. C., and Zwelling, L. A. (1981). Measurement of strand breaks and cross-links by alkaline elution. In E. C. Friedberg and P. C. Hanawalt (Eds.), *DNA Repair: A Laboratory Mannual of Research Procedures.* Dekker, New York, Vol. 1, Part B, pp. 379–401.

Lesko, S. A., Lorentzen, R. J., and Tso, P. O. P. (1980). Role of superoxide in deoxyribonucleic acid strand scission. *Biochemistry* **19**, 3023–3028.

Li, J. H., and Rossman, T. G. (1991). Comutagenesis of sodium arsenite with ultraviolet radiation in Chinese hamster V79 cells. *Biol. Met.* **4**, 197–200.

Marafante, E., and Vahter, M. (1984). The effect of methyltransferase inhibition on the metabolism of (^{74}As) arsenite in mice and rabbits. *Chem.-Biol. Interact.* **50**, 49–57.

Marafante, E., Vahter, M., and Envoll, J. (1985). The role of the methylation in the detoxication of arsenate in the rabbit. *Chem.-Biol. Interact.* **56**, 225–238.

McBride, B. C., and Wolfe, R. S. (1971). Biosynthesis of dimethylarsine by methanobacterium. *Biochemistry* **10**, 4312–4317.

Mello Filho, A. C., Hoffmann, M. E., and Meneghini, R. (1984) Cell killing and DNA damage by hydrogen peroxide are mediated by intracellular iron. *Biochem. J.* **218**, 273–275.

Nakano, M., Yamanaka, K., Hasegawa, K., Sawamura, R., and Okada, S. (1992). Preferential increase of heterochromatin in venular endothelium of lung in mice after administration of dimethylarsinic acid, a major metabolite of inorganic arsenics. *Carcinogenesis* **13**, 391–392.

Nishioka, H. (1975). Mutagenic activities of metal compounds in bacteria. *Mutat. Res.* **31**, 185–198.

Okada, S., Yamanaka, K., Ohba, H., and Kawazoe, Y. (1983). Effect of inorganic arsenics on cytotoxicity and mutagenicity of ultraviolet light on *Escherichia coli* and the mechanism involved. *J. Pharm. Dyn.* **6**, 496–504.

Pramila, S., and Costa, M. (1985). Induction of chromosomal damage in Chinese hamster ovary cells by soluble and particulate nickel compounds: Preferential fragmentation of the heterochromatic long arm of the X-chromosome by carcinogenic crystalline NiS particles. *Cancer Res.* **45**, 2320–2325.

Regelson, W., Kim, U., Ospina, J., and Halland, J. F. (1968). Hemangioendothelial sarcoma of liver from chronic arsenic intoxication by Fowler's solution. *Cancer (Philadelphia)* **21**, 514–522.

Robinson, S. H., and Costa, M. (1982). The induction of DNA strand breakage by nickel compounds in cultured Chinese hamster ovary cells. *Cancer Lett.* **15**, 35–40.

Rossman, T. G. (1981a). Effect of metals on mutagenesis and DNA repair. *Environ. Health Perspect.* **40**, 189–195.

Rossman T. G. (1981b). Enhancement of UV-mutagenesis by low concentrations of arsenite in *E. coli.* 11. *Mutat. Res.* **91**, 207–211.

Rossman, T. G., Meyn, M. S., and Troll, W. (1977). Effects of arsenite on DNA repair in *Escherichia coli.* 11. *Environ. Health Perspect.* **19**, 229–233.

Rossman, T. G., Molina, M., and Klein, C. B. (1986). Comutagens in *E. coli* and Chinese hamster cells with special attention to arsenite. *Genetic Toxicology of Environmetal Chemicals, Part A: Basic Principles and Mechanisms of Action.* Alan R. Liss, pp. 403–408.

Roth, F. (1957). Arsen-leber-tumoren (Hamangioendotheliom). *Z. Krebsforsch.* **61**, 468–508.

Rowley, D. A., and Halliwell, B. (1983). DNA damage by superoxide-generating systems in relation to the mechanism of action of the anti-tumour antibiotic adriamycin. *Biochim. Biophys. Acta* **761**, 86–93.

Tam, G. K. H., Charbonneau, S. M., Bryce, F., and Lacroix, G. (1978). Separation of arsenic metabolites in dog plasma and urine following intravenous injection of ^{74}As. *Anal. Biochem.* **86**, 505–511.

Tam, G. K. H., Charbonneau, S. M., Bryce, F., Pomroy, C., and Sandi E. (1979). Metabolism of inorganic arsenic (^{74}As) in humans following oral ingestion. *Toxicol. Appl. Pharmacol.* **50**, 319–322.

Tezuka, M., Hanioka, K., Yamanaka, K., and Okada, S. (1993). Gene damage induced in human alveolar type II cells by exposure to dimethylarsinic acid. *Biochem. Biophys. Res. Commun.* **191**, 1178–1183.

Vahter, M., and Marafante, E. (1983). Intracellular interaction and metabolic fate of arsenite and arsenate in mice and rabbits. *Chem.-Biol. Interact.* **47**, 29–44.

Vahter, M., and Marafante, E. (1987). Effects of lower dietary intake of methionine, choline or proteins on the biotransformation of arsenite in the rabbit. *Toxicol. Lett.* **37**, 41–46.

Vahter, M., Marafante, E., and Dencker, L (1984). Tissue distribution and retention of (^{74}As) dimethylarsinic acid in mice and rats. *Arch. Environ. Contam. Toxicol.* **13**, 259–264.

Woolson, E. A. (1977). Fate of arsenicals in different environmental substrates. *Environ. Health Perspect.* **19**, 73–81.

World Health Organization (WHO) (1981). *Environmental Health Criteria 18: Arsenic.* WHO, Geneva, pp. 122–127.

Yamanaka, K., Ohba, H., Hasegawa, A., Sawamura, R., and Okada, S. (1989a). Mutagenicity of dimethylated metabolites of inorganic arsenics. *Chem. Pharm. Bull.* **37**, 2753–2756.

Yamanaka, K., Hasegawa, A., Sawamura, R., and Okada, S. (1989b). DNA strand breaks in mammalian tissues induced by methylarsenics. *Biol. Trace Elem. Res.* **21**, 413–417.

Yamanaka, K., Hasegawa, A., Sawamura, R., and Okada, S. (1989c). Dimethylated arsenics induce DNA strand breaks in lung via the production of active oxygen in mice. *Biochem. Biophys. Res. Commun.* **165**, 43–50.

Yamanaka, K., Hoshino, M., Okamoto, M., Sawamura, R., Hasegawa, A., and Okada, S. (1990). Induction of DNA damage by dimethylarsine, a metabolite of inorganic arsenics, is for the major part likely due to its peroxyl radical. *Biochem. Biophys. Res. Commun.* **168**, 58–64.

Yamanaka, K., Hasegawa, A., Sawamura, R., and Okada, S. (1991). Cellular response to oxidative damage in lung induced by the administration of dimethylarsinic acid, a major metabolite of inorganic arsenics, in mice. *Toxicol. Appl. Pharmacol.* **108**, 205–213.

Yamanaka, K., Tezuka, M., Kato, K., Hasegawa, A, and Okada, S. (1993). Crosslink formation between DNA and nuclear proteins by in vivo and in vitro exposure of cells to dimethylarsinic acid. *Biochem. Biophys. Res. Commun.* **191**, 1184–1191.

Yamauchi, H., and Yamamura, Y. (1984). Metabolism and excretion of orally administered dimethylarsinic acid in the hamster. *Toxicol. Appl. Pharmacol.* **74**, 134–140.

9

CHRONIC ARSENISM FROM DRINKING WATER IN SOME AREAS OF XINJIANG, CHINA

Wang Lianfang and Huang Jianzhong

Xinjiang Institute for Endemic Disease Control and Research, Urumqi 830002, China

Arsenic in the Environment, Part II: Human Health and Ecosystem Effects,
Edited by Jerome O. Nriagu.
ISBN 0-471-30436-0

---

1. INTRODUCTION

Arsenic is a common environmental agent whose toxic properties have been known for centuries. Since ancient times, arsenicals have been known as agents of suicide and murder. In modern times, arsenicals are used as agents for suicide and murder. Human poisonings and deaths from arsenic have occurred as a result of drinking water contaminated with arsenic in Chile (Borgoño et al., 1977), Taiwan (Tseng et al., 1968), and other parts of the world. In 1980, we found a village where there was a high level of arsenic in the drinking-water source in the Dzungaria Basin, in the Xingjiang Uighur Autonomous Region of China. In the same area, we carried out a series of investigations the next year to determine the geographical distribution of arsenic in the environment. We found a zone with a high level of arsenic in the southwestern part of the basin, where underground water from an artesian well had arsenic concentrations of 0.04 to 0.75 mg/L. This zone, which covers many small communities, covers the Tianshan Mountains from the Manas River valley as far west as Aibi Lake. The whole distance is more than 200 km. Some cases of human poisoning from arsenic were found in villages where the drinking water had arsenic concentrations of more than 0.2 mg/L. In 1983, two projects for improving the quality of the drinking water, from which 100,000 inhabitants would benefit, began in the reclamation area of Kuitun and the Chepaiz subcounty of the Usum County, in the Xingjiang Uighur Autonomous Region of China. These were completed in 1984. Since 1982, we and others have made several studies of the endemic areas of arsenism. In addition, a scientific expedition led by Professor Tan, of the Institute of Geography, Acadamy of Sciences of China, consisting of scientists from eight institutes and two medical colleges, carried out a series of investigations along the Kuntun River in order to elucidate the correlations between endemic diseases and environmental chemical factors. The results of these investigations were in accordance with those of investigations carried out by us between 1980 and 1981. On the other hand, no cases of human poisoning from arsenic were found in areas along the Manas River valley and around Aibi Lake, where the drinking-water source had arsenic concentrations of less than 0.18 mg/L. Between 1980 and 1989, we made specific investigations of the geography, environment, epidemiology, and clinical and preventive medicine associated with endemic arsenism. The results of these investigations will be described in this chapter.

2. GEOGRAPHICAL ENVIRONMENT

2.1. General Environmental Conditions

The Tianshan mountain range, of which the parts rising above 3800 m are covered by glacier and accumulated snow all year round, is situated in the Xinjiang Uighur Autonomous Region of China. It traverses the middle part of Xinjiang and extends from west to east for more than 2000 km, dividing Xinjiang into two parts (South and North Xinjiang). It is well known that the main ores of arsenic are the arsenopyrites (FeAsS)–the pyrites of copper, gold, and zinc. Geological surveys have shown that the Tianshan mountain range, which is considered to have risen from the ancient Tianshan Sea during the Permian Period, consists mainly of granite and contains some mineral deposits rich in arsenic, such as the arsenopyrites, chalcopyrite, pyrite, and gold. However, these mineral deposits have not been mined so far. In 1983, we found a natural lake with a high level of arsenic in this mountainous district.

The rainfall is usually much higher in the mountainous district than in the plain. Four rivers originate from the Tianshan mountain range in this area. The altitude tends to decline gradually from the northern slope of the Tianshan mountain range as far north as the lowest land of the alluvial plain, and to increase gradually from there, as far as the hilly land within the limits of Toli County. The land in the higher part of the plain is mainly covered by cobble stone and gravel, and the land in the lower part consists of sandy and clay soils. Since the 1950s, local people have built up artificial reservoirs to retain the river water in order to ensure a supply of irrigation water in all seasons. The alluvial plain is in an arid area where the annual mean precipitation and evaporation capacity are 160 to 185 mm and 1800 mm, respectively. Most of the inhabitants of this area are farmers. Before 1962, the drinking water available to the local people was from shallow wells and rivers. Since 1962, many artesian wells have been drilled locally. The high-arsenic zone is located at the lowest part of the mountain apron of the Tianshan mountain range, ranging from Aibi Lake as far east as the Manas River, a distance of about 250 km. The northern part of the area of Kuitun City and Usum County, is situated in the middle section of the zone, is an area of endemic severe arsenism. Human poisonings from arsenic have occurred as a result of drinking high-arsenic water in some villages and towns in the area.

2.2. Distribution of Environmental Arsenic

2.2.1. Soil

A total of 11 samples were collected from soil 10 to 20 cm below the surface within the area ranging from the northern slope of the Tianshan Mountains. The results showed water-soluble arsenic concentrations of 0 to 0.4 mg/kg. Furthermore, no regularity was found in the distribution of environmental arsenic. In 1984, 100 samples were collected from the soil in different sections of the alluvial plain and again no regularity in the distribution of environmental arsenic was found.

2.2.2. Surface Water

2.2.2.1. River and Reservoir. During the period from 1980 to 1984, 22 samples were collected from the rivers and reservoirs in different sections of the alluvial plain and tested for water-soluble arsenic. The results showed that the water samples had an average arsenic concentration of 0.01 mg/L, with a range of 0 to 0.03 mg/L. No regularity was found in the distribution of environmental arsenic.

2.2.2.2. Lake. Arsenic found in a water sample collected from a natural lake at an altitude of 1860 m in the Tianshan Mountains had a concentration of 0.37 mg/L. In addition, one water sample collected from Aibi Lake, a saltwater lake at an altitude of 189 m in the lowest part of the Dzungaria Basin, had an arsenic concentration of 0.175 mg/L.

2.2.3. Shallow-Well Water

Arsenic found in 41 samples collected from various wells at a depth of 2 to 30 m had a mean concentration of 0.018 mg/L, with a range of 0 to 0.068 mg/L. However, no regularity in the distribution of arsenic in shallow-well water from various areas has been observed so far.

2.2.4 Deep-Well Water

Artesian wells have been sunk in some areas deeper than 660 m. Our findings have shown that arsenic concentrations in the water samples collected from artesian wells at a depth of 70 to 400 m may increase with depth. In the upper section of the alluvial plain, arsenic was found in deep-well water at concentration of less than 0.04 mg/L, while in the lower section of the alluvial plain, most of the samples collected from artesian wells had an arsenic concentration of more than 0.2 mg/L. Based on arsenic levels in deep-well water, the alluvial plain may be divided geographically into six belts, and the areas of endemic arsenism are situated in the fourth and fifth belts (see Table 1 and Fig. 1).

In the high-arsenic areas, it has been found that there are two rules by which arsenic is present in deep-well water. First, arsenic concentratoins in deep-well

Table 1 Relationship between Altitude and Arsenic Level in Deep Wells

Geographical Belt:	I	II	III	IV	V	VI
Number of samples:	48	66	37	42	20	9
Mean arsenic content:	0.007	0.021	0.053	0.27	0.08	0.008
(mg/L) SE:	0.001	0.001	0.005	0.017	0.012	0.003
Depth (m):	450–660	350–450	290–350	270–290	270–290	270–300

Note: $r_s = -0.9286$; $P < 0.05$.

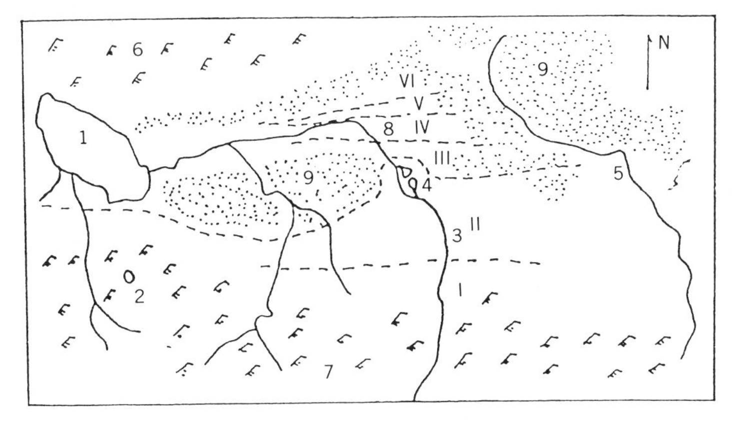

Figure 1. The distribution of arsenic in deep-well water on the alluvial plain: 1, Aibi Lake; 2, a nature pool with a high level of arsenic; 3, Kuntun River; 4, reservoir; 5, Mamas River; 6, the hills of Toli; 7, the Tianshan Mountains; 8, area of endemic arsenism; 9, desert.

water tend to go up with increasing depth of the well (2–400 m), with a range from 0.007 ± 0.005 mg/L (at a depth of 2–10 m, $n = 11$) to 0.251 ± 0.077 mg/L (at a depth of 316–400 m, $n = 5$), indicating that arsenic concentrations in the deep layer of the earth's crust might be greater than in the surface layer (see Table 2). Second, arsenic concentrations in deep wells around an artificial reservoir may be less than those in wells 5 km away (0.023 ± 0.009 mg/L versus 0.053 ± 0.026 mg/L), suggesting that there might be a belt of low arsenic in the groundwater around a reservoir.

Table 2 Relationship between Depth of Well and Arsenic level

Depth (m):	2–10	76–156	156–236	236–316	316–400
Number of wells:	11	7	12	37	5
As(mg/L)[a]:	0.007 ± 0.005	0.095 ± 0.05	0.157 ± 0.02	0.217 ± 0.028	0.251 ± 0.077

[a] Mean ± SE.

3. EPIDEMIOLOGY

3.1. Morbidity Rate

In the period from 1980 to 1984, a total of 31,141 inhabitants were examined for arsenic poisoning in 77 villages and towns situated in the alluvial plain, and 523 cases of human poisoning from arsenic were found among those subjects. Most of the cases were found in the fourth belt of the alluvial plain and only a few in the fifth belt, while no cases of human poisoning from arsenic occurred in the first, second, third, and sixth belts. In order to define quantitatively the correlation of water arsenic level with the morbidity rate from human arsenic poisonings, we carried out a series of investigations in 17 villages and towns, each of which had only one well for drinking water, in the Kuntun River valley. The results showed that the morbidity rate from human arsenic poisonings may increase with an increase in arsenic level in the drinking-water source. No cases of human poisonings from arsenic were found where the arsenic concentration in the drinking-water source was less than 0.1 mg/L. In addition, it was observed that cases of mild human poisoning occurred when arsenic concentration in the drinking-water source reached a range of 0.1 to 0.2 mg/L, the morbidity rate from human poisonings being no more than 1%. On the other hand, it was also observed that the morbidity rate from human poisonings might increase gradual-

Table 3 Correlation between Arsenic Level in Drinking Water and Human Morbidity Rate from Poisoning

Number of Villages	Water Arsenic (mg/L)	Subjects Examined	Cases Diagnosed	Rate(%)
1	0.035	409	0	0
2	0.066	271	0	0
3	0.11	180	0	0
4	0.17	188	0	0
5	0.21	322	1	0.3
6	0.135	247	2	0.81
7	0.13	285	2	0.7
8	0.23	543	8	1.5
9	0.24	241	3	1.2
10	0.26	1427	26	1.8
11	0.29	225	14	6.2
12	0.34	291	19	6.5
13	0.34	944	35	3.7
14	0.39	211	29	13.7
15	0.46	744	83	11.2
16	0.56	283	89	31.4
17	0.75	362	168	46.4

Note: $r = 0.912$; $r_{0.001(15)} = 0.725$; $P < 0.001$.

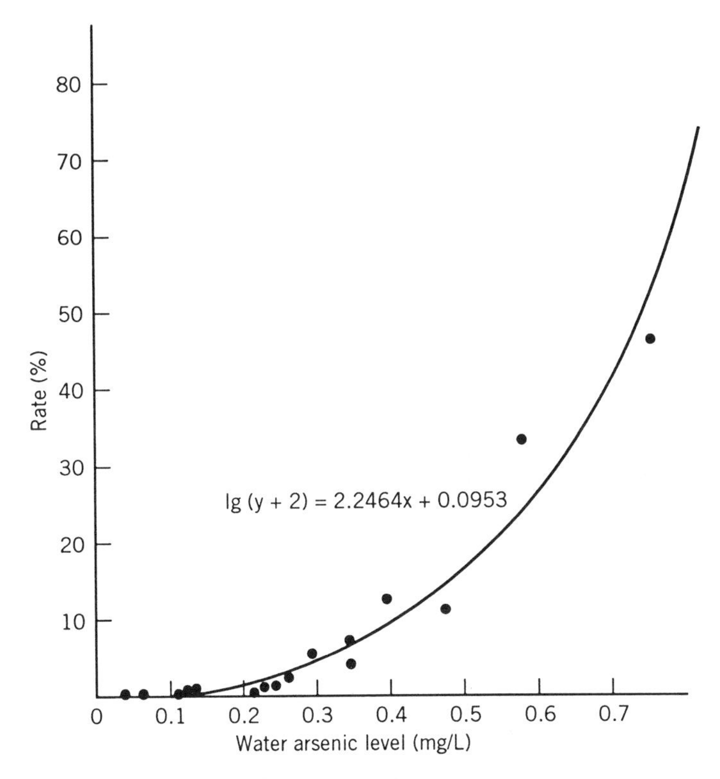

Figure 2. Relationship between the morbidity rate of human poisoning and arsenic levels in drinking water.

ly with increasing concentrations of arsenic in the drinking-water source when the arsenic concentration reached a range of 0.2 to 0.4 mg/L, and that the rate rose sharply when arsenic concentrations were more than 0.4 mg/L. It has thus been found that arsenic concentrations in drinking water may correlate positively with the morbidity rate from human arsenic poisonings. This positive correlation (see Table 3 and Fig. 2) may be expressed as a regression equation:

$$\log(y + 2) = 2.2464x + 0.0953$$

3.2. Epidemiological Studies on Endemic Arsenism

In August 1980, 362 inhabitants were examined for signs of arsenism in a small village where a well drilled in 1975 had an arsenic concentration of 0.75 mg/L. The cases of human poisoning among the inhabitants examined numbered 168, the morbidity rate being 46.4%. At the same time, in another village where arsenic was found in the drinking-water source at concentrations of 0.035 to 0.04 mg/L, 1267 inhabitants were examined and no cases of human poisoning were found.

3.2.1. Age and Sex

A total of 168 cases of human poisoning were found among the individuals examined (183 males and 179 females). Of the cases aged 3 to 67 years, 94 (51.4%) were males and 74 (41.3%) females. The morbidity rate from human poisonings may rise with increasing age, as shown in Table 4. This suggests that all the local inhabitants might be exposed to the disease, but that the younger might be more tolerant than the older inhabitants.

3.2.2. Occupation and Nationality

The village is in an agricultural area. We found that the morbidity rate from human arsenic poisoning was more than 70% throughout the different occupation communities (farmer, herdsman, etc.). It was observed that 148 cases of human poisoning (45.5%) occurred among the 325 inhabitants of the Han nationality, and 20 (54.1%) cases occurred among the 37 inhabitants of the Hui nationality. On the other hand, no significant difference was observed between the different occupations and nationalities in the morbidity rate from human arsenic poisoning.

3.2.3. Household

We carried out investigations using the household as a basic unit. It was found that of the 85 households studied, 81 had cases of human poisoning from arsenic. Most of the households had more than one case and in one household all six members suffered from arsenism.

3.2.4. Course of Arsenic Poisoning and Duration of High-Arsenic Water Ingestion

Our findings have shown that the morbidity rate from human arsenic poisoning may rise with increasing duration of high-arsenic water ingestion. It was observed that 65.6% of the cases of human poisoning had an attack 3 years after drinking high-arsenic water, and some cases had an attack only 6 months after drinking the water. No cases of poisoning were found among people who had drunk the surface water containing low arsenic.

3.2.5. Stages of Arsenic Poisoning

Based on the degrees of human poisoning observed, the disease can be divided into three stages: stage I, mild; stage II, moderate; stage III, serious. Among the

Table 4 Relationship between Morbidity Rate and Age of Human Cases

Age(yr):	0–9	10–19	20–39	40–49	50 and up	Total
Number examined:	98	103	56	59	46	362
Number of cases:	14	31	41	49	33	168
Incidence(%):	14.3	30.1	73.2	83.1	71.7	46.4

168 cases of human poisoning in the village investigated, 81 cases (48.2%) were found to have stage I arsenic poisoning, 60 (35.7%) had stage II, and 27 (16.1%) had stage III.

4. CLINICAL MANIFESTATIONS

4.1. Clinical Symptoms

We found that clinical symptoms occurring in the early stage of human arsenic poisoning were unspecific. Among the people who had drunk high-arsenic water, early symptoms included palpitations or chest discomfort, fatigue, headache, dizziness, insomnia, nightmare, and numbness in the extremities. However, we also found that some of those people had cutaneous changes, such as a dirty piebald skin, especially on the trunk, as a result of depigmentation, and keratinized eruptions on the skin of the palm or sole several months or years after drinking high-arsenic water. As is well known, these cutaneous changes are the specific manifestations of the disease.

4.2. Skin Signs

4.2.1. Arsenical Keratosis

The clinical findings suggest that arsenical keratosis might be the main sign of the disease. Of the 523 cases of human poisoning studied, 447 (85.5%) had keratinized eruptions characterized by many light-yellow and brown cutaneous lesions, papuloid or verrucous, present symmetrically on the palm and sole. Furthermore, no inflammatory reaction or painful sensation was present around the cutaneous lesions. Some cracks were observed on top of cutaneous lesions. These are considered to be the primary lesions.

4.2.2. Hyperpigmentation and Depigmentation

During our investigations, two other types of cutaneous lesions were found, one hyperpigmented and the other depigmented. Of the 523 cases of human poisoning from arsenic, 208 cases (39.8%) had lesions characterized by diffuse gray-black pigmentation or brown spots, obvious on the trunk but tending to become gradually less obvious on the extremities, and 306 (58.5%) had lesions characterized by dirty piepald skin or colorless spots the size of millet grains or soybeans on the trunk.

4.2.3. Other Cutaneous Changes

We noted in our practices that some cases of human arsenic poisoning, especially young females, had cyanosis on their extremities with a cold feeling in both hands and feet. Qian (1985) reported that malignant changes in the skin occurred in 5 of 317 cases of human arsenic poisoning in this area.

5. LABORATORY EXAMINATION

5.1. Biochemistry

Samples of both blood and urine were collected from 55 cases of human poisoning in a village where arsenic poisoning was prevalent and analyzed. The results showed that all the samples of blood and urine were normal. On the other hand, biochemical examination revealed that the concentration of albumin in sera increased in 10 cases, the concentration of globulin increased in 14 cases, and the ratio of albumin to globulin in sera was reversed in 19 cases (34.5%). Furthermore, the concentration of γ-globulin in sera rose in 21 cases, while that of α1- and α2- as well as β-globulin in sera declined.

5.2. Determination of Arsenic Content

During our investigation, urine, hair, and fingernail samples collected from poisoned and healthy people were examined for arsenic contents. The results showed that the mean value and standard error of the measured arsenic contents in the urinary samples from 109 poisoned cases were 297.0 μg/L and 33.0 μg/L respectively. Those of the controls were 61.0 μg/L, and 7.1 μg/L, respectively, or 4.9 times as high ($P < 0.05$). These results were similar to those reported from Taiwan (Lin et al., 1985), in which the arsenic contents in urinary samples collected from healthy people and from cases with blackfoot disease and Bowen's disease were 63.4, 75.7, and 201 μg/L, respectively. Our findings also revealed that the arsenic content in hair samples collected from 43 poisoned cases was 455 ± 47 μg/100 g, while that in the hair of 15 controls was 65.0 ± 18.3 μg/100 g. The arsenic content in hair samples from the poisoned cases was seven times as high as that in the controls. Furthermore, arsenic was found in the fingernails of 13 cases at levels of 18.9 ± 6.09 μg/g, 4.7 times as high as that in the controls (0.087–4.0 μg/g).

6. ELECTROCARDIOGRAPHIC CHANGES

Arsenic toxicity may result in impairment of the heart. A dose-response relationship between arsenic level in well water and cardiovascular disease in some areas of endemic arsenic toxicosis has been reported in Taiwan (Wu et al., 1989). Among our cases of arsenic poisoning, some showed heart symptoms such as palpitations. Of 80 cases of human poisoning in a village where arsenic toxicosis was prevalent, six showed arrhythmia. In addition, abnormal electrocardiograms were noted in 18(31.0%) of 58 cases of human poisoning resulting from drinking water contaminated with arsenic. Wang (1985) reported that abnormal electrocardiograms were noted in 48(78.69%) of 61 cases of human poisoning caused by drinking water contaminated with arsenic for about 13 years. Such abnor mal electrocardiograms were manifested mainly as a prolonged Q-T interval (31 cases), sinus arrhythmia (17 cases), and various cardiac blocks (15 cases).

Furthermore, a lasting prolongation of Q-T interval has been noted at the acute stage of arsenic poisoning in human beings (Barry and Herndon, 1962). However, we think that a prolonged Q-T interval might be a significant abnormality in electrocardiograms at the chronic as well as at the acute stage of arsenic poisoning in human beings.

7. HISTOPATHOLOGY

A histopathological examination was made in 25 cases of arsenic poisoning. Tissue specimens were collected from the areas of keratinized lesions and brown pigmentation, fixed in formalin solution (10%), and stained with hematoxylin-eosin for microscopic examination. The microscopic examination of tissue specimens from certain a typical cases revealed an acanthokeratodermia with hyperplasia in the prickle cell layer and vacuolar degeneration in some spine cells. In typical cases of arsenic poisoning, however, such histopathological changes could be seen as a thickened corneal layer, into which some papillary epiderm was herniating, as a thickened stratum granulasum without granular cells in certain areas where a great deal of pyknotic nucleuses were present in the adjacent corneal layer, and as a thickened layer of prickle cells with an abnormal structure and a great number of polynuclear cells or megakaryocytes. In some cases, moreover, degenerated vacuolar cells were revealed under the microscope, but the cytodesma could be seen clearly. It was also observed that keratincytes or keratinized cysts were obvious, with the basal layer proliferated but the membrane of basal layer intact. In cases of severe poisoning, the superficial dermis showed edema and chronic infiltration with inflammatory cells, and malignant changes of dermal tissue occurred in two of those cases. In our practice, however, the specific histopathological changes that could be used as evidence for a diagnosis have so far never been found. On the other hand, since arsenic toxicity tends to cause carcinoma, the possibility should be considered that cutaneous carcinoma might occur in cases of human poisoning.

8. PREVENTION OF ARSENIC POISONING IN HUMANS

It is obvious that high-arsenic drinking water may be a factor in arsenic toxicosis in human beings. It seems to be important in the control of the disease to consider how to prevent arsenic intake from drinking water. The village where the first cases of arsenic poisoning in humans occurred in August 1980 was deserted in December of the same year. The inhabitants moved to other villages where arsenic was found in the drinking-water source at concentrations of 0 to 0.37 mg/L. In some cases, the symptoms and signs of arsenic poisoning were reduced three years after the quality of drinking water improved; the morbidity rate from arsenic poisoning also declined from 46.4% to 34.6%. In contrast, in another village, where a water-quality improvement project is not yet complete, the symptoms of and morbidity from arsenic poisoning have remained the same. As shown in Table 5, the rate of improvement in the symptoms and signs of arsenic poisoning in human beings may increase with a decrease in arsenic level in the drinking-water source, suggesting a correlation. Furthermore, it was observed that new cases of human poisoning occurred only when arsenic concentrations in the drinking-water source exceeded 0.15 mg/L(see Table 5). At the same time, it was also found that arsenic levels in the urinary samples from cases of human poisoning also declined with a decrease in the arsenic levels in water source for drinking (see Table 6).

In 1984, two projects for improving the quality of the drinking water of 100,000 people were completed in the Xinjiang Uighur Autonomous Region of China, an area of endemic arsenic toxicosis. Based on our findings, an improvement in the symptoms and signs of arsenic poisoning in human beings has occurred as a result of the new water sources. Thus, it may be essential for the control of the disease to improve water quality in areas of endemic arsenic toxicosis.

Table 5 Relationship between the Rate of Improvement in Symptoms and Signs and Arsenic Level in Drinking Water

Water Arsenic (mg/L)	Improvement in Symptoms and Signs			New Cases		
	Number of Cases	Number Improved		Number Observed	New Cases	
		Cases	Rate(%)		Cases	Rate(%)
0–0.04	13	10	76.9	26	0	0
0.05–0.14	19	11	57.9	43	0	0
0.15–0.25	61	35	57.4	47	3	6.4
0.26–0.4	81	21	25.9	154	25	16.2
0.5 and up	7	0	0	4	3	

Table 6 Correlation between Arsenic Level in New Water Source and Arsenic Level in Urine

Water Arsenic Level (mg/L)	Cases Observed	Urine Arsenic Level (mg/L)	
		Mean	SE
0–0.04	32	0.03	0.005
0.05–0.14	9	0.059	0.013
0.15–0.25	10	0.127	0.024
0.26–0.3	7	0.152	0.014
0.4 and up	14	0.228	0.037

Note: $2r_s = 1.000$; $P < 0.05$.

9. REMOTE EFFECTS OF ARSENIC

In April and May 1989, we made a follow-up survey among 228 inhabitants of the area of endemic arsenic toxicosis where the project for improving the quality of drinking water was completed in 1984. We found that of the 141 original cases of human poisoning, 101 (71.63%) had improved symptoms and signs of arsenic poisoning, and 58 people who had suffered from mild and moderate arsenic toxicosis had recovered. In addition, only two cases of mild arsenic poisoning were observed among the 87 inhabitants who had not suffered originally from arsenic toxicosis. It was also noted that blackfoot disease, which is considered to be prevalent in an area of endemic arsenic toxicosis in Taiwan, occurred in six cases of human poisoning. In our cases of blackfoot disease, the clinical symptoms and signs of the disease were mild, manifested only as a feeling of numbness or cold in one or two feet or cyanosis of the skin of the feet, especially the toes. Moreover, vasculitis occurred in some of our cases of blackfoot disease, but no amputation or deaths occurred as a result of the disease. Our findings have shown that blackfoot disease in this region of China is less prevalent than in Taiwan. Tseng (1989) reviewed a 30-year follow-up survey of the cases of blackfoot disease in Taiwan and observed that amputation and death had occurred in 68% of the cases, the mortality rate being 4.84 per 100 cases each year.

Why are there such differences between the problem of blackfoot disease in the two endemic areas? We think that other factors that might promote the development of blackfoot disease exist in the area of endemic arsenic toxicosis in Taiwan. Moreover, Lu et al. (1975) and Yu (1984) made a study of a fluorescent substance in the well water in certain endemic areas in Taiwan and considered that it might be a causative agent of blackfoot disease. In addition, it was inferred that a fluorescent humic substance might be correlated closely with the development of the disease (Lu et al., 1990). The carcinogenicity of arsenic has been reported in many parts of the world (Brown et al., 1989; Chen et al., 1985; Chen and Wang, 1990; Tsuda et al., 1990; and others). It was found in Taiwan and Japan that there was a dose-response correlation between arsenic intake and the development of certain carcinomas. We reviewed 23 deaths that occurred in an area of endemic arsenic toxicosis with a population of 600 from 1981 to 1989 and observed that of the fatal cases, six (26.1%) had died of malignant neoplasms. Two died of lung cancer and the four others died of nasopharyngeal carcinoma, leukemia, lymphosarcoma, and osteosarcoma. Furthermore, Wu et al. (1989) noted that concentrations of arsenic in well water might be related to the development of cardiovascular disease. Of our group, 52.2% died of cardiovascular disease.

10. ARSENIC PROBLEMS IN OTHER AREAS OF XINJIANG

In some other areas of Xinjiang, such as Jinhe, Yengisar, Yopurga, Jiashi, and Shihezi, arsenic has been found in well water at concentrations of 0.05 to 0.2 mg/L. However, typical cases of arsenic poisoning in human beings have not been observed so far in those areas. Most of the wells that have been found to have high concentrations of arsenic have not been used as a source of drinking water, however.

REFERENCES

Barry, K. G., and Herndon, E. G., Jr. (1962). Electrocardiographic changes associated with acute arsenic poisoning. *Med. Ann. D. C.* **31**, 65–66.

Borgoño, J. M., Vicent, P., Venturino, H., and Infante, A. (1977). Arsenic in the drinking water of the city of Antofagasta: Epidemiological and clinical study before and after the installation of the treatment plant. *Environ. Health Perspect.* **19**, 103–105.

Brown, K. G., Boyle, K. E., Chen, C. W., and Gibb, H. J. (1989). A dose-response analysis of skin cancer from inorganic arsenic in drinking water. *Risk Anal.* **9**, 519–538.

Chen, C. J., and Wang, C. J. (1990). Ecological correlation between arsenic level in well water and age-adjusted mortality from malignant neoplasms. *Cancer Res.* **50**, 5470–5474.

Chen, C. J., Chuang, Y. C., Lin, T. M., and Wu, H. Y. (1985). Malignant neoplasms among residents of a blackfoot disease-endemic area in Taiwan: High-arsenic artesian well water and cancers. *Cancer Res.* **45**(11 pt. 2), 5895–5899.

Fowber, B. A. (1977). Toxicology of environmental arsenic. In R. A. Goyer and M. A. Mehlman (Eds.), *Toxicology of Trace Elements.* Wiley, New York, pp. 79–110.

Lin, S. M., Ching, C. H., and Yang, M. H. (1985). Arsenic concentration in the urine of patients with blackfoot disease and Bowen's disease. *Biol. Trace Elem. Res.* **8**, 11–18.

Lu, F. J., Yang, C. K., and Ling, K. H. (1975). Physio-chemical characteristics of drinking water in blackfoot disease endemic areas in China-I and Tainan Hsiens. *J. Formosan Med. Assoc.* **79**, 596–605.

Lu, F. J., Shih, S. R., Liu, T. M., and Shown, S. H. (1990). The effect of fluorescent humic substances existing in the well water of blackfoot disease endemic aresa in Taiwan on prothrombin time and activated partial thromboplastin time in vitro. *Thromb. Res.* **57**, 747–753.

Park, M. J., and Currier, M. (1991). Arsenic exposures in Mississippi, a review of cases. *South. Med. J.* **84**, 461–464.

Qian, X. C. (1985). Skin manifestations of endemic arsenism: With clinical analysis of 317 cases and pathological analysis of 46 cases. *Acta Acad. Med. Xinjiang* **8**, 103–107.

Tseng, W. P., Chu, M. M., How, S. W., Fong, J. M., Lin, C. S., and Yeh, S. (1968). Prevalence of skin cancer in an endemic area of chronic arsenism in Taiwan. *J. Natl. Cancer Inst. (U.S.)* **40**, 453–463.

Tsuda, T., Nagira, T., Yarnamoto, M., and Kume, Y. (1990). An epidemiological study on cancer in certified arsenic poisoning patients in Toroku. *Ind. Health* **28**, 53–62.

Wang, L. F., Sun, X. Z., and Liu, H. D. (1987). The value of skin affections with keratinization in palmaris et plantaris etc. on diagnosis of chronic endemic arsenism. *Endemic Dis. Bull.* **2**, 6–10.

Wang, S. Z. (1985). Electrocardiographic changes in arsenism. *Acta Acad. Med. Xinjiang* **8**, 117–120.

Wu, M. M., Kuo, T. L., Hwang, Y. H., and Chen, C. J. (1989). Dose-response relation between arsenic concentration in well water and mortality from cancers and vascular diseases. *Am. J. Epidemiol.* **130**, 1123–1132.

Yu, H. S. (1984). Blackfoot disease and chronic arsenism in southern Taiwan *Int. J. Dermatol.* **23**, 258–260.

10

ESTIMATION OF HUMAN EXPOSURE TO AND UPTAKE OF ARSENIC FOUND IN DRINKING WATER

Hao Xu

Groundwater Contamination Project, National Water Research Institute, Burlington, Ontario L7R 4A6, Canada

Anders Grimvall and Bert Allard

Department of Water and Environmental Studies, Linköping University, S-581 83 Linköping, Sweden

Arsenic in the Environment, Part II: Human Health and Ecosystem Effects,
Edited by Jerome O. Nriagu.
ISBN 0-471-30436-0

1. INTRODUCTION

Human exposure to arsenic can be estimated by direct or indirect methods. The direct methods are primarily based on chemical analysis of food, water, and air; the indirect methods include analysis of arsenic in urine and hair. Uptake and excretion models have been developed to relate direct exposure estimates to indirect estimates (Tam et al., 1979; Pomroy et al., 1980). However, the universal applicability of such models is not clear, and this may create considerable uncertainty about the contribution of arsenic from different sources.

Several studies have shown that, in certain areas, naturally occurring arsenic in drinking water can account for a considerable fraction of the total arsenic exposure (e.g., Harrington et al., 1978; Valentine et al., 1979). It is also known that high exposure to arsenic can occur in certain industrially polluted environments (Morse et al., 1979; Norin and Vahter, 1981) and that seafood can contain considerable amounts of the less toxic organic forms of arsenic (Vahter and Lind, 1986). The possibility of as yet unidentified sources can be illustrated by the results of a population study performed in Sweden (Vahter and Lind, 1986). Even though there were no known sources of arsenic exposure, the average excretion of inorganic and methyl arsenic was approximately 10 μg/g creatinine, and for certain individuals the excretion was significantly higher.

The uncertainty about uptake and excretion can be illustrated by the variability of the results obtained in different experimental studies. Pomroy et al. (1980) studied the fate of ingested low doses of radioactively labeled As(V) and concluded that 94% of the arsenic was excreted via urine. Other investigators, using much higher doses and different forms of arsenic, have reported excretions ranging from 50% (Crecelius, 1977) to 60 or 70% (Mappes, 1977; Butchet et al., 1981a, b).

The present report focuses on the estimation of exposure to arsenic via drinking water. The applicability of existing exposure–excretion models was investigated and the role of drinking water in relation to other sources was evaluated. This report consists of three parts:

1. Application of different exposure–excretion models to previously published data regarding arsenic exposure and urinary excretion of arsenic;
2. Experimental studies of absorption and excretion of different forms of arsenic;
3. A field and population study of arsenic in urine and hair in relation to arsenic in drinking water.

2. MATERIALS AND METHODS

2.1. Exposure–Excretion Model

Let A denote the fraction of ingested arsenic that is first absorbed and then excreted via urine. Arsenic retention in the human body was then assumed to be described by a first-order model

$$R(t) = A \sum_{i=1}^{m} a_i \exp(-\lambda_i t)$$

where t denotes time and $a_1, a_2, \ldots, a_m$ are nonnegative weight-factors summing to 1 and are excretion rates corresponding to different pathways of arsenic. The proposed model implies that the amount of arsenic retained in the body at time t is equal to

$$R(t) = A \sum_{x_j \leqslant t} c_j \sum_{i=1}^{m} a_i \exp[\lambda_i (t - x_j)]$$

if the person has been exposed to $c_1, c_2, \ldots, c_p$ units of arsenic at times $x_1, x_2, \ldots, x_p$, respectively. Furthermore, the amount $E(s, t)$ of arsenic excreted during the time interval (s, t) is equal to

$$\begin{aligned} E(s,t) &= R(t) - R(s) \\ &= A \sum_{i=1}^{m} a_i \sum_{j=1}^{p} c_j \{\exp[-\lambda_i \max(0, s - x_j)] - \exp[-\lambda_i \max(0, t - x_j)]\} \end{aligned} \quad (1)$$

Models assuming one, two or three pathways ($p = 1, 2$, or 3) were tested. The model subsequently referred to as Pomroy's model (Pomroy et al., 1980) is a three-term model with

$$a_1 = 0.659 \qquad a_2 = 0.304 \qquad a_3 = 0.037$$

$$\lambda_1 = 0.332 \qquad \lambda_2 = 0.0728 \qquad \lambda_3 = 0.0180$$

Parameter estimation was performed using a nonlinear regression procedure in the software package StatGraphics.

2.2. Experimental Studies

In one study, three healthy adult volunteers ingested 20 μg of As(V) (sodium arsenate dissolved in water) on 4 consecutive evenings. Morning urine was collected for a period of 14 days, starting 5 days prior to the first ingestion. In a second study, one of the volunteers ingested 50 μg of As(V) per day for 5 days.

Urine (12-hr night samples) was collected 3 days prior to the first ingestion for 15 consecutive days. Seafood was avoided during the entire study period.

The three volunteers also took part in a study of exposure to organic arsenic in seafood. Such food was first avoided for one week. Thereafter, about 200 g of shrimps were ingested, and seafood was then avoided for another week. Morning urine was collected during the entire study period.

2.3. Field and Population Study

A field and population study was performed in the municipality of Smedjebacken in central Sweden, which is one of the areas in the country where elevated arsenic concentrations have been found in groundwater (Allard et al., 1991). Urine and hair samples were collected from 38 persons (12 families) who used water from drilled or dug wells, and water samples were collected from the kitchen taps. Thirty-four persons who consumed water from a municipal water plant were chosen to make up a control group not exposed to arsenic via drinking water. All the participants were asked not to consume any seafood three days before the collection of urine. Data on dietary habits (consumption of seafood), smoking, and occupation were collected by interviews.

2.4. Analytical Procedures

Arsenic concentrations in water, urine, and hair samples were measured using an atomic absorption spectrophotometer (AAS, Perkin Elmer 1100) connected to an arsine generation system (MHS-10). The detection limit for water samples was $< 0.1\ \mu g$ As L^{-1} (Allard et al., 1991).

Urine samples were analyzed both with respect to total arsenic (As_{tot}) and the sum of inorganic and methylated arsenic ($As_{in + methyl}$). The latter species were measured directly by MHS–AAS after the addition of an antifoaming solution. As_{tot} was determined after digestion of the samples. More precisely, the samples were heated to 105 °C for 8 hr and then to 450 °C for 12.5 hr in the presence of 2 mL of an ashing aid solution [35 g MgO + 52.5 g $Mg(NO_3)_2 6H_2O$ + 500 mL deionized water]. The resulting ash was dissolved in 15 mL of 3 M HCl to allow analysis of arsenic by MHS–AAS.

Hair samples were washed with acetone (CH_3COCH_3), sodium lauryl sulfate [$CH_3(CH_2)_{10}CH_2OSO_3Na$], HCl, and, finally, hydroxyquinoline (HOC_6H_3N: CHCH:CH). The samples were then digested with a mixture of HNO_3 and H_2SO_4 (1:1) and analyzed for arsenic by MHS–AAS (Curatola et al., 1978).

Creatinine concentrations in the urine samples were analyzed using a procedure described by Dunicz (1963).

3. RESULTS

3.1. Reanalysis of Previously Published Exposure–Excretion Data

The excretion data shown in Figure 1 were observed in an experimental study of urinary arsenic excretion after ingestion of 500 μg As(III) per day for five con-

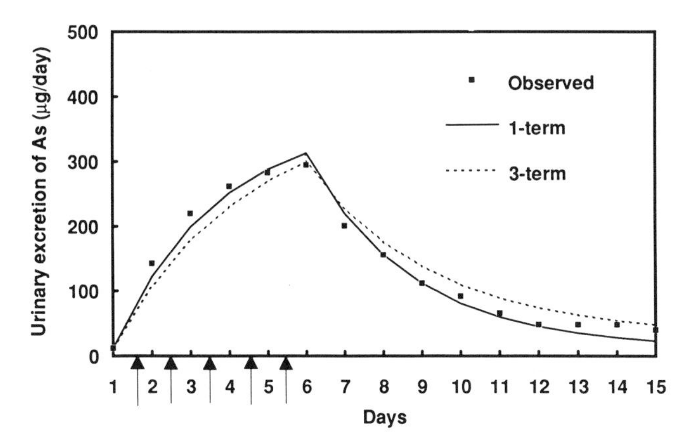

Figure 1. Predicted and observed urinary excretion of arsenic after a daily intake of 500 μg As(III) per day for five consecutive days (marked with arrows). Observed data from Buchet et al. (1981b). Predictions were made by using a one-term exponential model and Pomroy's three-term exponential model (Pomroy et al., 1980).

secutive days (Buchet et al., 1981b). The solid curve corresponds to a one-term exponential model ($m = 1$) of the form described in Section 2.1; the absorption (A) and the excretion rate were both treated as unknown parameters and estimated from observed data after subtracting a background excretion rate that was set at 12 μg/day. The broken-line curve corresponds to Pomroy's three-term model ($m = 3$) when using the weight factors and excretion rates listed in Section 2.1; the absorption (A) was estimated from the data of Buchet et al. after subtraction of the background excretion rate. As shown in Figure 1, both models produced a good fit with observed data. Despite this, the estimated absorption differed substantially between the two models (77.7% according to the one-term model and 91.5% according to the three-term model). Thus, the choice of exposure–excretion model is crucial in the estimation of arsenic absorption.

In their study of arsenic excretion, Pomroy et al. (1980) collected data for six individuals. A three-term exponential model was fitted to each of these data sets, and the estimated weight factors and excretion rates were presented in a table. In the present study, the excretion data observed by Buchet et al. (1981b) were matched against the six models proposed by Pomroy et al. (1980). More precisely, the absorption A in (1) was fitted to observed data, whereas the weight factors and excretion rates were taken from Pomroy et al. (1980). Figure 2 illustrates the calculated and observed arsenic excretion following ingestion of 500 μg arsenic per day for five consecutive days. Apparently, there was no contradiction between the two studies. The values observed by Buchet et al. (1981b) were within the variability between individuals observed by Pomroy et al. (1980).

The results in Table 1 show that the estimated arsenic absorption was different for different doses. However, there was no obvious trend in the observed data. Both the highest and the lowest dose resulted in a lower estimated absorption than the two middle doses. For each dose, the estimated absorption was strongly

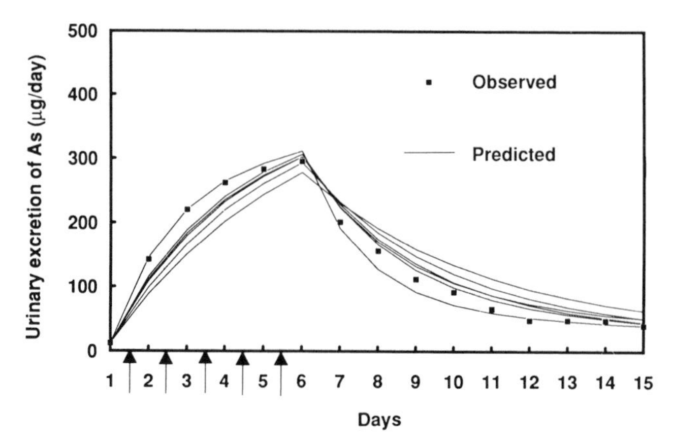

Figure 2. Predicted and observed urinary excretion of arsenic after a daily intake of 500 µg As(III) per day for five consecutive days (marked with arrows). Observed data from Buchet et al. (1981b). Predicted data obtained using Pomroy's three-term exponential model with parameter values corresponding to six different subjects (Pomroy et al., 1980).

Table 1 Estimated Absorption and Observed Urinary Excretion of Arsenic Following Daily Intake of As(III) for Five Consecutive Days[a]

	Calculated Absorption (%)		
Daily Dose of As(III) (µg)	One-Term Exponential Model	Three-Term Exponential Model	Observed Urinary Excretion (% in 14 days)
125	56	72	54
250	69	93	73
500	78	92	74
1000	65	78	64

[a] Observed excretion data from Buchet et al. (1981b); calculated data based on a one-term exponential model or on Pomroy's three-term exponential model.

Table 2 Estimated Absorption and Observed Urinary Excretion of Arsenic Following a Single Intake of 500 µg Arsenic[a]

	Calculated Absorption (%)		
Arsenic Form	One-Term Exponential Model	Three-Term Exponential Model	Observed Urinary Excretion (% in 14 days)
As(III)	51	75	44
MMA	71	90	78
DMA	74	102	75

[a] Observed data from Buchet et al. (1981a); calculated data based on a one-term exponential model or Pomroy's three-term exponential model.

dependent on the choice of exposure–excretion model. With a three-term exponential model, a considerable fraction of the arsenic excretion was predicted to occur more than two weeks after the exposure, whereas the one-term model predicted that practically all excretion would occur within two weeks.

The results in Table 2 show that the absorption of inorganic As(III) is considerably lower than that of methylated arsenic forms. In addition, the table illustrates that different exposure–excretion models can result in very different absorption estimates.

3.2. Experimental Studies

Studies of arsenic exposure and uptake from drinking water have usually been based on analyses of total arsenic (As_{tot}), although the sum of inorganic arsenic and different methylated arsenic forms ($As_{in + methyl}$) has also been used as an exposure indicator. The results presented in Figure 3 show how As_{tot} and $As_{in + methyl}$ in urine are affected by a predetermined exposure to moderate doses of inorganic As(V). Apparently, for $As_{in + methyl}$, the base line is more stable and the concentration peak more pronounced.

Shrimps and certain other types of seafood naturally contain considerable quantities of organic arsenic and possibly also some inorganic arsenic. The results in Figure 4 show that, after ingestion of shrimps, excretion of As_{tot} in urine is strongly elevated. The impact of the ingestion on $As_{in + methyl}$, however, was not large enough to be distinguished from the normal temporal variability of this parameter.

Previously performed experimental studies of arsenic uptake and excretion have been restricted to very low doses of As(V) and moderate to high doses of As(III) or methyl arsenic. The curves in Figure 5 illustrate the excretion of

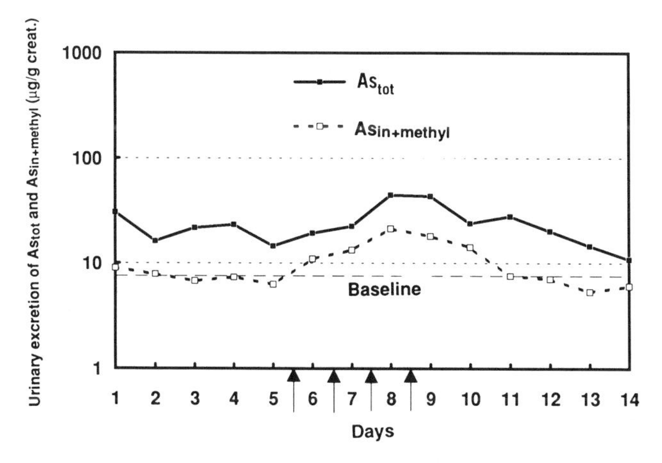

Figure 3. Urinary excretion of As_{tot} and $As_{in+methyl}$ after a daily intake of 20 μg As(V) for four consecutive days (marked with arrows); mean values for three subjects.

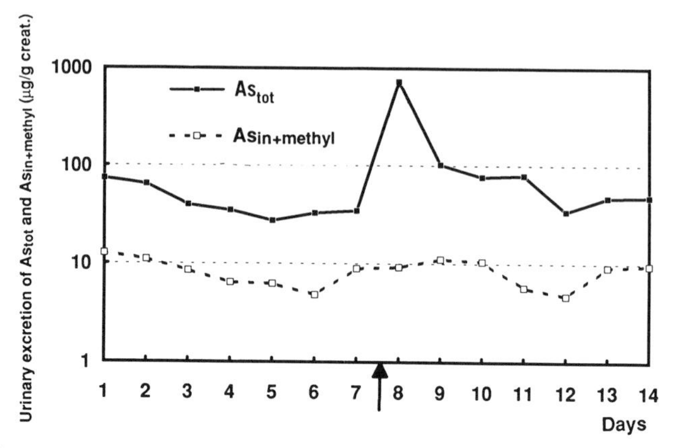

Figure 4. Urinary excretion of As_{tot} and $As_{in+methyl}$ after a single intake of about 200 g shrimps; mean values for three subjects.

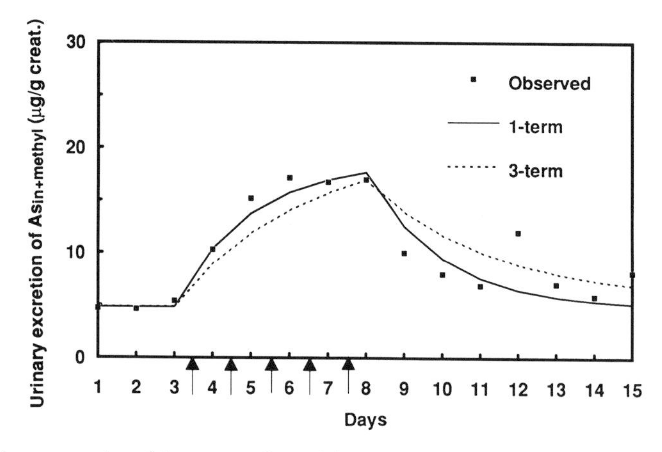

Figure 5. Urinary excretion of $As_{in+methyl}$ after a daily intake of 50 µg As(V) for five consecutive days (marked with arrows). A one-term exponential model and Pomroy's three-term model have been fitted to observed data.

$As_{in+methyl}$ following intake of 50 µg As(V) per day for five consecutive days. Because of the short duration of the experiment, the fit with observed data was not markedly improved when a simple one-term model was extended to a three-term exponential model. The estimated As(V) absorption, 49.6% in the one-term model and 71.8% in the three-term model, was almost identical to the As(III) absorption shown in Table 2.

3.3. Population Study

The diagram in Figure 6 shows that the urinary excretion of $As_{in+methyl}$ was markedly elevated among people exposed to arsenic via drinking water. The

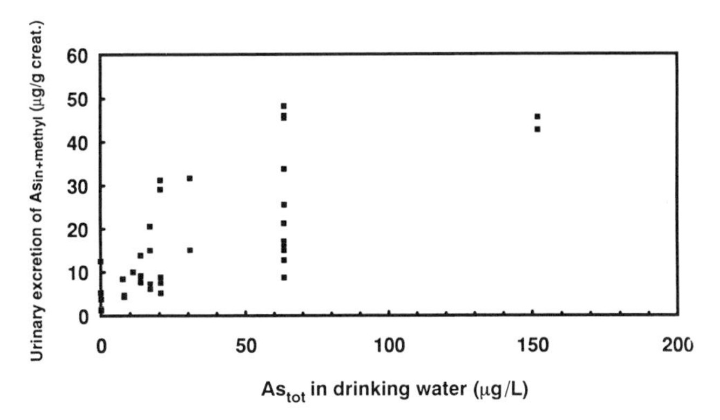

Figure 6. Relationship between $As_{in+methyl}$ in urine and As_{tot} in drinking water in the municipality of Smedjebacken in central Sweden.

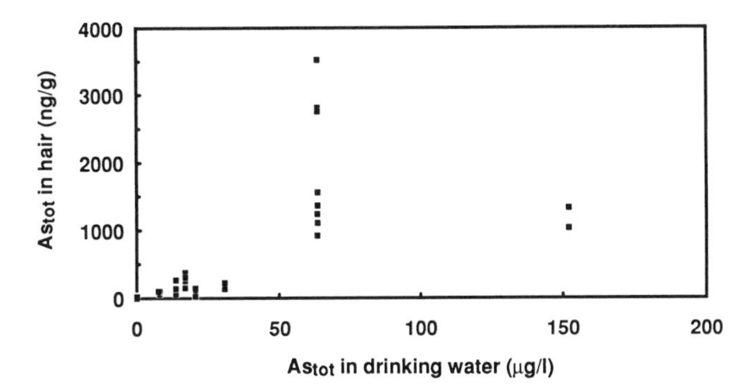

Figure 7. Relationship between As_{tot} in hair and As_{tot} in drinking water in the municipality of Smedjebacken in central Sweden.

average $As_{in+methyl}$ excretion in the unexposed control group was only 2.9 μg As/g creatinine, as compared to almost 50 μg As/g creatinine for some of the exposed individuals. A simple linear regression of $As_{in+methyl}$ excretion versus As_{tot} concentration in the drinking water showed that the relationship was statistically significant ($r = 0.785$; $p < 0.0001$). An almost identical relationship ($r = 0.792$; $p < 0.0001$) was observed between As_{tot} in hair and As_{tot} in water (Fig. 7). According to the interviews, there was no significant difference between the exposed group and the control group in terms of dietary habits (consumption of seafood) or smoking. None of the participants had an occupation in which arsenic exposure could be suspected.

4. DISCUSSION

The reanalysis of previously published exposure–uptake–excretion data provided at least two explanations for the seemingly contradictory absorption estimates that have been reported by different investigators (Bettley and O'Shea, 1975; Crecelius, 1977; Mappes, 1977; Pomroy et al., 1980; Buchet et al., 1981a, b). The choice of uptake–excretion model was found to be crucial in the estimation of arsenic absorption (Fig. 1 and Tables 1 and 2). Furthermore, it was clearly shown that variability between individuals must be taken into account when results from different studies are compared (Fig. 2).

The absorption and excretion rate estimates that were obtained for different doses and forms of arsenic indicated that methylated arsenic forms are more readily absorbed than inorganic As(III) (Table 2), but no difference was found between inorganic As(III) and As(V). The dose dependence of absorption and excretion rates was moderate, and the lack of clear trends in observed data indicated that variability between individuals was a more important factor (Table 1 and Fig. 2). In fact, if variability between individuals was taken into account, none of the results of the present study contradicted the applicability of Pomroy's three-term exponential model to a very wide range of arsenic doses and different inorganic arsenic forms. Less than 10^{-4} μg was ingested in the investigations by Pomroy et al. (1980), whereas up to 5000 μg was ingested in the experiments performed by Buchet et al. (1981b).

The population study of arsenic exposure via drinking water showed that the observed excretion of $As_{in + methyl}$ agreed well with the observed concentrations in water and the estimated absorption of the dominating arsenic species [inorganic As(V) and As(III)] in well water. The highest observed arsenic excretion surpassed excretion in the unexposed control group by approximately 40 μg As/g creatinine. If the excretion of creatinine is 1.6 g/day and the average absorption is 75%, this corresponds to an excess exposure of nearly 90 μg/day or, equivalently, 0.6 L/day of the most contaminated well water (150 μg/L) examined in the present study. The regression line fitted to observed As_{tot} concentrations in water and $As_{in + methyl}$ in urine showed that the excretion of $As_{in + methyl}$ was doubled at a drinking-water arsenic concentration as low as 15 μg/L. In addition, it is noteworthy that the background level was considerably lower than previously reported by other investigators (e.g., Vahter and Lind, 1986).

ACKNOWLEDGMENTS

Information and assistance by Mr. B. Jernberg and Colleagues, Smedjebacken, are gratefully acknowledged. This study was partly financed by the Swedish National Environmental Protection Board (SNV).

REFERENCES

Allard, B., Xu, H., and Grimvall, A. (1991). Concentration and speciation of arsenic in groundwaters—a case study. In H. Xu, Ph.D. Dissertation, Linköping University, Sweden, ISBN 91-7870-783-8.

Bettley, F. R., and O'Shea, J. A. (1975). The absorption of arsenic and its relation to carcinoma. *Br. J. Dermatol.* **92**, 563–568.

Buchet, J. P., Lauwerys, R., and Roels, H. (1981a). Comparison of the urinary excretion of arsenic metabolites after a single oral dose of sodium arsenite, monomethylarsonate, or dimethylarsinate in man. *Int. Arch. Occup. Environ. Health* **48**, 71–79.

Buchet, J. P., Lauwerys, R., and Roels, H. (1981b). Urinary excretion of inorganic arsenic and its metabolites after repeated ingestion of sodium metaarsenite by volunteers. *Int. Arch. Occup. Environ. Health* **48**, 111–118.

Crecelius, E. A. (1977). Changes in the chemical speciation of arsenic following ingestion by man. *Environ. Health Perspect.* **19**, 147–150.

Curatola, C. J., Grunder, F. I., and Moffett, A. E. (1978). Hydride generation atomic absorption spectrophotometry for determination of arsenic in hair. *Am. Ind. Hyg. Assoc. J.* **39**, 933–938.

Dunicz, B. L. (1963). Simple and accurate method for the determination of creatine and creatinine. *Clin. Chim. Acta* **9**, 203–209.

Harrington, J. M., Middaugh, J. P., Morse, D. L., and Housworth, J. (1978). A survey of a population exposed to high arsenic in well water in Fairbanks, Alaska. *Am. J. Epidemiol.* **108**, 377–385.

Mappes, R. (1977). Experiments on excretion of arsenic in urine. *Occup. Environ. Health* **40**, 267–272.

Morse, D. L., Harrington, J. M., Housworth, J., and Landrigan, P. J. (1979). Arsenic exposure in multiple environmental media in children near a smelter. *Clin. Toxicol.* **14**, 389–399.

Norin, H., and Vahter, M. (1981). A rapid method for the selective analysis of total urinary metabolites of inorganic arsenic. *Scand. J. Work Environ. Health* **7**, 38–44.

Pomroy, C., Charbonneau, S. M., McCullough, R. S., and Tam, G. K. H. (1980). Human retention studies with ^{74}As. *Toxicol. Appl. Pharmacol.* **53**, 550–556.

Tam, G. K. H., Charbonneau, S. M., Bryce, F., Pomroy, C., and Sandi, E. (1979). Metabolism of inorganic arsenic (^{74}As) in humans following oral ingestion. *Toxicol. Appl. Pharmacol.* **50**, 319–322.

Vather, M., and Lind, B. (1986). Concentrations of arsenic in urine of the general population in Sweden. *Sci. Total Environ.* **54**, 1–12.

Valentine, J. L., Kang, H. K., and Spivey, G. (1979). Arsenic levels in human blood, urine, and hair in response to exposure via drinking water. *Environ. Res.* **20**, 24–32.

11

A REVIEW OF ARSENIC HAZARDS TO PLANTS AND ANIMALS WITH EMPHASIS ON FISHERY AND WILDLIFE RESOURCES

Ronald Eisler

U.S. National Biological Survey, Patuxent Wildlife Research Center, Laurel, Maryland 20708

Arsenic in the Environment, Part II: Human Health and Ecosystem Effects,
Edited by Jerome O. Nriagu.
ISBN 0-471-30436-0

---

1. INTRODUCTION

Anxiety over arsenic (As) is understandable and frequently justifiable. Arsenic compounds were the preferred homicidal and suicidal agents during the Middle Ages, and arsenicals have been regarded largely in terms of their poisonous characteristics in the nonscientific literature [National Academy of Sciences (NAS), 1977]. Data collected on animals, including human beings, indicate that inorganic arsenic can cross the placenta and produce mutagenic, teratogenic, and carcinogenic effects in offspring (Nagymajtenyi et al., 1985). Correlations between elevated atmospheric arsenic levels and mortalities from cancer, bronchitis, and pneumonia were established in an epidemiological study in England and Wales, where deaths from respiratory cancer increased at air concentrations $> 3 \mu g$ As/m^3 [National Research Council of Canada (NRCC), 1978]. Chronic arsenical poisoning, including skin cancer and a gangrenous condition of the hands and feet called blackfoot disease, has occurred in people from communities in Europe, South America, and Taiwan who were exposed to elevated concentrations of arsenic in drinking water [Environmental Protection Agency (EPA), 1980]. More recently, about 12,000 Japanese infants were poisoned (128 deaths) after consuming dry milk containing 15 to 24 mg inorganic As/kg, which originated from contaminated sodium phosphate used as a milk stabilizer. Fifteen years after exposure, the survivors sustained an elevated frequency of severe hearing loss and brain-wave abnormalities (Pershagen and Vahter, 1979).

Many reviews on the ecotoxicological aspects of arsenic in the environment are available; particularly useful are those by Woolson (1975), NAS (1977), NRCC (1978), Pershagen and Vahter (1979), EPA (1980, 1985), Hood (1985), Andreae (1986), Eisler (1988), and Phillips (1990). These authorities agree on six points: (1) arsenic is a relatively common element and is present in air, water, soil, plants, and all living tissues; (2) arsenicals have been used in medicine as chemotherapeutics since 400 B.C.E., and organoarsenicals were used extensively for this purpose until about 1945, with no serious effects when judiciously administered; (3) large quantities of arsenicals are released into the environment as a result of industrial and especially agricultural activities, and these may pose potent ecological dangers; (4) exposure of humans and wildlife to arsenic may occur through air (emissions from smelters, coal-fired power plants, herbicide sprays), water (mine tailings runoff, smelter wastes, natural mineralization), and food (especially seafoods); (5) chronic exposure to arsenicals by way of the air, diet, and other routes is associated with liver, kidney, and heart damage, hearing loss, brain-wave abnormalities, and impaired resistance to viral infections; and (6) exposure to arsenic has been associated with different types of human cancers, such as respiratory cancers and epidermoid carcinomas of the skin, as well as precancerous dermal keratosis. Only recently (Deknudt et al., 1986) has the epidemiological evidence of human carcinogenicity been confirmed by carcinogenesis in experimental animals.

This account briefly reviews the ecological and toxicological aspects of arsenic in the environment, with emphasis on fish, wildlife, and invertebrates, and updates my earlier report (Eisler, 1988) on this subject.

2. SOURCES, FATE, AND USES

Global production of arsenic is estimated to be 75,000 to 100,000 tons annually, of which the United States produces about 21,000 tons and uses about 44,000 tons; major quantities are imported from Sweden, the world's leading producer (NAS, 1977; EPA, 1980). Almost all (97%) of the arsenic made worldwide enters end-product manufacture in the form of arsenic trioxide (As_2O_3), and the rest is used as additives in producing special lead and copper alloys (NAS, 1977). More than 80% of the As_2O_3 is used to manufacture products with agricultural applications, such as insecticides, herbicides, fungicides, algicides, sheep dips, wood preservatives, dyestuffs, and medicines for the eradication of tapeworm in sheep and cattle (NAS, 1977). The sole producer and refiner of As_2O_3 in the United States is a copper smelter in Tacoma, Washington (NAS, 1977).

Arsenic occurs naturally as sulfides and as complex sulfides of iron, nickel, and cobalt (Woolson, 1975). In one form or another, arsenic is present in rocks, soils, water, and living organisms at concentrations of parts per billion to parts per million (NAS, 1977). Soil arsenic levels are normally elevated near arseniferous deposits and in mineralized zones containing gold, silver, and sulfides of lead and zinc (Dudas, 1984). Secondary iron oxides formed from the weathering of pyrite act as scavengers of arsenic (Dudas, 1984). Pyrite is a known carrier of arsenic and may contain up to 5600 mg/kg; for example, total arsenic is 10 times above normal background levels in soils derived from pyritic shale (Dudas, 1984). Natural weathering of rocks and soils adds about 40,000 tons of arsenic to the oceans yearly, accounting for < 0.01 mg/L input to water on a global basis (NRCC, 1978). Many species of marine plants and animals often contain naturally high concentrations of arsenic (NAS, 1977), but it is usually present in a harmless organic form (Woolson, 1975). Anthropogenic input of arsenic to the environment is substantial and exceeds the amount contributed by natural weathering processes by a factor of about three (NRCC, 1978).

The most important factor in arsenic cycling in the environment is constant change. Arsenic is ubiquitous in living tissue and is constantly being oxidized, reduced, or otherwise metabolized. In soils, insoluble or slightly soluble arsenic compounds are constantly being resolubilized, and the arsenic is being presented for plant uptake or reduction by organisms and chemical processes. Human beings reportedly have modified the arsenic cycle only by causing localized high concentrations (NAS, 1977). The speciation of arsenic in the environment is affected partly by indiscriminate biological uptake, which consumes about 20% of the dissolved arsenate pool and results in measurable concentrations of reduced and methylated arsenic species. The overall arsenic cycle is similar to the phosphate cycle; however, regeneration time for arsenic is much slower—on the order of several months (Sanders, 1980). The ubiquity of arsenic in the environment is evidence of the redistribution processes that have been operating since early geologic time (Woolson, 1975). The prehuman steady-state global arsenic cycle (Austin and Millward, 1984) indicates that major reservoirs of arsenic (in kilotons) are magma (50 billion), sediments (25 billion), oceanic deep waters (1.56 million), land (1.4 million), and ocean mixed layers (270,000); minor amounts occur in ocean particulates (100) and in continental (2.5) and marine tropospheres (0.069). Arsenic is significantly mobilized from the land to the troposphere by natural and anthropogenic processes. Industrial emissions account for about 30% of the present-day burden of arsenic in the troposphere (Austin and Millward, 1984). Agronomic ecosystems, for example, may receive arsenic from agricultural sources such as organic herbicides, irrigation waters, and fertilizers, and from nonagricultural sources such as fossil fuels and industrial and municipal wastes (Woolson, 1975). Arsenic is mobile and nonaccumulative in the air, plant, and water phases of agronomic ecosystems; arsenicals sometimes accumulate in soils, but redistribution mechanisms usually preclude hazardous accumulations (Woolson, 1975).

Arsenic compounds have been used in medicine since the time of Hippocrates, ca. 400 B.C.E. (Woolson, 1975). Inorganic arsenicals have been used for centuries, and organoarsenicals have been used for at least a century in the treatment of syphylis, yaws, amoebic dysentery, and trypanosomiasis (NAS, 1977). During the period 1200 to 1650, arsenic was used extensively in homicides (NRCC, 1978). In 1815, the first accidental death was reported from arsine (AsH_3) poisoning, and in 1900 to 1903 accidental poisonings from consumption of arsenic-contaminated beer were widely reported (NRCC, 1978). In 1938, arsenic was found to counteract selenium toxicity (NRCC, 1978). The introduction of arsphenamine, an organoarsenical, to control venereal disease earlier in this century gave rise to intensive research by organic chemists, which resulted in the synthesis of at least 32,000 arsenic compounds. The advent of penicillin and other newer drugs, however, nearly eliminated the use of organic arsenicals as human therapeutic agents (EPA, 1980). Arsenical drugs are still used in treating certain tropical diseases, such as African sleeping sickness and amoebic dysentery, and in veterinary medicine to treat parasitic diseases, including filariasis in dog (*Canis familiaris*) and blackhead in turkeys (*Meleagris gallopavo*) and chickens (*Gallus* spp.) (NAS, 1977). Today, abnormal sources of arsenic that can enter the food chain from plants or animals include arsenical pesticides such as lead arsenate; arsenic acid, $HAsO_3$; sodium arsenite, $NaAsO_2$; sodium arsenate, Na_2AsO_4; and cacodylic acid, $(CH_3)_2As(OH)$ (NAS, 1977).

The major uses of arsenic are in the production of herbicides, insecticides, desiccants, wood preservatives, and growth stimulants for plants and animals. Much smaller amounts are used in the manufacture of glass (nearly all of which contains 0.2 to 1.0% arsenic as an additive—primarily as a decolorizing agent) and textiles, and in medical and veterinary applications (NAS, 1977; EPA, 1980). Arsenic is also an ingredient in lewisite, a blistering poison gas developed (but not used) during World War I, and in various police riot-control agents (NAS, 1977). The availability of arsenic in certain local areas has been increased by various human activities: smelting and refining of gold, silver, copper, zinc, uranium, and lead ores; combustion of fossil fuels, such as coal and gasoline; burning of vegetation from cotton gins treated with arsenical pesticides; careless or extensive use of arsenical herbicides, pesticides, and defoliants; dumping of land wastes and sewage sludge (1.1 mg As/L) in areas that allow leaching into groundwater; use of domestic detergents in wash water (2.5 to 1000 mg As/L); manufacture of glass; and the sinking of drinking-water wells into naturally arseniferous rock (NRCC, 1978; EPA, 1980). There are several major anthropogenic sources of environmental arsenic contamination: industrial smelters—the effluent from a copper smelter in Tacoma, Washington, contained up to 70 tons of arsenic discharged yearly into nearby Puget Sound (NRCC, 1978); coal-fired power plants, which collectively emit about 3000 tons of arsenic annually in the United States (EPA, 1980); and the production and use of arsenical pesticides, coupled with careless disposal of used pesticide containers (NAS, 1977). Elevated levels of arsenic have been reported in soils near smelters, in acid mine spoils, and in orchards receiving heavy applications of lead arsenate (NAS, 1977; Dudas, 1984). Air concentrations of arsenic are elevated near metal smelters, near sources of coal burning, and wherever arsenical pesticides are applied (NAS, 1977). Atmospheric deposition of arsenic has steadily increased for at least 30 years, on the basis of sedimentary evidence from lakes in upstate New York (Smith et al., 1987). Arsenic is introduced into the aquatic environment through atmospheric deposition of combustion products and through runoff from fly ash-storage areas near power plants and nonferrous smelters (Smith et al., 1987). Elevated arsenic concentrations in water were recorded near mining operations, and from mineral springs and other natural waters—usually alkaline with high sodium and bicarbonate contents (NAS, 1977). In the United States, the most widespread and frequent increases in dissolved arsenic concentrations in river waters have been in the northern Midwest; all evidence indicates that increased atmospheric deposition of fossil-fuel combustion products is the predominant cause of the trend (Smith et al., 1987).

Agricultural applications are the largest anthropogenic sources of arsenic in the environment (Woolson, 1975). Inorganic arsenicals (arsenic trioxide; arsenic acid; arsenates of calcium, copper, lead, and sodium; and arsenites of sodium and potassium) have been used widely for centuries as insecticides, herbicides, algicides, and desiccants. Paris green (cuprous arsenite) was successfully used in 1867 to control the Colorado potato beetle (*Leptinotarsa decemlineata*) in the eastern United States. Arsenic trioxide has been applied widely as a soil sterilant. Sodium arsenite has been used for aquatic weed control, as a defoliant to kill potato vines before tuber harvest, as a weed killer along roadsides and railroad rights-of-way, and for control of crabgrass (*Digitaria sanguinalis*). Calcium arsenates have been applied to cotton and tobacco fields to protect against the boll weevil (*Anthonomus grandis*) and other insects. Lead arsenate has been used to control insect pests on fruit trees, and for many years it was the only insecticide that controlled the codling moth (*Carpocapsa pomonella*) in apple orchards and the horn worm larva (Sphyngidae) on tobacco. Much smaller quantities of lead arsenate are used in orchards now that fruit growers rely primarily on carbamate and organophosphorus compounds to control insect pests; however, lead arsenate is still being used by some growers to protect orchards from certain chewing insects. The use of inorganic arsenicals has decreased in recent years due to the banning of sodium arsenite and some other arsenicals for most purposes, although they continue to be used on golf greens and fairways in certain areas to control annual bluegrass (*Poa annua*). In recent decades, inorganic arsenicals have been replaced by organoarsenicals for herbicidal application, and by carbamate and organophosphorus compounds for insect control (Woolson, 1975). By the mid-1950s, organoarsenicals were used extensively as desiccants, defoliants, and herbicides (NRCC, 1978). Organoarsenicals marketed in agriculture today, and used primarily for herbicidal application, include cacodylic acid (also known as dimethylarsinic acid) and its salts, monosodium and disodium methanearsonate (Woolson, 1975; NAS, 1977). Organoarsenicals are used as selective herbicides for weedy grasses in turf, and around cotton and noncrop areas for weed control; at least 1.8 million ha (4.4 million acres) have been treated with more than 8000 tons of organoarsenicals (NAS, 1977). In 1945, one organoarsenical (3-nitro-4-hydroxyphenylarsonic acid) was found to control coccidiosis and promote growth in domestic chickens (Woolson, 1975). Since that time, other substituted phenylarsonic acids have been shown to have both therapeutic and growth-promoting properties as feed additives for poultry and swine (*Sus* spp.), and they are used for this purpose today under existing regulations (Woolson, 1975; NAS, 1977), although the use of arsenicals in poultry food was banned in France in 1959 (NRCC, 1978).

3. CHEMICAL AND BIOCHEMICAL PROPERTIES

Elemental arsenic is a gray, crystalline material. Its atomic number is 33, its atomic weight is 74.92, its density is 5.727, its melting point is 817 °C, and its sublimation point is 613 °C; its chemical properties are similar to those of phosphorus (Woolson, 1975; NAS, 1977; NRCC, 1978; EPA, 1980, 1985). Arsenic has four valence states: $-3, 0, +3$, and $+5$. Arsines and methylarsines, which are characteristic of arsenic in the -3 oxidation state, are generally unstable in air. Elemental arsenic (As^0) is formed by the reduction of arsenic oxides. Arsenic trioxide (As^{+3}) is a product of smelting operations and is the material used in synthesizing most arsenicals. It is oxidized catalytically or by bacteria to arsenic pentoxide (As^{+5}) or orthoarsenic acid (H_3AsO_4). Arsenic in nature is rarely in its free state. Usually, it is a component of sulfidic ores, occurring as arsenides and arsenates, along with arsenic trioxide, which is a weathering product of arsenides. Most arsenicals degrade or weather to form arsenate, although arsenite may form under anaerobic conditions. Biotransformations may occur, resulting in volatile arsenicals that normally are returned to the land, where soil adsorption, plant uptake, erosion, leaching, reduction to arsines, and other processes occur. This natural arsenic cycle entails a constant shifting of arsenic between environmental compartments.

Arsenic species in flooded soils and water are subject to chemically and microbiologically mediated oxidation or reduction and methylation reactions. At high Eh values (i.e., high oxidation–reduction potential) typical of those encountered in oxygenated waters, pentavalent As^{+5} tends to exist as H_3AsO_4, $H_2AsO_4^-$, $HAsO_2$, and AsO_4^{3-}. At lower Eh, the corresponding trivalent arsenic species can be present, as well as AsS_2^- (Thanabalasingam and Pickering, 1986). In aerobic soils, the dominant arsenic species is As^{+5}, and small quantities of arsenite and monomethylarsonic acid are present in mineralized areas; in anaerobic soils, As^{+3} is the major soluble species (Haswell et al., 1985). Inorganic arsenic is more mobile than organic arsenic and thus poses greater problems by leaching into surface waters and groundwater (NRCC, 1978). The trivalent arsenic species are generally considered to be more toxic, more soluble, and more mobile than As^{+5} species (Thanabalasingam and Pickering, 1986). Soil microorganisms metabolize arsenic into volatile arsine derivatives. Depending on conditions, 17 to 60% of the total arsenic present in soil may be volatilized (NRCC, 1978). Estimates of the half-life of arsenic in soil vary from 6.5 years for arsenic trioxide to 16 years for lead arsenate (NRCC, 1978).

In water, arsenic occurs in both inorganic and organic forms and in dissolved and gaseous states (EPA, 1980). The form of arsenic in water depends on Eh, pH, organic content, suspended solids, dissolved oxygen, and other variables (EPA, 1985). Arsenic in water exists primarily as a dissolved ionic species; particulates account for $< 1\%$ of the total measurable arsenic (Maher, 1985a). Arsenic is rarely found in water in the elemental state (0), and it is found in the -3 state only at extremely low Eh values (Lima et al., 1984). Common forms of arsenic encountered in water are arsenate, arsenite, methanearsonic acid, and dimethylarsinic acid (EPA, 1985). The formation of inorganic pentavalent arsenic, the most common species in water, is favored under conditions of high dissolved oxygen, basic pH, high Eh, and reduced organic material content; the opposite conditions usually favor the formation of arsenites and arsenic sulfides (NRCC, 1978; Pershagen and Vahter, 1979; EPA, 1980), although some arsenite is attributed to biological activity (Maher, 1985a). Water temperature seems to affect arsenic species composition in the estuary of the River Beaulieu in the United Kingdom, where reduced and methylated species predominate during the warmer months and inorganic As^{+5} predominates during the colder months; the appearance of methylated arsenicals during the warmer months is attributed to bacterial and abiotic release from decaying plankton and to grazing by zooplankton (Howard et al., 1984). Also contributing to higher water or mobile levels are the natural levels of polyvalent anions, especially phosphate species. Phosphate, for example, displaces arsenic held by humic acids, and it sorbs strongly on the hydrous oxides of arsenates (Thanabalasingam and Pickering, 1986).

Physical processes play a key role in governing arsenic bioavailability in aquatic environments. For example, arsenates are readily sorbed by colloidal humic material under conditions of high organic content, low pH, low phosphate, and low mineral content (EPA, 1980; Thanabalasingam and Pickering, 1986). Arsenates also coprecipitate with, or adsorb on, hydrous iron oxides and form insoluble precipitates with calcium, sulfur, aluminium, and barium compounds (EPA, 1980). Removal of arsenic from seawater by iron hydroxide scavenging seems to be the predominant factor in certain estuaries. The process involves both As^{+3} and As^{+5} and results in a measurable increase in arsenic level of particulate matter, especially at low salinities (Sloot et al., 1985; Tremblay and Gobeil, 1990). Arsenic sulfides are comparatively insoluble under conditions prevalent in anaerobic aqueous and sedimentary media containing hydrogen sulfide; accordingly, these compounds may accumulate as precipitates and thus remove arsenic from the aqueous environment. In the absence of hydrogen sulfide, these sulfides decompose within several days to form arsenic oxides, sulfur, and hydrogen sulfide (NAS, 1977).

In reduced environments such as sediments, arsenate is reduced to arsenite and methylated to methylarsinic acid or dimethylarsenic acids. These compounds may be further methylated to trimethylarsine or reduced to dimethylarsine, and they may volatilize to the atmosphere, where oxidation reactions result in the formation of dimethylarsinic acid (Woolson, 1975). Arsenates are more strongly adsorbed to sediments than are other arsenic forms, the adsorption processes depending strongly on arsenic concentrations, sediment characteristics, pH, and ionic concentration of other compounds (EPA, 1980). An important mechanism of arsenic adsorption onto lake sediments involves the interaction of anionic arsenates and hydrous iron oxides. Evidence indicates that arsenic is incorporated into sediments at the time of hydrous oxide formation, rather than by adsorption onto existing surfaces (Aggett and Roberts, 1986). Arsenic concentrations in lake sediments are also correlated with manganese; hydrous manganese oxides—positively charged for the adsorption of Mn^{+2} ions—play a significant role in arsenic adsorption onto the surface of lake sediments (Takamatsu et al., 1985). The mobility of arsenic in lake sediments and its release to the overlying water is related partly to seasonal changes. In areas that become stratified in summer, arsenic released from sediments accumulates in the hypolimnion until turnover, when it is mixed with epilimnetic waters; this mixing may result in a 10 to 20% increase in arsenic concentration (Aggett and O'Brien, 1985). Microorganisms (including four species of fungi) in lake sediments oxidized inorganic As^{+3} to As^{+5} and reduced inorganic As^{+5} to As^{+3} under aerobic conditions; under anaerobic conditions, only reduction was observed (Freeman et al., 1986). Inorganic arsenic can be converted to organic alkyl arsenic acids and methylated arsines under anaerobic conditions by fungi, yeasts, and bacteria, although biomethylation may also occur under aerobic conditions (EPA, 1980).

Most arsenic investigators now agree on the following points: (1) arsenic may be absorbed by ingestion, inhalation, or through permeation of the skin or mucous membranes; (2) cells accumulate arsenic by using an active transport system normally used in phosphate transport; (3) arsenicals are readily absorbed after ingestion, most being rapidly excreted in the urine during the first few days, or at most a week (the effects seen after long-term exposure are probably a result of continuous daily exposure rather than accumulation); (4) the toxicity of arsenicals conforms to the following order, from greatest to least toxicity: arsines > inorganic arsenites > organic trivalent compounds (arsenoxides) > inorganic arsenates > organic pentavalent compounds > arsonium compounds > elemental arsenic; (5) solubility in water and body fluids appears to be directly related to toxicity (the low toxicity of elemental arsenic is attributed to its virtual insolubility in water and body fluids, whereas the highly toxic arsenic trioxide, for example, is soluble in water to 12.0 g/L at 0 °C, 21.0 g/L at 25 °C, and 56.0 g/L at 75 °C); and (6) the mechanisms of arsenical toxicity differ considerably among arsenic species, although signs of poisoning appear to be similar for all arsenicals (Woolson, 1975; NRCC, 1978; Pershagen and Vahter, 1979).

The primary toxicity mode of inorganic As^{+3} is through reaction with sulfhydryl groups of proteins and subsequent enzyme inhibition; inorganic pentavalent arsenate does not react as readily as As^{+3} with sulfhydryl groups, but may uncouple oxidative phosphorylation (Howard et al., 1984; EPA, 1985). Inorganic As^{+3} interrupts oxidative metabolic pathways and sometimes causes morphological changes in liver mitochondria. Arsenite in vitro reacts with protein-SH groups to inactivate enzymes such as dihydrolipoyl dehydrogenase and thiolase, producing inhibited oxidation of pyruvate and beta-oxidation of fatty acids (Belton et al., 1985). Inorganic As^{+3} may also exert toxic effects by the reaction of arsenous acid (HAsO) with the sulfhydryl (SH) groups of enzymes. In the first reaction, arsenous acid is reduced to arsonous acid ($AsOH_2$), which then condenses to either monothiols or dithiols to yield dithioesters of arsonous acid. Arsonous acid may then condense with enzyme-SH groups to form a binary complex (Knowles and Benson, 1984a, b).

Methylation to methylarsonic acid $[(CH_3)_2AsO_3H_2]$ and dimethylarsinic acid $[(CH_3)_2AsO_2H]$ is usually the major detoxification mechanism for inorganic pentavalent arsenates and trivalent arsenites in mammals. Methylated arsenicals rapidly clear from all tissues, except perhaps the thyroid (Marafante et al., 1985; Vahter and Marafante, 1985; Yamauchi et al., 1986). Methylated arsenicals are probably common in nature. Methylation of arsenic (unlike methylation of mercury) greatly reduces toxicity and is a true detoxification process (Woolson, 1975). Before methylation (which occurs largely in the liver), As^{+5} is reduced to As^{+3} the kidney being an important site for this transformation (Belton et al., 1985). Arsenate reduction and subsequent methylation is rapid. Arsenite and dimethylarsinate were present in hamster (*Cricetus* sp.) plasma only 12 minutes after injection of inorganic As^{+5} (Hanlon and Ferm, 1986c). Demethylation of methylated arsenicals formed in vivo has not yet been reported (EPA, 1980).

The toxic effects of organoarsenicals are exerted by initial metabolism to the trivalent arsonoxide form, and then by reaction with sulfhydryl groups of tissue proteins and enzymes to form an arylbis (organylthio) arsine (NAS, 1977). This form, in turn, inhibits oxidative degradation of carbohydrates and decreases cellular ATP, the energy-storage molecule of the cell (NRCC, 1978). Among the organoarsenicals, those most injurious physiologically are methylarsonous acid [$CH_3As(OH)_2$] and dimethylarsinous acid [$(CH_3)_2AsOH$] (Knowles and Benson, 1984b). The enzyme inhibitory forms of organoarsenicals (arsonous acid) are formed from arsenous acid, and the corresponding arsonic acids are formed by a wide variety of enzymes and subcellular particles (Knowles and Benson, 1984a). Organoarsenicals used as growth promoters and drugs are converted to more easily excretable (and sometimes more toxic) substances, although most organoarsenicals are eliminated without being converted to inorganic arsenic or to demethylarsinic acids (Pershagen and Vahter, 1979).

4. ESSENTIALITY, SYNERGISM, AND ANTAGONISM

Arsenic apparently behaves more like an environmental contaminant than a nutritionally essential mineral (NAS, 1977). Nevertheless, low doses ($< 2\ \mu g$/day) of arsenic stimulated growth and metamorphosis in tadpoles and increased viability and cocoon yield in silkworm caterpillars (NAS, 1977). Arsenic deficiency has been observed in rats: signs include rough haircoat, low growth rate, decreased hematocrit, increased fragility of red cells, and enlarged spleen (NAS, 1977). Similar results have been documented in goats and pigs fed diets containing less than 0.05 mg As/kg. In these animals, reproductive performance was impaired, neonatal mortality was increased, birth weight was lower, and weight gain in second-generation animals was decreased; these effects were not evident in animals fed diets containing 0.35 mg As/kg (NAS, 1977).

The use of phenylarsonic feed additives to promote growth in poultry and swine and to treat specific diseases does not seem to constitute a hazard to the animal or its consumers. Animal deaths and elevated tissue arsenic residues occur only when the arsenicals are fed at excessive dosages for long periods (NAS, 1977). Arsenic can be detected at low levels in tissues of animals fed organoarsenicals, but it is rapidly eliminated when the arsenicals are removed from the feed for the required 5-day period before marketing (Woolson, 1975).

Selenium and arsenic are antagonists in several animal species. In rats, dogs, swine, cattle, and poultry, the arsenic protects against selenium poisoning if arsenic is administered in the drinking water and selenium in the diet (NAS, 1977; NRCC, 1978; Pershagen and Vahter, 1979). Inorganic arsenic compounds decrease the toxicity of inorganic selenium compounds by increasing biliary excretion (NRCC, 1978). However, in contrast to the antagonism shown by inorganic arsenic–inorganic selenium mixtures, the toxic effects of naturally methylated selenium compounds (trimethylselenomium chloride and dimethyl selenide) are markedly enhanced by inorganic arsenicals (NRCC, 1978).

The toxic effects of arsenic can be counteracted with (1) saline purgatives, (2) various demulcents that coat irritated gastrointestinal mucous membranes, (3) sodium thiosulfate (NAS, 1977), and (4) mono- and dithiol-containing compounds and 2,3-dimercaptopropanol (Pershagen and Vahter, 1979). Arsenic uptake in rabbit intestine is inhibited by phosphate, casein, and various metal chelating agents (EPA, 1980). Mice and rabbits are significantly protected against sodium arsenite intoxication by *N*-(2,3-dimercaptopropyl) phthalamidic acid (Stine et al., 1984). Conversely, the toxic effects of arsenite are potentiated by excess dithiols, cadmium, and lead, as evidenced by reduced food efficiency and disrupted blood chemistry in rodents (Pershagen and Vahter, 1979).

Arsenic effectively controls filariasis in cattle; new protective uses are under investigation. The control of parasitic nematodes (*Parafilaria bovicola*) in cattle was successful after 30 weekly treatments in plungement dips containing 1600 mg As_2O_3/L; however, the muscle of treated cattle contained up to 1.3 mg As/kg, or 12 times the amount in controls (Nevill, 1985). Existing anionic organic arsenicals used to control tropical nematode infections in humans have sporadic and lethal side effects. Cationic derivatives have been synthesized in an attempt to avoid the side effects and have been examined for their effect on adult nematodes (*Brugia pahangi*) in gerbils (*Meriones unguiculatus*). All arsenicals were potent filaricides; the most effective compounds tested killed 95% of adult *B. pahangi* after five daily subcutaneous injections of 3.1 mg As/kg body weight (Denham et al., 1986).

Animals previously exposed to sublethal levels of arsenic may develop tolerance to arsenic on reexposure. Although the mechanism of this process is not fully understood, it probably includes the efficiency of in vivo methylation processes (EPA, 1980). For example, resistance to toxic doses of As^{+3} or As^{+5} increases in mouse fibroblast cells pretreated with a low As^{+3} concentration (Fischer et al., 1985). Also, growth is better in arsenic-conditioned mouse cells in the presence of arsenic than in previously unexposed cells, and inorganic arsenic is more efficiently methylated. In vivo biotransformation and excretion of inorganic arsenic as monomethylarsonic acid (MMA) and dimethylarsinic acid (DMA) have been demonstrated in a number of mammalian species, including human beings. Cells may adapt to arsenic by increasing the biotransformation rate of the element to methylated forms, such as MMA and DMA (Fischer et al., 1985). Pretreatment of Chinese hamster (*Cricetus* spp.) ovary cells with sodium arsenite provided partial protection against the adverse effects of methyl methanesulfonate (MMS), and may even have benefited the MMS-treated cells; however, posttreatment dramatically increased the cytotoxic, clastogenic, and mitotic effects of MMS (Lee et al., 1986b).

Although arsenic is not an essential plant nutrient, small yield increases have sometimes been observed at low soil-arsenic levels, especially for tolerant crops such as potatoes, corn, rye, and wheat (Woolson, 1975). Arsenic phytotoxicity of soils is reduced with increasing lime, organic matter, iron, zinc, and phosphates (NRCC, 1978). In most soil systems, the chemistry of arsenic becomes the chemistry of arsenate; the estimated half-time of arsenic in soils is about 6.5 years, although losses of 60% in 3 years and 67% in 7 years have been reported (Woolson, 1975). Additional research is warranted on the role of arsenic in crop production and nutrition, with special reference to essentiality for aquatic and terrestrial wildlife.

5. BACKGROUND CONCENTRATIONS

In abundance, arsenic ranks twentieth among the elements in the earth's crust (1.5 to 2 mg/kg), fourteenth in seawater, and twelfth in the human body (Woolson, 1975). It occurs in various forms, including inorganic and organic compounds and trivalent and pentavalent states (Pershagen and Vahter, 1979). In aquatic environments, higher arsenic concentrations are reported in hot springs, in groundwaters from areas of thermal activity or in areas containing rocks with high arsenic content, and in some waters with high dissolved-salt content (NAS, 1977). Most of the other elevated values reported in lakes, rivers, and sediments are probably due to anthropogenic sources, which include smelting and mining operations, combustion of fossil fuels, arsenical grasshopper baits, synthetic detergent and sewage sludge wastes, and arsenical defoliants, herbicides, and pesticides (NAS, 1977). Most living organisms normally contain measurable concentrations of arsenic, but except for marine biota, these are usually less than 1 mg/kg fresh weight. Marine organisms, especially crustaceans, may contain more than 100 mg As/kg dry weight, usually as arsenobetaine, a water-soluble organoarsenical that poses little risk to the organism or its consumer. Plants and animals collected from naturally arseniferous areas or near anthropogenic sources may contain significantly elevated tissue residues of arsenic. Additional and more detailed information on background concentrations of arsenic in abiotic and living resources can be found in NAS (1977), Hall et al. (1978), NRCC (1978), EPA (1980), Jenkins (1980), Eisler (1981), and Phillips (1990).

5.1. Nonbiological Samples

Arsenic is a major constituent of at least 245 mineral species, of which arsenopyrite is the most common (NAS, 1977). In general, background concentrations of arsenic are 0.2 to 15 mg/kg in the lithosphere, 0.005 to 0.1 $\mu g/m^3$ in air, $< 10\ \mu g/l$ in water, and < 15 mg/kg in soil (NRCC, 1978). Commercial use and production of arsenic compounds have raised local concentrations in the environment far above natural background concentrations (Table 1).

Weathering of rocks and soils adds about 45,000 tons of arsenic to the oceans annually, accounting for < 0.01 mg/L on a global basis (NRCC, 1978). However, arsenic inputs to oceans increased during the past century from natural sources and as a result of industrial use, agricultural and deforestation activities, emissions from coal and oil combustion, and loss during mining of metal ores. If present activities continue, arsenic concentrations in oceanic surface waters may increase overall by about 2% by the year 2000; most of the increased burden will be in estuaries and coastal oceans, for example, Puget Sound, Washington; the Tamar, England; and the Tejo, Portugal (Sanders, 1985). Estimates of residence times of arsenic are 60,000 years in oceans and 45 years in freshwater lakes (NRCC, 1978). In the hydrosphere, inorganic arsenic occurs predominantly as As^{+5} in surface water, and significantly as As^{+3} in groundwater containing high levels of total arsenic. The main organic species in freshwater are methylarsonic

Table 1 Total Arsenic Concentrations in Selected Nonbiological Materials

Material (concentration unit)	Concentration [mean, (range), max.]	Reference[a]
AIR ($\mu g/m^3$)		
Remote areas	< 0.021	1
Urban areas	(0.0–0.16)	1
Near smelters		
Former USSR	(0.5–1.9)	2
Texas	Max. 1.4	2
Tacoma, Washington	Max. 1.5	2
Romania	Max. 1.6	2
Germany	(0.9–1.5)	2
Coal-fired power plant, Czechoslovakia	(19–69)	3
Orchard spraying of Pb arsenate	Max. 260,000	3
Near U.S. cotton gin burning vegetation treated with arsenic	Max. 400	3
DRINKING WATER ($\mu g/L$)		
Nationwide, USA	2.4 (0.5–214)	4
Fairbanks, Alaska	224 (1–2450)	4
Bakersfield, California	(6–393)	4
Nevada, 3 communities	(51–123)	4
Mexico, from plant producing As_2O_3	(4000–6000)	2
Japan, near factory producing arsenic sulfide	3000	2
Ghana, near gold mine	1400	2
Minnesota, contaminated by residual arsenical grasshopper bait	(11,800–21,000)	1
Methylated arsenicals, USA	Usually < 0.3 (0.01–1)	5
DUST (mg/kg)		
Tacoma, Washington		
Near smelter	1300	1
Remote from smelter	70	1
FOSSIL FUELS (mg/kg)		
Coal		
Canada	4 (0.3–100)	3
USA	5	2
Czechoslovakia	Max. 1500	2
Worldwide	13 (0.0–2000)	1
Coal ash	(< 20–8000)	3
Fly ash	(2.8–200)	3
Petroleum	0.2 (2.8–200)	3
Petroleum ash	Max. 100,000	3
Automobile particulates	298	3
GROUNDWATER ($\mu g/L$)		
Near polymetallic sulfide deposits	Max. 400,000	3
Near gold mining activities	> 50	3

Table 1 *(Continued)*

Material (concentration unit)	Concentration [mean, (range), max.]	Reference[a]
USA	Usually < 10	2
USA	17.9 (0.01–800)	3
LAKE WATER (μg/L)		
Dissolved solids		
< 2,000 mg/L	(0.0–100)	6
> 2,000 mg/L	(0.0–2,000)	6
Lake Superior	(0.1–1.6)	6
Japan, various	(0.2–1.9)	6
Germany, Elbe River	(20–25)	6
Searles Lake, California	(198,000–243,000)	1, 4
California, other lakes	(0.0–100)	1, 4
Michigan	Max. 2.4	1, 4
Wisconsin	(4–117)	1, 4
Florida	3.6	1, 4
Lake Chautauqua, New York	(3.5–35.6)	1, 4
Lake Ohakuri, New Zealand	(30–60)	7
Finfeather Lake, Texas	Max. 240,000	8
Thermal waters, worldwide	(20–3800 usually), Max. 276,000	1–3, 9
RAIN (μg/L)		
Canada	(0.01–5)	3
Rhode Island	0.8	1
Seattle, Washington	17.0	1
RIVER WATER (μg/L)		
Polluted, USA	Max. 6,000	4
Nonpolluted, USA	Usually < 5	4
Nationwide, USA, 1974–1981		
25th percentile	< 1	10
50th percentile	1	10
75th percentile	3	10
ROCK (mg/kg)		
Limestone	1.7 (0.1–20)	1
Sandstone	2 (0.6–120)	1
Shale and clay	14.5 (0.3–490)	1
Phosphate	22.6 (0.4–188)	1
Igneous, various	1.5–3 (0.06–113)	1
SEAWATER (μg/L)		
Worldwide	2 (0.15–6)	6
Pacific Ocean	(1.4–1.8)	11
Atlantic Ocean	(1.0–1.5)	11
South Australia		
Total dissolved As	1.3 (1.1–1.6)	12
As^{+5}	1.29	12

Table 1 *(Continued)*

Material (concentration unit)	Concentration [mean, (range), max.]	Reference[a]
As^{+3}	0.03	12
Particulate As	< 0.0006	12
UK, Beaulieu Estuary		
Water temperature < 12 °C		
Inorganic arsenic	(0.4–0.9)	13
Suspended arsenic	(0.02–0.24)	13
Organoarsenicals	(0.19–0.75)	13
Water temperature > 12 °C		
Inorganic arsenic	(0.6–1.1)	13
Suspended arsenic	(0.2–0.6)	13
Organoarsenicals	ND	13
SEDIMENTS (mg/kg dry weight)		
Near sewer outfall	35	3
From areas contaminated by smelters, arsenical herbicides, or mine tailings		
Surface	(198–3500)	1, 7, 9, 14, 15
Subsurface	(12–25)	1, 7, 9, 14, 15
Upper Mississippi River	2.6 (0.6–6.2)	16
Lake Michigan	(5–30)	1
Naturally elevated	> 500	1, 9
Oceanic	33.7 (< 0.4–455)	3
Lacustrine	(Usually 5–26.9), max. 13,000	3
SNOW (mg/kg)		
Near smelter	> 1000	3
SOIL POREWATERS (μg/L)		
Mineralized areas		
Arsenate	(79–210)	17
Arsenite	(2–11)	17
Monomethyl arsonic acid (MMAA)	(4–22)	17
Total arsenic	(93–240)	17
Unmineralized areas		
Arsenate	(18–49)	17
Arsenite	(1–7)	17
MMAA	< 1	17
Total arsenic	(13–59)	17
SOILS (mg/kg dry weight)		
USA, uncontaminated	7.4	18
Worldwide, uncontaminated	7.2	18
Canada		
Near gold mine		
Air levels 3.9 μg As/m^3	21,213	3
80 km distant	(10–25)	3

Table 1 *(Continued)*

Material (concentration unit)	Concentration [mean, (range), max.]	Reference[a]
Near smelter		
Japan	Max. 2470	2
Tacoma, Washington	Max. 380	2
Treated with arsenical pesticides		2
USA	165 (1–2553)	6
Canada	121	6
SYNTHETIC DETERGENTS (mg/kg)		
Household, heavy duty	(1–73)	1,2

[a]1, NAS (1977); 2, Pershagen and Vahter (1979); 3, NRCC (1978); 4, EPA (1980); 5, Hood (1985); 6, Woolson (1975); 7, Freeman et al. (1986); 8, Sorensen et al. (1985); 9, Farmer and Lovell (1986); 10, Smith et al. (1987); 11, Sanders (1980); 12, Maher (1985a); 13, Howard et al. (1984); 14, Hallacher et al. (1985); 15, Takamatsu et al. (1985); 16, Wiener et al. (1984); 17, Haswell et al. (1985); 18, Dudas (1984).

acid and dimethylarsinic acid, and these are usually present in lower concentrations than inorganic arsenites and arsenates (Pershagen and Vahter, 1979). Total arsenic concentrations in surface water and groundwater are usually $< 10\ \mu g/L$; in certain areas, however, levels above 1 mg/L have been recorded (Pershagen and Vahter, 1979).

In air, most arsenic particulates consist of inorganic arsenic compounds, often as As^{+3}. Burning of coal and arsenic-treated wood and smelting of metals are major sources of atmospheric arsenic contamination (i.e., $> 1\ \mu g/m^3$); in general, atmospheric arsenic levels are higher in winter, due to increased use of coal for heating (Pershagen and Vahter, 1979).

The main carrier of arsenic in rocks and in most types of mineral deposits is iron pyrite (FeS_2), which may contain > 2000 mg/kg of arsenic (NRCC, 1978). In localized areas, soils are contaminated by arsenic oxide fallout from smelting of ores (especially sulfide ores) and combustion of arsenic-rich coal (Woolson, 1975).

Arsenic in lacustrine sediment columns is subject to control by diagenetic processes and adsorption mechanisms, as well as anthropogenic influences (Farmer and Lovell, 1986). For example, elevated levels of arsenic in or near surface sediments may have several causes (Farmer and Lovell, 1986), including natural processes (Loch Lomond, Scotland) and human activities such as smelting (Lake Washington, Washington; Kelly Lake, Ontario, Canada), manufacture of arsenical herbicides (Brown's Lake, Wisconsin), and mining operations (Northwest Territories, Canada). Elevated levels of arsenic in sediments of the Wailoa River, Hawaii, are the result of As_2O_3 applied as an antitermite agent between 1932 and 1963 (Hallacher et al., 1985). These elevated levels are found mainly in anaerobic sediment regions where the chemical has been relatively undisturbed by activity. Low levels of arsenic in the biota of the Wailoa River estuary suggest that arsenic is trapped in the anaerobic sediment layers.

Arsenic geochemistry in Chesapeake Bay, Maryland, depend on anthropogenic inputs and phytoplankton species composition (Sanders, 1985). Inputs of anthropogenic arsenic into Chesapeake Bay are estimated at 100 kg daily or 39 tons/year—probably from sources such as unreported industrial discharges, arsenical herbicides, and wood preservatives (Sanders, 1985). The chemical form of the arsenic in solution varies seasonally and along the axis of the bay. Arsenic is present only as arsenate in winter, but substantial quantities of reduced and methylated forms are present in summer in different areas. The forms and distribution patterns of arsenic during summer suggest that separate formation processes exist. Arsenite, present in low-salinity regions, may have been formed by chemical reduction in anoxic, subsurface waters and then mixed into the surface layer, Methylated arsenicals are highly correlated with standing crops of algae. One particular form, methylarsonate, is significantly correlated with the dominant alga *Chroomonas*. Since arsenic reactivity and toxicity are altered by transformation of chemical form, observed variations in arsenic speciation have considerable geochemical and ecological significance (Sanders, 1985).

5.2. Biological Samples

Background arsenic concentrations in living organisms are usually < 1 mg/kg fresh weight in terrestrial flora and fauna, birds, and freshwater biota. These levels are higher, sometimes markedly so, in biota collected from mine waste sites and arsenic-treated areas, and near smelters and mining areas, areas with high geothermal activity, and manufacturers of arsenical defoliants and pesticides (Table 2). For example, bloaters (*Coregonus hoyi*) collected in Lake Michigan near a facility that produced arsenical herbicides consistently had the highest (1.5–2.9 mg As/kg fresh weight whole body) arsenic concentrations measured in freshwater fishes in the United States between 1976 and 1984 (Schmitt and Brumbaugh, 1990). Marine organisms, however, normally contain arsenic residues of several to more than 100 mg/kg dry weight (Lunde, 1977); however, as will be discussed later, these concentrations present little hazard to the organism or its consumers.

Shorebirds (seven species) wintering in the Corpus Christi, Texas, area contained an average of only 0.3 mg As/kg fresh weight in their livers (maximum of 1.5 mg/kg), despite the presence of smelters and the heavy use of arsenical herbicides and defoliants; these values probably reflect normal background concentrations (White et al., 1980). Similar arsenic levels were reported in the livers of brown pelicans (*Pelecanus occidentalis*) collected from South Carolina (Blus et al., 1977). The highest arsenic concentration recorded in seemingly unstressed coastal birds was 13.2 mg/kg fresh weight lipids (Table 2). This observation tends to corroborate the findings of others that arsenic concentrates in lipid fractions of marine plants, invertebrates, and higher organisms. An abnormal concentration of 16.7 mg As/kg fresh weight was recorded in the liver of an osprey (*Pandion haliaetus*) from the Chesapeake Bay region (Wiemeyer et al., 1980). This bird was alive but weak, with serious histopathology including

Table 2 Arsenic Concentrations in Field Collections of Selected Species of Flora and Fauna

Ecosystem, Species, and Other Variables	Concentration (mg As/kg FW or DW[a]), in ppm [mean, (range), max.]	Reference[b]
TERRESTRIAL PLANTS		
Colonial bentgrass, *Agrostis tenuis*		
On mine waste site	1480 DW, max. 3470 DW	1
On low-arsenic soil	(0.3–3) DW	1
Scotch heather, *Calluna vulgaris*		
On mine waste site	1260 DW	1
On low-arsenic soil	0.3 DW	1
Coontail, *Ceratophyllum demersum*		
From geothermal area, New Zealand	(20–1060) DW	1
Cereal grains		
From arsenic-treated areas	Usually < 3 DW, max. 252 DW	2
Nontreated areas	Usually < 0.5 DW, max. 5 DW	2
Grasses		
From arsenic-treated areas	(0.5–60,000) DW	2
Nontreated areas	(0.1–0.9) DW	2
Apple, *Malus sylvestris*		
Fruit	< 0.1 FW, < 1.8 DW	1
Alfalfa, *Medicago sativa*		
USA	1.6 FW	1
Montana, smelter area	(0.4–5.7) FW	1
White spruce, *Picea alba*		
Arsenic-contaminated soil		
Branch	(2.8–14.3) DW	1
Leaf	(2.1–9.5) DW	1
Trunk	(0.3–55) DW	1
Root	(45–130) DW	1
Control site		
All samples	< 2.4 DW	1
Pine, *Pinus silvestrus*, needles		
Near USSR metals smelter; soil levels 120 mg As/kg	22 FW	3
Trees		
Nontreated areas	Usually < 1 DW	2
Lowbush blueberry, *Vaccinum angustifolium*		
Maine, leaf		
Arsenic-treated soil	(6.8–15) DW	1
Control	0.8 DW	1
Various species		1
From uncontaminated soils	(< 0.01–5) DW	2
From arsenic-impacted (80 mg/kg) soils	1.2 (< 0.2–5.8) DW	4

[a,b]See page 213 for footnotes.

Table 2 *(Continued)*

Ecosystem, Species, and Other Variables	Concentration (mg As/kg FW or DW[a]), in ppm [mean, (range), max.]	Reference[b]
Vegetables		
From arsenic-treated areas	Usually < 3 DW, max. 145 DW	2
Nontreated areas	Usually < 1 DW, max. 8 DW	2
Vegetation		
Near gold mine, Canada, air levels up to 3.9 μg As/m^3	Max. 11,438 DW	5
80 km distant	(12–20) DW	5
FRESHWATER FLORA		
Aquatic plants		
Arsenic-treated areas	(20–1450) DW	2
Untreated areas	(1.4–13) DW	2
Irish moss, *Chondrus crispus*		2
Whole	(5–12) DW	1
Pondweeds, *Potamogeton* spp.		1
Whole		
Near geothermal area	(11–436) DW	1
Control site	< 6 DW	1
Widgeongrass, *Ruppia maritima*, from Kern National Wildlife Refuge, California, contaminated by agricultural drainwater of 12–190 μg As/L	Max. 430 DW	6
FRESHWATER FAUNA		
Alewife, *Alosa pseudoharengus*		
Whole, Michigan	0.02 FW	1
Muscle, Wisconsin	0.0 FW	1
White sucker, *Catostomus commersoni*		1
Muscle	(0.03–0.13) FW	1
Whole	(0.05–0.16) FW	1
Common carp, *Cyprinus carpio*		
Upper Mississippi River, 1979		
Whole	0.4 (0.2–0.6) DW	7
Liver	0.4 (0.3–1) DW	7
Nationwide, USA		
Whole	0.05 FW	1
Muscle	(0.0–0.2) FW	1
Northern pike, *Esox lucius*		
Muscle		
Canada	(0.05–0.09) FW	1
Great Lakes	< 0.05 FW	1
Sweden	0.03 FW	1
New York	< 0.1 FW	1
Wisconsin	< 0.01 FW	1

Table 2 *(Continued)*

Ecosystem, Species, and Other Variables	Concentration (mg As/kg FW or DW[a]), in ppm [mean, (range), max.]	Reference[b]
Fish, various species		
Whole	Max. 1.9 FW	2
Whole	(0.04–0.2) FW	8
Netherlands, 1977–1984, muscle	(0.04–0.2) FW	9
Nationwide, USA, whole fish		
1976–1977	0.27 FW, max. 2.9 FW	10, 11
1978–1979	0.16 FW, max. 2.1 FW	11
1980–1981	0.15 FW, max. 1.7 FW	11
1984	0.14 FW, max. 1.5 FW	11
Near smelter (water arsenic 2.3–2.9 μg/L)		
Muscle, 3 species		
Total arsenic	(0.05–0.24) FW	12
Inorganic arsenic	(0.01–0.02) FW	12
Liver, 2 species		
Total arsenic	0.15 FW	12
Inorganic arsenic	0.01 FW	12
Control location (water arsenic < 0.5 μg/L)		
Muscle		
Total arsenic	(0.06–0.09) FW	12
Inorganic arsenic	< 0.03 FW	12
Liver		
Total arsenic	0.09 FW	12
Inorganic arsenic	< 0.01 FW	12
Channel catfish, *Ictalurus punctatus*		
Muscle		
Native	(0.0–0.3) FW	1
Cultured	(0.2–3.1) FW	1
Whole, nationwide	(< 0.05–0.3) FW	1
Green sunfish, *Lepomis cynellus,* liver		
Polluted waters (from manufacturer of arsenical defoliants and pesticides), Texas. Mean water concentration 13.5 mg As/L; sediment content of 4700 mg/kg		
Age 1 to 2	(19.7–64.2) DW	13
Age 3	15 DW	13
Age ≥ 4	(6.1–11.5) DW	13
Bluegill, *Lepomis macrochirus*		
From pools treated with arsenic		
Muscle	1.3 FW	1
Skin and scales	2.4 FW	1

Table 2 *(Continued)*

Ecosystem, Species, and Other Variables	Concentration (mg As/kg FW or DW[a]), in ppm [mean, (range), max.]	Reference[b]
Gills and GI tract	17.6 FW	1
Liver	11.6 FW	1
Kidney	5.9 FW	1
Ovary	8.4 FW	1
Control locations		
All tissues	< 0.2 FW	1
Whole		
Nationwide, USA	(< 0.05–0.15) FW	1
Upper Mississippi River, 1979	0.3 (0.2–0.4) DW	7
Smallmouth bass, *Micropterus dolomieui*		
Muscle		
Wisconsin	< 0.13 FW	1
Lake Erie	0.22 FW	1
New York	(0.03–0.51) FW	1
Whole, nationwide, USA	(< 0.05–0.28) FW	1
Largemouth bass, *Micropterus salmoides*		
Whole, nationwide, USA	(< 0.05–0.22) FW	1
Muscle		
Wisconsin	(0.0–0.12) FW	1
New York	(0.03–0.16) FW	1
Striped bass, *Morone saxatilis*		
Muscle	(0.2–0.7) FW	1
Coho salmon, *Oncorhynchus kisutch*		
Muscle		
Wisconsin	< 0.15 FW	1
Lake Erie	(< 0.07–0.17) FW	1
New York	< 0.5 FW	1
USA	0.09 FW	1
Rainbow trout, *Oncorhynchus mykiss*		
All tissues	< 0.4 FW	1
Yellow perch, *Perca flavescens*		
All tissues	< 0.16 FW	1
Atlantic salmon, *Salmo salar*		
Oil		
Liver	6.7 FW	1
Muscle	(0.8 –3.1) FW	1
Lake trout, *Salvelinus namaycush*		
Whole, nationwide, USA	(0.06–0.68) FW	1
MARINE FLORA		
Algae		
Green	(0.05–5) DW	2

Table 2 *(Continued)*

Ecosystem, Species, and Other Variables	Concentration (mg As/kg FW or DW[a]), in ppm [mean, (range), max.]	Reference[b]
Brown	Max. 30 DW	2
11 species	(2–58) DW	14
Various species	(10–100) DW	15
Seaweed, *Chondrus crispus*	5.2 DW	2
Alga, *Fucus* spp.		
Oil	(6–27) FW	2
Fatty acid	(5–6) FW	2
Brown alga, *Fucus vesiculosus*		
Whole	(35–80) DW	1
Brown alga, *Laminaria digitata*		
Whole	94 DW	2
Whole	(42–50) DW	1
Oil	(155–221) DW	2
Fatty acid	(8–36) DW	2
Alga, *Laminaria hyperborea*		
Total arsenic	142 DW	14
Organic arsenic	139 DW	14
Periphyton, Louisana, USA	8.4 DW	16
Sargassum weed, *Sargassum fluitans*		
Total arsenic	19.5 FW	8
As^{+3}	1.8 FW	8
As^{+5}	17.7 FW	8
Organoarsenicals	0.2 FW	8
Seaweed, *Sargassum* sp.		
Total arsenic	(4.1–8.7) FW	5
As^{+3}	(0.14–0.35) FW	5
As^{+5}	(1.9–7.3) FW	5
Organoarsenicals	Max. 0.1 FW	5
Seaweeds		
Whole	(4–94) DW	2
Whole	(10–109) DW	14
Oil fraction	(6–221) FW	14
MARINE MOLLUSKS		
Bivalves, California, 1984–1986, soft parts		
Clam, *Corbicula* sp.	5.4–11.5 DW	17
Clam, *Macoma balthica*	7.6–12.1 DW	17
Ivory shell, *Buccinum striatissimum*		
Muscle		
Total arsenic	38 FW	18
Arsenobetaine	24.2 FW	18
Midgut gland		18
Total arsenic	18 FW	18
Arsenobetaine	10.8 FW	18

Table 2 *(Continued)*

Ecosystem, Species, and Other Variables	Concentration (mg As/kg FW or DW[a]), in ppm [mean, (range), max.]	Reference[b]
Oysters, *Crassostrea* spp.		
Soft parts	(1.3–10) DW, (0.3–3.4) FW	14
American oyster, *Crassostrea virginica*		
Soft parts	2.9 FW	1
Soft parts	10.3 DW	19
Spindle shells, *Hemifusus* spp.		
Hong Kong 1984, muscle		
Total arsenic	Max. 500 FW	20
Inorganic arsenic	< 0.5 FW	20
Limpet, *Littorina littorea*		
Soft parts		
Near arsenic source	11.5 DW	14
Offshore	4 DW	14
Squid, *Loligo vulgaris*		
Soft parts	(0.8–7.5) FW	1
Hardshell clam, *Mercenaria mercenaria*		
Soft parts		
Age 3 years	3.8 DW	14
Age 4 years	4.7 DW	14
Age 10 years	9.3 DW	14
Age 15 years	8.4 DW	14
Mollusks, edible tissues		
Hong Kong, 1976–1978		
Bivalves	(3.2–39.6) FW	21
Gastropods	(19–176) FW	21
Cephalopods	(0.7–5.5) FW	21
USA		
6 species	(2–3) FW	22
8 species	(3–4) FW	22
3 species	(4–5) FW	22
4 species	(7–20) FW	22
Yugoslovia, northern Adriatic Sea, summer 1986		
6 species	21–31 FW	23
Mussel, *Mytilus edulis*		
Soft parts	2.5 (1.4–4.6) FW	9
Soft parts	(1.6–16) DW	14
Mussels, Louisiana, USA		
Soft parts	1.4–4.5 DW	16
Scallop, *Placopecten magellanicus*		
Soft parts	1.6 (1.3–2.4) FW	1

Table 2 *(Continued)*

Ecosystem, Species, and Other Variables	Concentration (mg As/kg FW or DW[a]), in ppm [mean, (range), max.]	Reference[b]
MARINE CRUSTACEANS		
Blue crab, *Callinectes sapidus*		
Florida, whole	7.7 FW	1
Maryland, soft parts	(0.5–1.8) FW	1
Dungeness crab, *Cancer magister*		
Muscle	6.5 (2.2–37.8) FW	1
Muscle	4 FW	24
Alaskan snow crab, *Chionocetes bairdii*		
Muscle	7.4 FW	24
Copepods, whole	(2–8.2) DW, (0.4–1.3) FW	14
Shrimp, *Crangon crangon*		
Netherlands, 1977–1984		
Muscle	3 (2–6.8) FW	9
Crabs and shrimps,		
Louisiana, USA		
Muscle	< 0.2 DW	16
Crustaceans, edible tissues		
Hong Kong, 1976–1978		
Crabs	(5.4–19.1) FW	21
Lobsters	(26.7–52.8) FW	21
Prawns and shrimps	(1.2–44) FW	21
USA		
6 species	(3–5) FW	22
3 species	(5–10) FW	22
4 species	(10–20) FW	22
2 species	(20–30) FW	22
1 species	(40–50) FW	22
American lobster, *Homarus americanus*		
Muscle	(3.8–7.6) DW, max. 40.5 FW	1
Hepatopancreas	22.5 FW	1
Whole	(3.8–16) DW, (1–3) FW	14
Lesser spider crab, *Maia crispata*		
Yugoslavia, 1986		
Digestive gland	25.4–37.1 FW	23
Muscle	24.5–28.0 FW	23
Stone crab, *Menippe mercenaria*		
Whole	(9–11.8) FW	1
Deep-sea prawn, *Pandalus borealis*		
Head and shell	68.3 DW	1

Table 2 *(Continued)*

Ecosystem, Species, and Other Variables	Concentration (mg As/kg FW or DW[a]), in ppm [mean, (range), max.]	Reference[b]
Muscle	61.6 DW	1
Oil	42 DW, 10.1 FW	1
Egg	3.7–14 FW	1
Prawns, *Pandalus* spp.		
Whole	(7.3–11.5) FW	14
Alaskan king crab, *Paralithodes camtschatica*		
Muscle	8.6 FW	24
Brown shrimp, *Penaeus aztecus*		
Muscle	(3.1–5.2) FW	1
Whole	0.6 DW	1
White shrimp, *Penaeus setiferus*		
Muscle		
Mississippi	(1.7–4.4) FW	1
Florida	(2.8–7.7) FW	1
Shrimp, *Sergestes lucens*		
Muscle		
Total arsenic	5.5 FW	25
Arsenobetaine	4.5 FW	25
Shrimps		
Exoskeleton	15.3 FW	8
Muscle, 2 species	(18.8–41.6) FW, max. 128 DW	2
MARINE FISHES AND ELASMOBRANCHS		
Whitetip shark, *Carcharhinus longimanus*		
Muscle	3.1 FW	26
Black sea bass, *Centropristis striata*		
Muscle	6.4 DW	1
Peacock wrasse, *Crenilabrus pavo*, Yugoslovia, 1986		
Liver	26.9–37.6 FW	23
Muscle	20.7–22.2 FW	23
Elasmobranchs		
Muscle		
Sharks	Max. 30 FW	14
Rays	Max. 16.2 FW	14
Roundnose flounder, *Eopsetta grigorjewi*		
Muscle	20.1 FW	27
Finfishes		
Near metal smelter,		

Table 2 *(Continued)*

Ecosystem, Species, and Other Variables	Concentration (mg As/kg FW or DW[a]), in ppm [mean, (range), max.]	Reference[b]
water concentration		
2.3–2.9 μg As/L		
Muscle, 6 species		
Total arsenic	(0.2–2.6) FW	12
Inorganic arsenic	(0.02–0.1) FW	12
Liver, 4 species		
Total arsenic	(0.4–1.8) FW	12
Inorganic arsenic	(0.02–0.07) FW	12
Control location, water concentration < 2.0 μg As/L		
Muscle, 5 species		
Total arsenic	(0.1–1.2) FW	12
Inorganic Arsenic	(0.02–0.15) FW	12
Liver, 4 species		
Total arsenic	(0.2–1.5) FW	12
Inorganic arsenic	(0.02–0.05) FW	12
Finfish, Hong Kong, 1976–1978		
Edible tissues	Max. 21.1 FW	21
Finfish, Netherlands, 1977–1984		
Muscle, 4 species	(2.8–10.9) FW	9
Finfish, North America		
Liver		
49 species	(0.7–5) FW	22
26 species	(5–20) FW	22
6 species	(20–50) FW	22
Muscle		
91 species	(0.6–4) FW	22
41 species	(4–8) FW	22
27 species	(8–30) FW	22
6 species	0.2 (0.18–0.30) DW	16
4 species		
Total arsenic	(1.4–10) FW	28
Inorganic arsenic	< 0.5 FW	28
Whole		
16 species	(1–8) FW	22
Finfish, worldwide		
Various tissues		
Total arsenic	Max. 142 FW	2
Inorganic arsenic	(0.7–3.2) FW	2
Organic arsenic	(3.4–139) FW	2
Atlantic cod, *Gadus morhua*		
Muscle	2.2 FW	2
Liver	9.8 FW	2

Table 2 *(Continued)*

Ecosystem, Species, and Other Variables	Concentration (mg As/kg FW or DW[a]), in ppm [mean, (range), max.]	Reference[b]
Blue pointer, *Isurus oxyrhinchus*		
Muscle	9.5 FW	26
Striped bass, *Morone saxatilis*		
Muscle	(0.3–0.5) FW, 1.8 DW	14
Liver	0.7 FW	14
Striped mullet, *Mugil cephalus*		
Viscera	Max. 1.3 FW	29
English sole, *Parophrys vetulus*		
Muscle	1.1. (0.6–11.5) FW	1
Skate, *Raja* sp.		
Muscle	16.2 FW	1
Windowpane flounder, *Scopthalmus aquosus*		
Muscle	(1.4–2.8) FW	1
Spiny dogfish, *Squalus acanthias*		
Muscle	10 DW	30
Liver	5.7 DW	30
Spleen	9.8 DW	30
Yolk sac	9.1 DW	30
Embryo	2.6 DW	30
AMPHIBIANS AND REPTILES		
Alligator, *Alligator mississipiensis*		
Egg	(0.05–0.2) FW	1
Crocodile, *Crocodylus acutus*		
Egg	0.2 FW	31
Frogs, *Rana* spp.		
All tissues	< 0.4 FW	1
Toads, 2 species		
All tissues	< 0.05 FW	1
BIRDS		
American black duck, *Anas rubripes*		
Egg	0.2 FW	14
Ducks, *Anas* spp.		
All tissues	< 0.4 FW	1
Scaup, *Aythya* spp.		
All tissues	< 0.4 FW	1
Gulls, 3 species		
Oil	(0.6–13.2) FW	14
Osprey, *Pandion haliaetus*		
Liver	Max. 16.7 FW	32
Brown pelican, *Pelecanus occidentalis*		
Egg		
South Carolina, 1971–1972	0.3 (0.08–0.8) FW	33

Table 2 *(Continued)*

Ecosystem, Species, and Other Variables	Concentration (mg As/kg FW or DW[a]), in ppm [mean, (range), max.]	Reference[b]
Florida, 1969–1970	0.1 (0.07–0.2) FW	33
Liver, 1972–1973, GA, FL, SC		
Found dead	(0.2–1) FW	33
Shot	(0.3–0.9) FW	33
Shorebirds		
Corpus Christi, Texas, 1976–1977		
Liver, 7 species	(0.05–1.5) FW	34
New Zealand, 5 species		
Feather	< 1 FW	14
Liver	Max. 2.6 FW	14
Starling, *Sturnus vulgaris*		
Whole, nationwide, USA, 1971	(<0.01–0.21) FW	2
Icelandic redshank, *Trinqa totanus robusta*		
Netherlands, 1979–1982		
Feather		
Juveniles	Max. 0.8 FW	35
Adults	(0.5–3.2) FW	35
MAMMALS		
Fin whale, *Balaenoptera physalis*		
Blubber oil	1.8 FW	1
Cow, *Bos bovis*		
Downwind from copper smelter		
16–21 km		
Hair	8.9 FW	1
Milk	0.013 FW	1
Blood	0.026 FW	1
60 km		
Hair	0.46 FW	1
Milk	0.002 FW	1
Blood	0.009 FW	1
Controls		
Milk	< 0.001 FW	36
Muscle	0.005 FW	36
Liver	(0.008–0.012) FW	36
Kidney	(0.017–0.053) FW	36
Domestic animals		
All tissues	< 0.3 FW	2
Livestock		
All tissues	< 0.6 FW	2
Marine mammals		14
Pinnipeds		
All tissues	Max. 1.7 FW	14

Table 2 *(Continued)*

Ecosystem, Species, and Other Variables	Concentration (mg As/kg FW or DW[a]), in ppm [mean, (range), max.]	Reference[b]
Cetaceans		
Muscle	0.4 DW	14
Oil	(0.6–2.8) FW	14
White-tailed deer, *Odocoileus virginianus* Tennessee, killed from arsenic herbicide		
Liver	19 FW	1
Kidney	17.8 FW	1
Rumen contents	22.5 FW	1
Harbor seal, *Phoca vitulina*		
UK, all tissues	< 0.3 FW	1
Fox, *Vulpes* sp.		
All tissues	< 0.7 FW	1
Wildlife		
All tissues	< 1 FW	2

[a] Abbreviations: DW = dry weight; FW = fresh weight.

[b] 1, Jenkins (1980); 2, NAS (1977); 3, Mankovska (1986); 4, Merry et al. (1986); 5, NRCC (1978); 6, Camardese et al. (1990); 7, Wiener et al. (1984); 8, Woolson (1975); 9, Vos and Hovens (1986); 10, Lima et al. (1984); 11, Schmitt and Brumbaugh (1990); 12, Norin et al. (1985); 13, Sorensen et al. (1985); 14, Eisler (1981); 15, Pershagen and Vahter (1979); 16, Ramelow et al. (1989); 17, Johns and Luoma (1990); 18, Shiomi et al. (1984a); 19, Zaroogian and Hoffman (1982); 20, Phillips and Depledge (1986); 21, Phillips et al. (1982); 22, Hall et al. (1978); 23, Ozretic et al. (1990); 24, Francesconi et al. (1985); 25, Shiomi et al. (1984b); 26, Hanaoka and Tagawa, (1985a); 27, Hanaoka and Tagawa, (1985b); 28, Reinke et al. (1975); 29, Hallacher et al. (1985); 30, Windom et al. (1973); 31, Hall (1980); 32, Wiemeyer et al. (1980); 33, Blus et al. (1977); 34, White et al. (1980); 35, Goede, (1985); 36, Vreman et al. (1986).

absence of subcutaneous fat, presence of serous fluid in the pericardial sac, and disorders of the lung and kidney. The bird died shortly after collection. Arsenic concentrations in the livers of other ospreys collected in the same area usually were < 1.5 mg As/kg fresh weight.

Arsenic concentrations in the tissues of marine biota show a wide range of values, being highest in lipids, liver, and muscle tissues, and varying with age of the organism, geographic locale, and proximity to anthropogenic activities (Table 2). In general, tissues with high lipid content contain high levels of arsenic. Crustacean tissues sold for human consumption and collected in U.S. coastal waters usually contain 3 to 10 mg As/kg fresh weight (Hall et al., 1978) or 1 to 100 mg/kg dry weight (Fowler and Unlu, 1978), and are somewhat higher than those reported for finfish and molluskan tissues. Marine finfish tissues usually contain 2 to 5 mg As/kg fresh weight (Table 2). However, postmortem reduction of As^{+5} to As^{+3} occurs rapidly in fish tissues (Reinke et al., 1975), suggesting a need for additional research in this area. Maximum arsenic values recorded in elasmobranchs (mg/kg fresh weight) were 30 in the muscle of a shark, *Mustelus antarcticus*, and 16.2 in the muscle of a ray, *Raja* sp. (Eisler, 1981). The highest arsenic concentration recorded in a marine mammal, 2.8 mg As/kg fresh weight lipid, was from a whale (Eisler, 1981).

Arsenic appears to be elevated in marine biota because of their ability to accumulate arsenic from seawater and food sources, not because of localized pollution (Maher, 1985b). The great majority of arsenic in marine organisms exists as water-soluble and lipid-soluble organoarsenicals, including arsenolipids, arsenosugars, arsenocholine, arsenobetaine [$(CH_3)_3AsCH_2COOH$], monomethylarsonate [$CH_3AsO(OH)_2$], and dimethylarsinate [$(CH_3)_2AsO(OH)$], as well as other forms. There is no convincing hypothesis to account for the existence of all the various forms of organoarsenicals found in marine organisms. One hypothesis is that each form involves a single metabolic pathway concerned with the synthesis and turnover of phosphatidylcholine (Phillips and Depledge, 1986). Arsenosugars (arsenobetaine precursors) are the dominant arsenic species in brown kelp, *Ecklonia radiata*; giant clam, *Tridacna maxima*; shrimp, *Pandalus borealis*; and ivory shell, *Buccinum striatissimum* (Shiomi et al., 1984a, b; Francesconi et al., 1985; Matsuto et al., 1986; Phillips and Depledge, 1986). For most marine species, however, there is general agreement that arsenic exists primarily as arsenobetaine, a water-soluble organoarsenical that has been identified in the tissues of western rock lobster (*Panulirus cygnus*), American lobster (*Homarus americanus*), octopus (*Paroctopus* sp.), sea cucumber (*Stichopus japonicus*), blue shark (*Prionace gluaca*), sole (*Limanda* sp.), squid (*Sepioteuthis australis*), prawn (*Penaeus latisulcatus*), scallop (*Pecten alba*), and many other species, including teleosts, mollusks, tunicates, and crustaceans (Shiomi et al., 1984b; Francesconi et al., 1985; Hanaoka and Tagawa, 1985a, b; Maher, 1985b; Norin et al., 1985; Matsuto et al., 1986; Ozretic et al., 1990; Phillips, 1990). The potential risks associated with consumption of seafoods containing arsenobetaine seem to be minor. The chemical was not mutagenic in the bacterial *Salmonella typhimurium* assay (Ames test), had no effect on metabolic inhibition of Chinese hamster ovary cells at 10,000 mg/L, and showed no synergism or antagonism on the action of other contaminants (Jongen et al., 1985). Arsenobetaine was not toxic to mice at oral doses of 10,000 mg/kg body weight during a 7-day observation period, and it was rapidly absorbed from the gastrointestinal tract and rapidly excreted in urine without metabolism, owing to its high polar and hydrophylic characteristics (Kaise et al., 1985).

6. LETHAL AND SUBLETHAL EFFECTS

As will be discussed later, most authorities agree on 10 points: (1) inorganic arsenicals are more toxic than organic arsenicals, and trivalent forms are more toxic than pentavalent forms; (2) episodes of arsenic poisoning are either acute or subacute; cases of chronic arsenosis are rarely encountered, except in humans; (3)

sensitivity to arsenic is greatest during the early developmental stages; (4) arsenic can traverse placental barriers; as little as 1.7 mg As^{+5}/kg body weight at critical stages of hamster embryogenesis, for example, can produce fetal death and malformation; (5) biomethylation is the preferred detoxification mechanism for inorganic arsenicals; (6) arsenic is bioconcentrated by organisms, but not biomagnified in the food chain; (7) in soils, depressed crop yields were recorded at 3 to 28 mg water-soluble As/L, or about 25 to 85 mg total As/kg soil; adverse effects on vegetation were recorded at concentrations in air $> 3.9\,\mu g$ As/m^3; (8) some aquatic species were adversely affected at water concentrations of 19 to 48 μg As/L, or 120 mg As/kg in the diet, or tissue residues of 1.3 to 5 mg As/kg fresh weight; (9) sensitive species of birds died following single oral doses of 17.4 to 47.6 mg As/kg body weight; and (10) adverse effects were noted in mammals at single oral doses of 2.5 to 33 mg As/kg body weight, at chronic oral doses of 1 to 10 mg As/kg body weight, and at feeding levels of 50 mg—and sometimes only 5 mg—As/kg in the diet.

The literature emphasizes that arsenic metabolism and toxicity vary greatly among species and that its effects are significantly altered by numerous physical, chemical, and biological modifiers. Adverse health effects, for example, may involve the respiratory, gastrointestinal, cardiovascular, and hematopoietic systems, and may range from reversible effects to cancer and death, depending in part on the physical and chemical forms of arsenic tested, the route of administration, and the dose.

6.1. Carcinogenesis, Mutagenesis, and Teratogenesis

Epidemiological studies show that an increased risk of cancers in the skin, lung, liver, lymph, and hematopoietic systems of humans is associated with exposure to inorganic arsenicals. These increased cancer risks are especially prevalent among smelter workers and those engaged in the production and use of arsenical pesticides where atmospheric levels exceed 54.6 μg As/m^3 (NRCC, 1978; Belton et al., 1985; Pershagen and Bjorklund, 1985). Skin tumors, mainly of low malignancy, have been reported after consumption of arsenic-rich drinking waters; a total dose of several grams, probably as As^{+3}, is usually required for the development of skin tumors (Pershagen and Vahter, 1979). High incidences of skin cancer and hyperpigmentation were noted among several population groups, especially Taiwanese and Chileans, who consumed water containing more than 0.6 mg As/L; the frequency of cancer was highest among people over age 60 who demonstrated symptoms of chronic arsenic poisoning (NRCC, 1978).

Arsenic reportedly inhibits cancer formation in species having a high incidence of spontaneous cancers (NRCC, 1978). In fact, arsenic may be the only chemical for which there is sufficient evidence of carcinogenicity in humans but not in animals (Woolson, 1975; Belton et al., 1985; Lee et al., 1985). In general, animal carcinogenicity tests with inorganic and organic arsenicals have been negative (Hood, 1985), even when the chemicals were administered at or near the highest tolerated dosages for long periods (NAS, 1977). Most studies of arsenic carcinogenesis in animals were presumably of insufficient duration to simulate conditions in long-lived species such as humans (NRCC, 1978). However, mice developed leukemia and lymphoma after 20 subcutaneous injections of 0.5 mg As^{+5}/kg body weight; 46% of the experimental group developed these signs, versus none of the controls (NRCC, 1978). Pulmonary tumorogenicity has been demonstrated in hamsters administered calcium arsenate intratracheally (Pershagen and Bjorklund, 1985). Cacodylic acid and other organoarsenicals are not carcinogenic, but may be mutagenic at very high doses (Hood, 1985).

Several inorganic arsenic compounds are weak inducers of chromosomal aberrations, sister-chromatid exchange, and in vitro transformation of mammalian cells; however, there is no conclusive evidence that arsenic causes point mutations in any cellular system (Pershagen and Vahter, 1979; Belton et al., 1985; Lee et al., 1985; Deknudt et al., 1986). Studies with bacteria suggest that arsenite is a comutagen, or that it may inhibit DNA repair (Belton et al., 1985).

Arsenic is a known teratogen in several classes of vertebrates, and it has been implicated as a cause of birth defects in humans. Specific developmental malformations have been produced experimentally in mammals with inorganic As^{+3} or As^{+5} through either a single dose or a continuous dose during embryogenesis (Hanlon and Ferm, 1986b). Teratogenic effects are initiated no later than 4 hours after administration of arsenic; fetal abnormalities are primarily neural tube defects (Hanlon and Ferm, 1986c) but may also include protruding eyes, incomplete development of the skull, abnormally small jaws, and other skeletal anomalies (NRCC, 1978). Inorganic As^{+3} and As^{+5}, but not organoarsenicals, cross placental barriers in many species of mammals, which results in fetal death and malformations (NRCC, 1978; EPA, 1980). Studies with hamsters, for example, showed that sodium arsenite can induce chromatid breaks and chromatid exchanges in Chinese hamster ovary cells in a dose-dependent manner (Lee et al., 1986b). In an earlier study (Lee et al., 1986a), As^{+3} was about 10 times more potent than As^{+5} in causing transformations. The birth defects were most pronounced in golden hamsters exposed to As^{+5} during the 24-hr period of critical embryogenesis—day 8 of gestation (Ferm and Hanlon, 1985, 1986)—when 1.7 mg As^{+5}/kg body weight induced neural tube defects in about 90% of the fetuses. Hanlon and Ferm (1986a) showed that hamsters exposed to As^{+5} and heat stress (39 °C for 50 minutes) on day 8 of gestation produced a greater percentage of malformed offspring (18 to 39%) than did hamsters exposed to As^{+5} alone (4 to 8%).

6.2. Terrestrial Plants and Invertebrates

In general, arsenic availability to plants is highest in coarse-textured soils having little colloidal material and little ion exchange capacity, and lowest in fine-textured soils high in clay, organic material, iron, calcium, and phosphate (NRCC, 1978). To be absorbed by plants, arsenic compounds must be in a mobile form in the soil solution. Except for locations where arsenic content is high (e.g. around smelters), the accumulated arsenic is distributed throughout the plant body in nontoxic amounts (NAS, 1977). For most plants, a significant depression in crop yield was evident at soil arsenic concentrations of 3 to 28 mg/L of water-soluble arsenic and 25 to 85 mg/kg of total arsenic (NRCC, 1978). Yields of peas (*Pisum sativum*), a sensitive species, were decreased at 1 mg/L of water-soluble arsenic or 25 mg/kg of total soil arsenic; rice (*Oryza sativum*) yields were decreased 75% at 50 mg/L of disodium methylarsonate in silty loam; and soybeans (*Glycine max*) grew poorly when residues exceeded 1 mg As/kg (Table 3) (NRCC, 1978). Forage plants grown in soils contaminated with up to 80 mg total As/kg from arsenical orchard sprays contained up to 5.8 mg As/kg dry weight; however, these plants were considered nonhazardous to grazing ruminants (Merry et al., 1986).

Attention was focused on inorganic arsenical pesticides after accumulations of arsenic in soils eventually became toxic to several agricultural crops, especially in former orchards and cotton fields. Once toxicity is observed, it persists for several years even if no additional arsenic treatment is made (Woolson, 1975). Poor crop growth was associated with the bioavailability of arsenic in soils. For example, alfalfa (*Medicago sativa*) and barley (*Hordeum vulgare*) grew poorly in soils containing only 3.4 to 9.5 mg As/kg, provided the soils contained excess moisture and were acidic, lightly textured, low in phosphorus and aluminum, and high in iron and calcium (Woolson, 1975). Use of inorganic arsenical herbicides, such as calcium arsenate, on golf-course turfs for control of fungal blight sometimes worsens the disease. The use of arsenicals on Kentucky bluegrass (*Poa pratensis*) is discouraged under conditions of high moisture and root stress induced by previous arsenical applications (Smiley et al., 1985).

Methylated arsenicals, as herbicides or defoliants, are sprayed on plant surfaces. They can reach the soil during application or can be washed from the plants. Additional arsenic enters soils by exchange from the roots or when dead plant materials decay (Hood, 1985). Cacodylic acid and sodium cacodylate are nonselective herbicides used in at least 82 products to eliminate weeds and grasses around trees and shrubs and to eradicate vegetation from rights-of-way and other noncrop areas (Hood, 1985). Normal application rates of various organoarsenicals for crop and noncrop purposes rarely exceed 5 kg/ha (Woolson, 1975). At recommended treatment levels, organoarsenical soil residues are not toxic to crops, and those tested (soybean, beet, wheat) were more resistant to organoarsenicals than to comparable levels of inorganic arsenicals (Woolson, 1975).

Air concentrations up to 3.9 μg As/m^3 near gold mining operations were associated with adverse effects on vegetation; higher concentrations of 19 to 69 μg As/m^3, near a coal-fired power plant in Czechoslovakia, produced measurable contamination in soils and vegetation in a 6-km radius (NRCC, 1978).

The phytotoxic actions of inorganic and organic arsenicals are different, and each is significantly modified by physical processes. The primary mode of action of arsenite in plants is inhibition of light activation, probably through interference with the pentose phosphate pathway (Marques and Anderson, 1986). Arsenites penetrate the plant cuticle to a greater degree than arsenates (NAS,

Table 3 Lethal and Sublethal Effects of Various Arsenic Compounds on Selected Species of Terrestrial Plants and Invertebrates

Ecosystem, Organism, and Other Variables	Arsenic Concentration and Effects	Reference[a]
TERRESTRIAL PLANTS		
Crops		
Total water-soluble As in soils	Depressed crop yields at 3 to 28 mg/L	1
Total soil As concentrations	Depressed crop yields at 25 to 85 mg/kg	1
Common Bermuda grass, *Cynodon dactylon*		
Arsenite	Plants grown on As-amended soils (up to 90 mg As^{+3}/kg) contained up to 17 mg As/kg dry weight (DW) in stems, 20 in leaves, and 304 in roots	2
Fruit orchards		
Inorganic arsenites and arsenates	Soils contain 31 to 94 mg As/kg DW (vs. 2.4 in untreated orchards); whole rodents contain $<$ 0.002 mg As/kg fresh weight (FW) vs. non-detectable in untreated orchards	3
Soybean, *Glycine max*		
Total As	Toxic signs at plant residues $>$ 1 mg total As/kg	1
Grasslands		
Cacodylic acid	Kill of 75 to 90% of all species at 17 kg/ha; recovery modest	3
Rice, *Oryza sativum*		
Disodium methylarsonate	75% decrease in yield at soil (silty loam) concentrations of 50 mg/kg	1
Scots pine, *Pinus sylvestris*		
Inorganic As^{+5}	Seedlings die when soil (sandy) concentrations exceed 250 mg/kg DW; maximum bioconcentration factors low: 0.6 for roots, 0.1 for shoots; residues $>$ 62 mg As/kg DW in shoots are toxic, and 3300 mg/kg DW usually fatal	4
Pea, *Pisum sativum*		
Sodium arsenite	15 mg/L inhibits light activation and photosynthetic CO_2 fixation in chloroplasts	5

Table 3 *(Continued)*

Ecosystem, Species, and Other Variables	Arsenic Concentration and Effects	Reference[a]
Sandhill plant communities		
Cacodylic acid	No lasting effect at 2.25 kg/ha; some species defoliated at 6.8 kg/ha; significant effect, including 75% defoliation of oaks and death of all pine trees, at 34 kg/ha	3
Cowpea, *Vigna* sp.		
Total water-soluble As in soils	Decreased yields at 1 mg/L	1
Total soil As concentrations (loamy sand)	Toxic at 25 mg/kg	1
Yeast		
Arsenate	At 75 mg/L, 60% reduction in phosphate transport and glucose metabolism in 30 minutes; at 375 mg/L, 100% reduction	1
TERRESTRIAL INVERTEBRATES		
Honeybee, *Apis mellifera*		
Inorganic arsenite	Following arsenic spray dusting, dead bees contained 20.8 to 31.2 mg/kg FW (adults) or 5 to 13 mg/kg FW (larvae)	6
Beetles		
Cacodylic acid	Dietary levels of 100 to 1000 mg/kg fatal to certain pestiferous species	3
Western spruce budworm, *Choristoneura occidentalis*, sixth instar stage		
Arsenic trioxide	Dietary levels of 99.5 mg/kg FW ration killed 10%, 2550 mg/kg killed 50%, and 65,300 mg/kg was fatal to 90%; newly molted pupae and adults of As-exposed larvae had reduced weight. Regardless of dietary level, concentrations of As ranged up to 2640 mg/kg DW in dead pupae and 1708 mg/kg DW in adults	7

[a] 1, NRCC (1978); 2, Wang et al. (1984); 3, Hood (1985); 4, Sheppard et al (1985); 5, Marques and Anderson (1986); 6, Jenkins (1980); 7, Robertson and McLean (1985).

1977). One of the first indications of plant injury by sodium arsenite is wilting caused by loss of turgor, whereas stress due to sodium arsenate does not involve rapid loss of turgor (NAS, 1977). Organoarsenicals such as cacodylic acid enter plants mostly by absorption of sprays; uptake from the soil contributes only a minor fraction (Hood, 1985). The phytotoxicity of organoarsenical herbicides is characterized by chlorosis, cessation of growth, gradual browning, dehydration, and death (NAS, 1977). In general, plants cease to grow and develop after the roots have absorbed much arsenic (NRCC, 1978). Plants can absorb arsenic through the roots and foliage, although translocation is species dependent. Concentrations of arsenic in plants correlate highly and consistently with water-extractable soil arsenic and usually poorly with total soil arsenic (NRCC, 1978). For example, concentrations of arsenic in corn (*Zea mays*) grown in calcareous soils for 25 days were significantly correlated with the soil water-extractable arsenic fraction, but not with other fractions; extractable phosphorus was correlated positively with arsenic in corn and the water-soluble arsenic fraction (Sadiq, 1986). In the moss *Hylocomium splendens*, arsenate accumulation from solution was through living shoots, optimum uptake being between pH 3 and 5 (Wells and Richardson, 1985). Beets (*Beta vulgaris*) accumulated arsenic more readily at elevated temperatures, but the addition of phosphate fertilizers markedly depressed uptake (Merry et al., 1986).

Soils amended with arsenic-contaminated plant tissues were not measurably affected in CO_2 evolution and nitrification, suggesting that the effects of adding arsenic to soils do not influence the decomposition rate of plant tissues by soil microorganisms (Wang et al., 1984). The half-time of cacodylic acid is about 20 days in untreated soils and 31 days in arsenic-amended soils (Hood, 1985). Estimates of the half-time of inorganic arsenicals in soils are much longer, ranging from 6.5 years for arsenic trioxide to 16 years for lead arsenate (NRCC, 1978).

Data on arsenic effects on soil biota and insects are limited. In general, soil microorganisms are capable of tolerating and metabolizing relatively high concentrations of arsenic (Wang et al., 1984). This adaptation seems to be due to decreased permeability of the microorganism to arsenic (NAS, 1977). Tolerant soil microbiota can withstand concentrations up to 1600 mg/kg; however, growth and metabolism were reduced in sensitive species at 375 mg As/kg, and at 150 to 165 mg As/kg soils were devoid of earthworms and showed diminished quantities of bacteria and protozoans (NRCC, 1978). Honeybees (*Apis mellifera*) that were killed accidentally by sprayed As^{+3} contained 4 to 5 μg arsenic per bee (NAS, 1977), equivalent to 21 to 31 mg/kg body weight (Table 3). Larvae of the western spruce budworm (*Choristoneura occidentalis*) continued to feed on As^{+3} contaminated vegetation until a threshold level of about 2300 to 3300 mg As/kg dry weight whole larvae was reached; death then sometimes occurred (Table 3) (Robertson and McLean, 1985). Larvae that had accumulated sufficient energy reserves completed the first stage of metamorphosis but developed into pupae of subnormal weight; larvae containing < 2600 mg As^{+3}/kg ultimately developed into adults of less than normal weight, and some containing > 2600 mg/kg dry weight died as pupae (Robertson and McLean, 1985).

6.3. Aquatic Biota

Adverse effects of arsenicals on aquatic organisms have been reported at concentrations of 19 to 48 μg/L in water, 120 mg/kg in diets, and 1.3 to 5 mg/kg fresh weight in tissues (Table 4). The most sensitive of the aquatic species tested that showed adverse effects were three species of marine algae, which showed reduced growth in the range of 19 to 22 μg As^{+3}/L; developing embryos of the narrow-mouthed toad (*Gastrophryne carolenensis*), of which 50% were dead or malformed in 7 days at 40 μg As^{+3}/L; and a freshwater alga (*Scenedesmus obliquus*), in which growth was inhibited by 50% in 14 days at 48 μg As^{+5}/L (Table 4). Studies of mass cultures of natural phytoplankton communities chronically exposed to low levels of arsenate (1.0 to 15.2 μg/L) showed that As^{+5} differentially inhibits certain plants, causing a marked change in species composition, succession, and predator–prey relations; the significance of these changes on carbon transfer between trophic levels is unknown (Sanders and Cibik, 1985; Sanders, 1986). Adverse biological effects have also been documented at water concentrations of 75 to 100 μg As/L. At 75 μg As^{+5}/L, growth and biomass in freshwater and marine algae were reduced; at 85 to 88 μg/L of As^{+5} or various methylated arsenicals, mortality was 10 to 32% in amphipods (*Gammarus pseudolimnaeus*) in 28 days; at 95 μg As^{+3}/L, marine red alga failed to reproduce sexually; and at 100 μg As^{+5}/L, marine copepods died, and goldfish behavior was impaired (Table 4). Rainbow trout (*Oncorhynchus mykiss*) fed diets containing up to 90 mg As^{+5}/kg were only slightly affected, but those given diets containing > 120 mg As/kg (as As^{+3} or As^{+5}) grew poorly, avoided food, and failed to metabolize food efficiently; no toxic effects were reported over 8 weeks of exposure to diets containing 1600 mg/kg as methylated arsenicals (Table 4). In bluegills (*Lepomis macrochirus*), tissue residues of 1.35 mg As/kg fresh weight in juveniles and 5 mg/kg in adults are considered elevated and potentially hazardous (NRCC, 1978).

Toxic and other effects of arsenicals on aquatic life are significantly modified by numerous biological and abiotic factors (Woolson, 1975; NAS, 1977; NRCC, 1978; EPA, 1980, 1985; Howard et al., 1984; Michnowicz and Weaks, 1984; Bryant et al., 1985; Sanders, 1986). The LC_{50} values, for example, are markedly affected by water temperature, pH, Eh, organic content, phosphate concentration, suspended solids, and the presence of other substances and toxicants, as well as arsenic speciation and duration of exposure. In general, inorganic arsenicals are more toxic than organoarsenicals to aquatic biota, and trivalent species are more toxic than pentavalent species. Early life stages are most sensitive, and large interspecies differences are recorded, even among species closely related taxonomically.

Arsenic is bioaccumulated from the water by many organisms; however, there is no evidence of biomagnification in aquatic food chains (Woolson, 1975; NAS, 1977; NRCC, 1978; Hallacher et al., 1985; Hood, 1985). In a marine ecosystem based on the alga *Fucus vesiculosus*, arsenate (7.5 μg As^{+5}/L) was accumulated by all biota. After 3 months, arsenic was concentrated most efficiently by *Fucus* (120 mg/kg dry weight in apical fronds) and filamentous algal species (30 mg/kg

Table 4 Lethal and Sublethal Effects of Various Arsenic Compounds on Selected Species of Aquatic Biota

Taxonomic Group, Species, Arsenic Compound,[a] and Other Variables	Arsenic Concentration or Dose	Effect	Reference[b]
FRESHWATER PLANTS			
Algae, various species			
As^{+3}	1.7 mg/L	Toxic	1
As^{+3}	4 mg/L	Decomposition	1
As^{+3}	2.3 mg/L	95% to 100% kill of 4 species in 2 to 4 weeks	2, 3
As^{+5}	0.075 mg/L	Decreased growth	3
Alga, *Ankistrodesmus falcatus*			
As^{+5}	0.26 mg/L	Growth reduced 50% in 14 days	3
Alga, *Scenedesmus obliquus*			
As^{+5}	0.048 mg/L	Growth reduced 50% in 14 days	3
Alga, *Selenastrum capricornutum*			
As^{+5}	0.69 mg/L	Growth reduced 50% in 96 hr	3
FRESHWATER INVERTEBRATES			
Cladoceran, *Bosmina longirostris*			
As^{+5}	0.85 mg/L	50% immobilization in 96 hr	4
Cladoceran, *Daphnia magna*			
As^{+3}	0.63–1.32 mg/L	MATC[c]	3
As^{+3}	0.96 mg/L	LC_5 (28 days)	5
As^{+3}			
Starved	1.5 mg/L	50% immobilization in 96 hr	6
Fed	4.3 mg/L	50% immobilization in 96 hr	6
As^{+5}	0.52 mg/L	Reproductive impairment of 16% in 3 weeks	3
As^{+5}	0.93 mg/L	LC_5 (28 days); maximum bioconcentration factor (BCF) of 219X	5
As^{+5}	7.4 mg/L	LC_{50} in 96 hr	2
DSMA	0.83 mg/L	No deaths in 28 days	5

[a,b,c] See page 229 for footnotes.

Table 4 *(Continued)*

Taxonomic Group, Species, Arsenic Compound,[a] and Other Variables	Arsenic Concentration or Dose	Effect	Reference[b]
SDMA	1.1 mg/L	No deaths in 28 days	5
Total As	1 mg/L	18% decrease in body weight in 3 weeks	1
Total As	1.4 mg/L	50% reproductive impairment in 3 weeks	1
Total As	2.8 mg/L	LC_{50} (21 days)	1
Total As	4.3–7.5 mg/L	Immobilization (21 days)	1
Cladoceran, *Daphnia pulex*			
As^{+5}	49.6 mg/L	50% immobilization in 48 hr	4
As^{+3}	1.3 mg/L	LC_{50} in 96 hr	2, 3
As^{+3}	3 mg/L	50% immobilization in 48 hr	7
Amphipod, *Gammarus pseudolimnaeus*			
As^{+3}	0.87 mg/L	50% immobilization in 96 hr	6
As^{+3}	0.088 mg/L	LC_{20} (28 days)	5
As^{+3}	0.96 mg/L	LC_{100} (28 days)	5
As^{+5}	0.97 mg/L	LC_{20} (28 days); no accumulations	5
DSMA	0.086 mg/L	LC_{10} (28 days)	5
DSMA	0.97 mg/L	LC_{40} (28 days)	5
SDMA	0.85 mg/L	LC_{0} (28 days)	5
Snail *Helisoma campanulata*			
As^{+3}	0.96 mg/L	LC_{10} (28 days)	5
As^{+5}	0.97 mg/L	LC_{0} (28 days); maximum BCF of 99X.	5
DSMA	0.97 mg/L	LC_{0} (28 days)	5
SDMA	0.085 mg/L	LC_{0} (28 days)	5
Red crayfish, *Procambarus clarkii*			
MSMA	Nominal concentration of 0.5 mg/L, equivalent to 0.23 mg As/L	Whole body arsenic concentration after 8-week exposure plus 8-week depuration was 0.3 mg/kg DW whole body vs. 0.4 for controls	8

Table 4 *(Continued)*

Taxonomic Group, Species, Arsenic Compound,[a] and Other Variables	Arsenic Concentration or Dose	Effect	Reference[b]
MSMA	Nominal concentration of 5 mg/L, equivalent to 2.3 mg As/L	Exposure and depuration as above; maximum As concentration was 4.3 mg/kg DW whole body during exposure, 0.6 at end of depuration	8
MSMA	Nominal concentration of 50 mg/L, equivalent to 23.1 mg As/L	Exposure and depuration as above; maximum As concentration during exposure was 9 mg/kg DW whole animal and 2.1 at end of depuration	8
MSMA	Nominal concentration of 100 mg/L, equivalent to 46.3 mg As/L	No effect on growth or survival during 24-week exposure, but hatching success reduced to 17% vs. 78% for controls	9
MSMA	1019 mg/L	LC_{50} (96 hr)	9
Stonefly, *Pteronarcys californica*			
As^{+3}	38 mg/L	LC_{50} (96 hr)	7
Stonefly, *Pteronarcys dorsata*			
As^{+3}	0.96 mg/L	LC_{0} (28 days)	5
As^{+5}	0.97 mg/L	LC_{20} (28 days); maximum BCF of 131X	5
DSMA	0.97 mg/L	LC_{0} (28 days)	5
SDMA	0.85 mg/L	LC_{0} (28 days)	5
Cladoceran, *Simocepalus serrulatus*			
As^{+3}	0.81 mg/L	LC_{50} (96 hr)	3
Zooplankton			
As^{+3}	0.4 mg/L	No effect	1
As^{+3}	1.2 mg/L	Population reduction	1

Table 4 *(Continued)*

Taxonomic Group, Species, Arsenic Compound,[a] and Other Variables	Arsenic Concentration or Dose	Effect	Reference[b]
FRESHWATER VERTEBRATES			
Marbled salamander, *Ambystoma opacum*			
As^{+3}	4.5 mg/L	50% mortality and malformations in 8 days in developing embryos	3
Goldfish, *Carassius auratus*			
As^{+5}	0.1 mg/L	15% behavioral impairment in 24 hr; 30% impairment in 48 hr	1
As^{+3}	24.6–41.6 mg/L	LC_{50} (7 days)	1
As^{+3}	0.49 mg/L	EC_{50} (7 days)	3
MSMA	5 mg/L	LC_{50} (96 hr)	3
Narrow-mouthed toad, *Gastrophryne carolinensis*			
As^{+3}	0.04 mg/L	50% death or malformations noted in developing embryos in 7 days	3
Channel catfish, *Ictalurus punctatus*			
As^{+3}	25.9 mg/L	LC_{50} (96 hr)	10
Flagfish, *Jordanella floridae*			
As^{+3}	14.4 mg/L	LC_{50} (96 hr)	6
As^{+3}	2.1–4.1 mg/L	MATC[c]	3
Bluegill, *Lepomis macrochirus*			
As^{+3}			
Juveniles	0.69 mg/L	Reduced survival 16 weeks after a single treatment	2, 3
Adults	0.69 mg/L	Histopathology after 16 weekly treatments	2
As^{+3}	4 mg/L	Population reduction of 42% after several monthly applications	10
As^{+3}	30–35 mg/L	LC_{50} (96 hr)	7, 10
MSMA	1.9 mg/L	LC_{50} (96 hr)	3

Table 4 *(Continued)*

Taxonomic Group, Species, Arsenic Compound,[a] and Other Variables	Arsenic Concentration or Dose	Effect	Reference[b]
Total As	Tissue residues of 1.35 mg/kg fresh weight in juveniles and 5 mg/kg in adults	Threshold acute toxic value	1
Spottail shiner, *Notropis hudsonius*			
As^{+3}	45 mg/L	LC_{50} (25 hr)	10
As^{+3}	29 mg/L	LC_{50} (48 hr); survivors with fin and scale damage	10
Chum salmon, *Oncorhynchus keta*			
As^{+3}	11 mg/L	LC_{50} (48 hr)	10
Rainbow trout, *Oncorhynchus mykiss*			
As^{+3}			
Embryos	0.54 mg/L	LC_{50} (28 days)	2
Adults	23 to 26.6 mg/L	LC_{50} (96 hr)	5
As^{+3}	0.96 mg/L	LC_{0} (28 days)	7, 10
As^{+3} or As^{+5}	Fed diets containing 120 to 1600 mg As/kg for 8 weeks	Growth depression, food avoidance, and impaired feed efficiency at all levels	11
As^{+5}	Fed diets containing 10 to 90 mg As/kg for 16 weeks	No effect at about 10 mg/kg diet; some adaptation to dietary As observed in trout fed 90 mg/kg diet, as initial negative growth gave way to slow positive growth over time	11
As^{+5}	0.97 mg/L	LC_{0} (28 days); no accumulations	5
DSMA	0.97 mg/L	LC_{0} (28 days)	5
SDMA	0.85 mg/L	LC_{0} (28 days)	5
SC	1000 mg/L	LC_{0} (28 days)	12

Table 4 *(Continued)*

Taxonomic Group, Species, Arsenic Compound,[a] and Other Variables	Arsenic Concentration or Dose	Effect	Reference[b]
DMA or ABA	Fed diet containing 120 to 1600 mg/kg for 8 weeks	No toxic response at any level tested	11
Minnow, *Phoxinus phoxinus*			
As^{+3}	20 mg/L	Equilibrium loss in 36 hr	10
As^{+5}	234–250 mg/L	Lethal	10
Fathead minnow, *Pimephales promelas*			
As^{+3}	14.1 mg/L	LC_{50} (96 hr)	6
As^{+3}	2.1–4.8 mg/L	MATC[c]	6
As^{+5}	25.6 mg/L	LC_{50} (96 hr)	3
As^{+5}	0.53–1.50 mg/L	MATC[c]	3
Brook trout, *Salvelinus fontinalis*			
As^{+3}	15 mg/L	LC_{50} (96 hr)	3
MARINE PLANTS			
Algae, 2 spp.			
As^{+3} or As^{+5}	1 mg/L	No effect	13
As^{+5}	1000 mg/L	No deaths	13
Algae, 3 spp.			
As^{+3}	0.019–0.022 mg/L	Reduced growth	3
Red alga, *Champia parvula*			
As^{+3}	0.065 mg/L	Normal sexual reproduction	13
As^{+3}	0.095 mg/L	Normal sexual reproduction	13
As^{+3}	0.30 mg/L	Death	13
As^{+5}	10 mg/L	Normal growth but no sexual reproduction	13
Phytoplankton			
As^{+5}	0.075 mg/L	Reduced biomass of populations in 4 days	4
Red Alga, *Plumaria elegans*			
As^{+3}	0.58 mg/L	Arrested sporeling development 7 days after exposure for 18 hr	2

Table 4 *(Continued)*

Taxonomic Group, Species, Arsenic Compound,[a] and Other Variables	Arsenic Concentration or Dose	Effect	Reference[b]
Alga, *Skeletonema costatum*			
As^{+5}	0.13 mg/L	Growth inhibition	3
Alga, *Thalassiosira aestivalis*			
As^{+5}	0.075 mg/L	Reduced chlorophyll a	3
MARINE INVERTEBRATES			
Copepod, *Acartia clausi*			
As^{+3}	0.51 mg/L	LC_{50} (96 hr)	3
Dungeness crab, *Cancer magister*			
As^{+5}	0.23 mg/L	LC_{50} (96 hr) for zoea	3
Amphipod, *Corophium volutator*			
As^{+5}			
Water temperature, °C			
5	8 mg/L	LC_{50} (230 hr)	14
10	8 mg/L	LC_{50} (150 hr)	14
15	8 mg/L	LC_{50} (74 hr)	14
15	4 mg/L	LC_{50} (140 hr)	14
15	2 mg/L	LC_{50} (192 hr)	14
Pacific oyster, *Crassostrea gigas*			
As^{+3}	0.33 mg/L	LC_{50} (96 hr) for embryos	3
American oyster, *Crassostrea virginica*			
As^{+3}	7.5 mg/L	LC_{50} (48 hr, eggs)	10
Copepod, *Eurytemora affinis*			
As^{+5}	0.025 mg/L	No effect	15
As^{+5}	0.1 mg/L	Reduced juvenile survival	15
As^{+5}	1 mg/L	Reduced adult survival	15
Clam, *Macoma balthica*			
As^{+5}			
Water temperature, °C			
5	220 mg/L	LC_{50} (192 hr)	14
10	60 mg/L	LC_{50} (192 hr)	14
15	15 mg/L	LC_{50} (192 hr)	14
Mysid, *Mysidopsis bahia*			
As^{+3}	0.63–1.27 mg/L	MATC[c]	3
As^{+5}	2.3 mg/L	LC_{50} (96 hr)	3
Blue mussel, *Mytilus edulis*			
As^{+3}	16 mg/L	Lethal in 3 to 16 days	10

Table 4 *(Continued)*

Taxonomic Group, Species, Arsenic Compound,[a] and Other Variables	Arsenic Concentration or Dose	Effect	Reference[b]
Mud snail, *Nassarius obsoletus*			
As^{+3}	2 mg/L	Depressed oxygen consumption in 72 hr	10
Oligochaete annelid, *Tubifex costatus*			
As^{+5}			
Water temperature, °C			
5	500 mg/L	LC_{50} (130 hr)	14
10	500 mg/L	LC_{50} (115 hr)	14
15	500 mg/L	LC_{50} (85 hr)	14
MARINE VERTEBRATES			
Grey mullet, *Chelon labrosus*			
As^{+3}	27.3 mg/L	LC_{50} (96 hr); some skin discoloration	16
Dab, *Limanda limanda*			
As^{+3}	8.5 mg/L	LC_{50} (96 hr); respiratory problems	16
Pink salmon, *Oncorhynchus gorbuscha*			
As^{+3}	2.5 mg/L	LC_0 (10 days)	10
As^{+3}	3.8 mg/L	LC_{54} (10 days)	3
As^{+3}	7.2 mg/L	LC_{100} (7 days)	3
Teleosts, 3 spp.			
As^{+3}	12.7–16 mg/L	LC_{50} (96 hr)	3

[a] As^{+3}, inorganic trivalent arsenite; As^{+5}, inorganic pentavalent arsenate; DMA, dimethylarsinic acid; ABA, *p*-aminobenzenearsonic acid; DSMA, disodium methylarsenate [$CH_3AsO(ONa)_2$]; SDMA, sodium dimethylarsenate [$(CH_3)_2AsO(ONa)$]; MSMA, monosodium methanearsonate; SC, sodium cacodylate.

[b] 1, NRCC (1978); 2, EPA (1980); 3, EPA (1985); 4, Passino and Novak (1984); 5, Spehar et al. (1980); 6, Lima et al. (1984); 7, Johnson and Finley, (1980); 8, Naqvi et al. (1990); 9, Naqvi and Flagge (1990); 10, NAS (1977); 11, Cockell and Hilton (1985); 12, Hood (1985); 13, Thursby and Steele (1984); 14, Bryant et al. (1985); 15, Sanders (1986); 16, Taylor et al. (1985).

[c] Maximum acceptable toxicant concentration. Lower value in each pair indicates highest concentration tested producing no measurable effect on growth, survival, reproduction, or metabolism during chronic exposure; higher value indicates lowest concentration tested producing a measurable effect.

dry weight); there was little or no bioaccumulation in mussels (Rosemarin et al., 1985). In a freshwater food chain composed of algae, daphnids, and fish, water concentrations of 0.1 mg cacodylic acid/L produced residues after 48 hr of 4.5 (mg As/kg dry weight) in algae and 3.9 in daphnids, but only 0.09 in fish (NAS, 1977). Microcosms of a Delaware cordgrass (*Spartina alterniflora*) salt marsh exposed to elevated levels of As^{+5} showed that virtually all arsenic was incorporated into plant tissue or strongly sorbed to cell surfaces (Sanders and Osman, 1985). Studies with radioarsenic and mussels (*Mytilus galloprovincialis*) showed that accumulation varied with nominal arsenic concentration, tissue, age of the mussel, and temperature and salinity of the medium (Unlu and Fowler, 1979). Arsenate uptake increased with increasing arsenic concentration in the medium, but the response was not linear, accumulation being suppressed at higher external arsenic concentrations. Smaller mussels took up more arsenic than larger ones. In both size groups, arsenic was concentrated in the byssus and digestive gland. In general, arsenic uptake and loss increased at increasing temperatures. Uptake was significantly higher at 1.9‰ salinity than at 3.8‰, but the loss rate was about the same at both salinities. Radioarsenic loss followed a biphasic pattern; biological half-life was 3 and 32 days for the fast and slow compartments, respectively; secretion via the byssal thread played a key role in elimination (Unly and Fowler, 1979). Factors known to modify rates of arsenic accumulation and retention in a marine shrimp (*Lysmata seticaudata*) include water temperature and salinity, arsenic concentration, age, and especially frequency of molting (Fowler and Unlu, 1978).

Bioconcentration factors (BCF) experimentally determined for arsenic in aquatic organisms are, except for algae, relatively low. The BCF values for inorganic As^{+3} in most aquatic invertebrates and fish exposed for 21 to 30 days did not exceed 17 times; the maxima were 6 times for As^{+5} and 9 times for organoarsenicals (EPA, 1980, 1985). Significantly higher BCF values were recorded in other aquatic organisms (NRCC, 1978), but they were based on mean arsenic concentrations in natural waters that seemed artificially high. A BCF of 350 times was reported for the American oyster (*Crassostrea virginica*) held in 5 μg As^{+5}/L for 112 days (Zaroogian and Hoffman, 1982). There was no relation between oyster body burden of arsenic and exposure concentration; however, diet seemed to contribute more to arsenic uptake than did seawater concentration (Zaroogian and Hoffman, 1982). An arsenic-tolerant strain of freshwater alga (*Chlorella vulgarus*) from an arsenic-polluted environment showed increasing growth up to 2000 mg As^{+5}/L, and it could survive at 10,000 mg As^{+5}/L (Maeda et al., 1985). Accumulations up to 50,000 mg As/kg dry weight were recorded (Maeda et al., 1985), suggesting a need for additional research on the extent of this phenomenon and its implications on food-web dynamics.

Some investigators have suggested that arsenic in the form of arsenite is preferentially utilized by marine algae and bacteria (Johnson, 1972; Bottino et al., 1978; Johnson and Burke, 1978). Arsenate reduction to arsenite in seawater depends on phosphorus in solution and available algal biomass (Johnson and Burke, 1978). During algal growth, as phosphate is depleted and the P^{+5}: As^{+5} ratio drops, the rate of As^{+5} reduction increases. The resultant As^{+3}, after an initial peak, is rapidly oxidized to As^{+5}, indicating the possibility of biological catalysis as well as mediation of As^{+5} reduction. Researchers generally agree that As^{+3} is more toxic than arsenates to higher organisms; however, As^{+5} has a more profound effect on the growth and morphology of marine algae than does As^{+3}.

Possibly, marine algae erect a barrier against the absorption of As^{+3} but not against As^{+5}. Within the cell, As^{+5} can then be reduced to the possibly more toxic As^{+3}. For example, the culture of two species of marine algae (*Tetraselmis chui, Hymenomonas carterae*) in media containing various concentrations of As^{+5} or As^{+3} showed that arsenic effects varied with oxidation state, concentration, and light intensity. Arsenate was incorporated and later partly released by both species. Differences between rates of uptake and release suggest that arsenic undergoes chemical changes after incorporation into algal cells (Bottino et al., 1978). When bacterial cultures from the Sargasso Sea and from marine waters off of Rhode Island were grown in As^{+3}-enriched media, the bacteria reduced all available As^{+5} and utilized As^{+3} during the exponential growth phase, presumably as an essential trace nutrient. The arsenate reduction rate per cell was estimated to be 75×10^{-11} mg As/minute (Johnson, 1972).

The ability of marine phytoplankton to accumulate high concentrations of inorganic arsenicals and to transform them into methylated arsenicals that are later effeciently transferred in the food chain is well documented (Irgolic et al., 1977; Benson, 1984; Matsuto et al., 1984; Freeman, 1985; Froelich et al., 1985; Maeda et al., 1985; Norin et al., 1985; Sanders, 1985; Yamaoka and Takimura, 1986). Algae constitute an important source of organoarsenic compounds in marine food webs. In the food chain composed of the alga *Dunaliella marina*, the grazing shrimp *Artemia salina*, and the carnivorous shrimp *Lysmata seticaudata*, organic forms of arsenic were derived from in vivo synthesis by *Dunaliella* and efficiently transferred, without magnification, along the food chain (Wrench et al., 1979). Laboratory studies with five species of euryhaline algae grown in freshwater or seawater showed that all species synthesized fat-soluble and water-soluble arsenoorganic compounds from inorganic As^{+3} and As^{+5}. The BCF values in the five species ranged from 200 times to about 3000 times—accumulations being highest in lipid phases (Lunde, 1973). In Charlotte Harbor, Florida, a region that has become phosphate-enriched due to agricultural activity, virtually all of the arsenic taken up by phytoplankton was biomethylated and returned to the estuary, usually as monomethylarsonic and dimethylarsenic acids (Froelich et al., 1985). The ability of marine phytoplankton to methylate arsenic and release the products to the surrounding environment varies between species, and even within a particular species, according to their possession of the necessary methylating enzymes (Sanders, 1985). The processes involved in detoxifying arsenate after its absorption by phytoplankton are not firmly established but seem to be nearly identical in all plants, suggesting a similar evolutionary development. Like phosphates and sulfates, arsenate may be fixed with ADP, reduced to the arsonous level, and successfully methylated and adenosylated, ultimately producing the 5-dimethylarsenosoribosyl derivatives accumulating in algae (Benson, 1984).

Sodium arsenite has been used extensively as an herbicide for the control of mixed submerged aquatic vegetation in freshwater ponds and lakes; concentrations of 1.5 to 3.8 mg As^{+3}/L have usually been effective and are considered safe for fish (NAS, 1977). However, As^{+3} concentrations considered effective for aquatic weed control may be harmful to several species of freshwater teleosts, including bluegills, flagfish (*Jordanella floridae*), fathead minnows (*Pimephales promelas*), and rainbow trout (*Oncorhynchus mykiss*) (Table 4). Fish exposed to 1 to 2 mg total As/L for 2 to 3 days may show one or more of several signs: hemorrhagic spheres on gills, fatty infiltration of the liver, and necrosis of heart, liver, and ovarian tissues (NRCC, 1978). In green sunfish (*Lepomis cyanellus*), hepatocyte changes parallel arsenic accumulation in the liver (Sorensen et al., 1985). Organoarsenicals are usually eliminated rapidly by fish and other aquatic fauna. Rainbow trout, for example, fed a marine diet containing 15 mg organic As/kg had only negligible tissue residues 6 to 10 days later, although some enrichment was noted in the eyes, throat, gills, and pyloric caeca (Pershagen and Vahter, 1979). Oral administration of sodium arsenate to estuary catfish (*Cnidoglanis macrocephalus*) and school whiting (*Sillago bassensis*) resulted in tissue accumulations of trimethylarsine oxide. Arsenobetaine levels, which occur naturally in these teleosts, were not affected by As^{+5} dosing. The toxicity of trimethylarsine oxide is unknown, but the ease with which it can be reduced to the highly toxic trimethylarsine is cause for concern (Edmonds and Francesconi, 1987).

6.4. Birds

Signs of inorganic trivalent arsenite poisoning in birds (muscular incoordination, debility, slowness, jerkiness, falling, hyperactivity, fluffed feathers, drooped eyelids, huddled position, unkempt appearance, loss of righting reflex, immobility, seizures) were similar to those induced by many other toxicants and did not seem to be specific for arsenosis. Signs occurred within 1 hr and death within 1 to 6 days after administration; remission took up to 1 month (Hudson et al., 1984). Internal examination suggested that the lethal effects of acute inorganic arsenic poisoning were due to the destruction of blood vessels lining the gut, which resulted in decreased blood pressure and subsequent shock (Nyström, 1984). For example, coturnix (*Coturnix coturnix*) exposed to acute oral doses of As^{+3} showed hepatocyte damage (swelling of granular endoplasmic reticulum); these effects were attributed to osmotic imbalance, possibly induced by direct inhibition of the sodium pump by arsenic (Nyström, 1984).

Arsenic, as arsenate, in aquatic plants (up to 430 mg As/kg plant dry weight) from agricultural drainwater areas can impair normal development of mallard ducklings (Camardese et al., 1990) (Table 5). Pen studies with ducklings showed that diets of 30 mg As/kg ration adversely affects growth and physiology, and 300 mg As/kg diet alters brain biochemistry and nesting behavior. Decreased energy levels and altered behavior can further decrease duckling survival in a natural environment (Camardese et al., 1990).

Western grasshoppers (*Melanophis* spp.) poisoned by arsenic trioxide were fed, with essentially no deleterious effects, to nestling northern bobwhites (*Colinus virginianus*), mockingbirds (*Mimus polyglottos*), American robins (*Turdus migratorius*), and other songbirds (NAS, 1977). Up to 134 poisoned grasshoppers, containing a total of about 40 mg arsenic, were fed to individual nestlings without any apparent toxic effect. Species tested that were most sensitive to various arsenicals

Table 5 Lethal and Sublethal Effects of Various Arsenicals on Selected Species of Birds

Species and Arsenic Compound	Effect	Reference[a]
Chukar, *Alectoris chukar*		
Silvisar-510 (mixture of cacodylic acid and triethanolamine cacodylate)	Single oral LD_{50} dose of about 2000 mg/kg body weight (BW); signs of poisoning evident within 10 minutes and mortalities within 1 to 2 days; remission took up to one month	1
Mallard, *Anas platyrhynchos*		
Sodium arsenate	Ducklings were fed 30, 100, or 300 mg As/kg diet for 10 weeks. All treatment levels produced elevated hepatic glutathione and ATP concentrations and decreased overall weight gain and rate of growth in females. Arsenic concentrations were elevated in brain and liver of ducklings fed 100 or 300 mg/kg diets; at 300 mg/kg, all ducklings had altered behavior, i.e., increased resting time; male ducklings had reduced growth	2
Sodium arsenite	323 mg/kg BW is LD_{50} acute oral value	1, 3, 4
Sodium arsenite	500 mg/kg diet is fatal to 50% in 32 days; 1000 mg/kg diet fatal to 50% in 6 days	3
Sodium cacodylate	1740 to 5000 mg/kg diet not measurably harmful to ducklings in 5 days	5
Silvisar 510	Single oral $LD_{50} > 2400$ mg/kg BW; regurgitation and excessive drinking noted	1
Lead arsenate	5000 mg/kg diet not fatal in 11 days	3
Copper acetoarsenite	5000 mg/kg diet fatal to 20% in 11 days	3
California quail, *Callipepla californica*		
Sodium arsenite	Single oral LD_{50} value of 47.6 mg/kg BW	1
Northern bobwhite, *Colinus virqinianus*		
Copper acetoarsenite	480 mg/kg in diet fatal to 50% in 11 days	3
Sodium cacodylate	1740 mg/kg in diet for 5 days produced no effect on behavior, no signs of intoxication, and negative necropsy	5

Table 5 *(Continued)*

Species and Arsenic Compound	Effect	Reference[a]
Monosodium methanearsonate, CH_4AsNaO_3	Single oral LD_{50} dose of 3300 mg/kg BW	5
Chicken, *Gallus gallus*		
Inorganic trivalent arsenite	Up to 34% dead embryos at dose range of 0.01–1 μg As^{+3}/embryo; threshold for malformations at dose range 0.03–0.3 μg/embryo	4
Inorganic pentavalent arsenate	Up to 8% dead at dose range 0.01–1 μg As^{+5}/embryo; threshold for malformations at dose range 0.3–3 μg/embryo	4
Disodium methylarsenate	Teratogenic to embryos when injected at 1 to 2 mg/egg	4, 5
Sodium cacodylate	Developmental abnormalities at embryonic injected doses of 1 to 2 mg/egg	5
Dodecylamine *p*-chlorophenylarsonate	At dietary levels of 23.3 mg/kg, liver residues were 2.9 mg/kg fresh weight (FW) at 9 weeks. No ill effects noted	6
3-Nitro-4-hydroxy-phenylarsonic acid	At 18.7 mg/kg diet for 9 weeks, liver residues of 2.4 mg/kg FW. Those fed diets containing 187 mg/kg for 9 weeks had no ill effects; liver content of 7.5 mg/kg FW	6
3-Nitro-4-hydroxy-phenylarsonic acid	LD_{50} dose of 33 mg/kg BW (single oral) or 9.7 mg/kg BW (intraperitoneal injection)	3
Arsanilic acid	Fed diets containing 45 mg/kg for 9 weeks; no effect except slightly elevated liver content of 1.2 mg/kg FW. At dietary levels of 455 mg/kg, liver residues were 6.4 mg/kg FW after 9 weeks; no other effects evident	6
Cacodylic acid	Dosed orally without effect at 100 mg/kg BW daily for 10 days	5
Chickens, *Gallus* spp.		
Arsanilic acid	50% excreted in 36 to 38 hr	4
Arsenate	50% excreted in 60 to 63 hr	4
Turkey, *Meleagris gallopavo*		
3-Nitro-4-hydroxy-phenylarsonic acid	Single oral LD_{50} dose of 17.4 mg/kg BW	3
Brown-headed cowbird, *Molothrus ater*		
Copper acetoarsenite	All survived 11 mg/kg diet for 6	

Table 5 *(Continued)*

Species and Arsenic Compound	Effect	Reference[a]
	months; maximum whole body residue of 1.7 mg As/kg dry weight (DW)	3
Copper acetoarsenite	All survived 33 mg/kg diet for 6 months (whole body content of 6.6 mg As/kg DW) or 7 months (8.6 DW)	3
Copper acetoarsenite	99.8 mg/kg in diet fatal to 50% in 11 days	3
Copper acetoarsenite	100 mg/kg in diet for 3 months fatal to 100%; tissue residues, in mg/kg DW, of 6.1 in brain, 40.6 in liver	3
Gray partridge, *Perdix perdix*		
Lead arsenate	300 mg/kg BW fatal in 52 hr	3
Ring-necked pheasant, *Phasianus colchicus*		
Sodium arsenite	Single oral dose of 386 mg/kg BW is LD_{50} value	1
Copper acetoarsenite	Single oral dose of 1403 mg/kg BW is LD_{50} value	4
Lead arsenate	4989 mg/kg in diet fatal	3

[a] 1, Hudson et al., 1984; 2, Camardese et al., 1990; 3, NAS, 1977; 4, NRCC, 1978; 5, Hood, 1985; 6, Woolson, 1975.

were the brown-headed cowbird (*Molothrus ater*), with an LD_{50} (11-day) value of 99.8 mg of copper acetoarsenite/kg diet; California quail (*Callipepla californica*), with an LD_{50} single oral dose value of 47.6 mg of sodium arsenite/kg body weight; and chicken with 33 and turkey with 17.4 mg/kg body weight of 3-nitro-4-hydroxyphenylarsonic acid as a single oral LD_{50} dose (Table 5).

Chickens rapidly excrete arsenicals; only 2% of dietary sodium arsenite remained after 60 hr (NAS, 1977), and arsanilic acid was excreted largely unchanged (Woolson, 1975). Excretion of arsanilic acid by chickens was affected by uptake route: excretion was more rapid when administration was by intramuscular injection than when it was oral (NRCC, 1978). Studies with inorganic As^{+5} and chickens indicated that (1) arsenates rapidly penetrated the mucosal and serosal surfaces of epithelial membranes, (2) As^{+5} intestinal absorption was essentially complete within 1 hr at 370 mg As^{+5}/kg body weight but only 50% complete at 3700 mg/kg body weight, (3) vitamin D_3 was effective in enhancing duodenal As^{+5} absorption in rachitic chicks, and (4) As^{+5} and phosphate did not appear to share a common transport pathway in the avian duodenum (Fullmer and Wasserman, 1985).

6.5. Mammals

Mammals are exposed to arsenic primarily through the ingestion of naturally contaminated vegetation and water or through human activity. In addition, feed additives containing arsonic acid derivatives are often fed to domestic livestock to promote growth and retard disease. Some commercial pet foods contain up to 2.3 mg As/kg dry weight (NRCC, 1978). Uptake may occur by ingestion (the most likely route), inhalation, or absorption through skin and mucous membranes. Soluble arsenicals are absorbed more rapidly and completely than the sparingly soluble arsenicals, regardless of the route of administration (NRCC, 1978).

Acute episodes of poisoning in warm-blooded organisms by inorganic and organic arsenicals are usually characterized by high mortality and morbidity over a period of 2 to 3 days (NAS, 1977; Selby et al., 1977). General signs of arsenic toxicosis include intense abdominal pain, staggering gait, extreme weakness, trembling, salivation, vomiting, diarrhea, fast and feeble pulse, prostration, collapse, and death. Gross necropsy shows a reddening of gastric mucosa and intestinal mucosa, a soft yellow liver, and red edematous lungs. Histopathological findings show edema of gastrointestinal mucosa and submucosa, necrosis and sloughing of mucosal epithelium, renal tubular degeneration, hepatic fatty changes and necrosis, and capillary degeneration in the gastrointestinal tract, vascular beds, skin, and other organs. In subacute episodes, in which animals remain alive for several days, signs of arsenosis include depression, anorexia, increased urination, dehydration, thirst, partial paralysis of the rear limbs, trembling, stupor, coldness of extremities, and subnormal body temperatures (NAS, 1977; Selby et al., 1977). In cases involving cutaneous exposure to arsenicals, a dry, cracked, leathery, and peeling skin may be a prominent feature (Selby et al., 1977). Nasal discharges and eye irritation were documented in rodents exposed to organoarsenicals in inhalation toxicity tests (Hood, 1985). Subacute effects in humans and laboratory animals include peripheral nervous disturbances, melanosis, anemia, leukopenia, cardiac abnormalities, and liver changes. Most adverse signs rapidly disappear after exposure ceases (Pershagen and Vahter, 1979).

Arsenic poisoning in most animals is usually manifested by acute or subacute signs; chronic poisoning is infrequently seen (NAS, 1977). The probability of chronic arsenic poisoning from continuous ingestion of small doses is rare because detoxication and excretion are rapid (Woolson, 1975). Chronic toxicity of inorganic arsenicals is associated with weakness, paralysis, conjunctivitis, dermatitis, decreased growth, and liver damage (NRCC, 1978). Arsenosis, produced as a result of chronic exposure to organic arsenicals, is associated with demyelination of the optic and sciatic nerves, depressed growth, and decreased resistance to infection (NRCC, 1978).

Research results on arsenic poisoning in mammals (Table 6) show general agreement on eight points: (1) arsenic metabolism and effects are significantly influenced by the organism tested, the route of administration, the physical and chemical form of the arsenical, and the dose; (2) inorganic arsenic compounds are

Table 6 Lethal and Sublethal Effects of Various Arsenicals on Selected Species of Mammals

Organism and Arsenical	Effect	Reference[a]
Cow, *Bos bovis*		
Arsenate	Cows fed 33 mg As^{+5} daily per animal for 3 months had slightly elevated levels in muscle [0.02 mg/kg fresh weight (FW) vs. 0.005 in controls] and liver (0.03 vs. 0.012) but normal levels in milk and kidney	1
Arsenite	Cows fed 33 mg As^{+3} daily per animal for 15 to 28 months had tissue levels, in mg/kg FW, of 0.002 for milk (vs. <0.001 for controls), 0.03 for muscle (vs. 0.005), 0.1 for liver (vs. 0.012), and 0.16 for kidney (vs. 0.053)	1
Cattle, *Bos* spp.		
Arsenic pentoxide (Wood ashes treated with As preservative)	Several deaths after eating wood ashes (780 mg/kg dry weight); tissue residues, in mg As/kg FW, of 13.9 in liver, 23.7 in kidney, and 25.8 in rumen contents (vs. control values of <0.5)	2
Arsenic trioxide	Single oral dose of 15 to 45 g/ animal was fatal	3
Arsenic trioxide	Toxic dose is 33 to 55 mg/kg body weight (BW), or 13.2 to 22 g for a 400-kg animal. Animals accidentally poisoned topically contained up to 15 mg As/kg FW liver, 23 in kidney, and 45 in urine (vs. <1 for all control tissues)	4
Cacodylic acid, $(CH_3)_2AsO(OH)$	Calves were anorexic in 3 to 6 days when fed diets containing 4700 mg/kg. Adult oral dosages of 10 mg/kg BW daily for 3 weeks followed by 20 mg/kg BW daily for 5 to 6 weeks were lethal. Adverse effects at 25 mg/kg BW daily for 10 days	5
Methanearsonic acid, $CH_3AsO(OH)_2$	Calves were anorexic in 3 to 6 days when fed diets containing 4000 mg/kg	5
Monosodium methanearsonate	10 mg/kg BW daily for 10 days was fatal	3
Sodium arsenite	Single oral dose of 1 to 4 g was fatal	3
Dog, *Canis familiaris*		
Cacodylic acid	Single oral LD_{50} value of 1000 mg/kg BW. Fed diets containing 30 mg/kg for 90 days with no ill effects	5
Methanearsonic acid	Fed diets containing 30 mg/kg for 90 days with no ill effects	5
Sodium arsenite	50 to 150 mg was fatal	3

Table 6 *(Continued)*

Organism and Arsenical	Effect	Reference[a]
Domestic goat, *Capra* sp.		
Arsenic acid	Single oral dose of 2.5 to 7.5 mg/kg BW (50 to 150 mg) was acutely toxic	3
Guinea pig, *Cavia* sp.		
Arsanilic acid	Dietary levels of 350 mg/kg resulted in blindness and optic disc atrophy in 25 to 30 days	6
Arsenic trioxide	Fed diets containing 50 mg/kg for 21 days; elevated As residues, in mg/kg FW, of 4 in blood, 15 in spleen, and 20 in heart (vs. < 1 for all control tissues)	7
Sodium arsanilate	Subcutaneous injection of 70 mg/kg BW caused degeneration of sensory walls of inner ear; elevated As residues in cochlea	6
Sodium arsenate	Intraperitoneal injection of 0.2 mg/kg BW at age 2 months caused deafness	6
Hamster, *Cricetus* sp.		
Arsenate	Maternal dose of 5 mg As^{+5}/kg BW caused some fetal mortality, but no malformations; higher dose of 20 mg/kg BW caused 54% fetal deaths and malformations	3
Calcium arsenate	Pulmonary tumorogenicity demonstrated 70 weeks after 15 intratracheal weekly injections of 3 mg/kg BW	8
Dimethylarsinate	50% growth reduction in Chinese hamster ovary cells (CHOC) at 90 to 112 mg/L	9
Gallium arsenide	Single oral dose of 100 mg/kg BW mostly (85%) eliminated in 5 days, usually in form of organoarsenicals; all tissue levels < 0.25 mg/kg	10
Sodium arsenate	Dosed intravenously on day 8 of gestation: 2 mg/kg BW had no measurable effect; 8 mg/kg BW produced increased incidence of malformation and resorption; 16 mg/kg BW killed all embryos.	6
Sodium arsenate	50% growth reduction in CHOC at 2.25 mg/L	9
Sodium arsenite	Chinese hamster ovary cells showed 50% growth reduction at 0.3 mg/L	9
Sodium cacodylate	Single intraperitoneal injection of 900 to 1000 mg/kg BW during midgestation resulted in some maternal deaths and increased incidence of fetal malformations	5
Horse, *Equus caballus*		
Sodium arsenite	Daily doses of 2 to 6 mg/kg BW (1 to 3 g) for 14 weeks were fatal	3

Table 6 *(Continued)*

Organism and Arsenical	Effect	Reference[a]
Cat, *Felis domesticus*		
Inorganic arsenate or arsenite	Chronic oral toxicity at 1.5 mg/kg BW	6
Human, *Homo sapiens*		
Arsenic trioxide	Fatal at 70 to 180 mg, equivalent to about 1 to 2.6 mg As/kg BW	6
Arsenic trioxide	LD_{50} dose of 7 mg/kg BW	3
Cacodylic acid	LD_{50} of 1,350 mg/kg BW	3
Lead arsenate	Some deaths at 7 mg/kg BW	3
Total arsenic	Accumulations of 1 mg/kg BW daily for 3 months in children, or 80 mg/kg BW daily for 3 years, produced symptoms of chronic arsenic poisoning	3
Total arsenic, daily oral dose	Prolonged dosages of 3 to 4 mg daily produced clinical symptoms of chronic arsenic intoxication	3
Total arsenic in drinking and cooking water	Prolonged use produced symptoms of chronic arsenic intoxication (0.6 mg/L) or skin cancer (0.29 mg/L)	3
Total arsenic , probably as arsenate	12,000 Japanese infants poisoned (128 deaths) from consumption of dry milk contaminated with arsenic; average exposure of 3.5 mg As daily for one month. Severe hearing loss, brain-wave abnormalities, and other central nervous system disturbances noted 15 years after exposure	6
Total inorganic arsenic	Daily dose of 3 mg for 2 weeks may cause severe poisoning in infants and symptoms of toxicity in adults	6
Cynomolgus monkey, *Macaca* sp.		
Fish arsenic meal (witch flounder, *Glyptocephalus cynoglossus*) containing 77 mg total As/kg	Given single meal at 1 mg/kg BW; tissue residues normal after 14 days	11
As above, except arsenic trioxide substituted for total arsenic	As above	11
Mammals, many species		
Calcium arsenate	Single oral LD_{50} range of 35 to 100 mg/kg BW	3
Lead arsenate	Single oral LD_{50} range of 10 to 50 mg/kg BW	3

Table 6 *(Continued)*

Organism and Arsenical	Effect	Reference[a]
Mammals, most species		
Arsenic trioxide	3 to 250 mg/kg BW lethal	12
Sodium arsenite	1 to 25 mg/kg BW lethal	12
Mouse, *Mus* spp.		
Arsenate	Maternal dose of 10 mg As^{+5}/kg BW resulted in some fetal deaths and malformations	3
Arsenic trioxide	Single oral LD_{50} (96 hr) value of 39.4 mg/kg BW; LD_0 (96 hr) of 10.4 mg/kg BW	12
Arsenic trioxide	"Adapted" group (50 mg As/L in drinking water for 3 months) had subcutaneous LD_{50} value of 14 mg/kg BW vs. 11 for nonadapted group	12
Arsenic trioxide	Air concentrations of 28.5 mg/m^3 for 4 hr daily on days 9 to 12 of gestation caused fetotoxic effects and chromosomal damage to liver cells by day 18; effects included reduced survival, impaired growth, retarded limb ossification, and bone abnormalities. At 2.9 mg/m^3, a 9.9% decrease in fetal weight was recorded; at 0.26 mg/m^3, a 3.1% decrease was measured	13
Cacodylic acid	Oral dosages of 400 to 600 mg/kg BW on days 7 to 16 of gestation produced fetal malformations (cleft palate), delayed skeletal ossification, and reduced fetal weight	5
Sodium arsenate	Maximum tolerated doses in terms of abortion or maternal death over 24 hr in 18-day pregnant mice were 20 mg As^{+5}/kg BW, intraperitoneal route, and 50 mg/kg BW administered orally. Residue half-life was about 10 hr regardless of route of administration	14
Sodium arsenite	Fed 5 mg/kg diet for three generations: reduced litter size, but outwardly normal	6
Sodium arsenite	LD_{50} of 9.6 mg/kg BW, subcutaneous route; LD_{90} (7 days after administration) of 11.3 mg/kg BW, subcutaneous route	15
Sodium arsenite	LD_{50} of 12 mg/kg BW, intraperitoneal route. At 10 mg/kg BW, damage to bone marrow and sperm	16
Sodium cacodylate	Single intraperitoneal injection of 1200 mg/kg BW during midgestation	

Table 6 *(Continued)*

Organism and Arsenical	Effect	Reference[a]
	resulted in increased rates of fetal skeletal malformations	5
Mule deer, *Odocoileus hemionus hemionus*		
Silvisar-510 (mixture of cacodylic acid and triethanolamine cacodylate)	Single oral LD_{50} dose > 320 mg/kg BW produced appetite loss	17
White-tailed deer, *Odocoileus virqinianus*		
Sodium arsenite (used to debark trees)	Lethal dose of 923 to 2770 mg, equivalent to about 34 mg/kg BW; liver residues of 40 mg/kg FW	12
Arsenic acid (herbicide to control Johnson grass)	23 deer killed from apparent misuse. Arsenic levels, in mg/kg FW, in deer found dead were 19 in liver, 18 in kidney, and 22.5 in rumen contents. Soils from area contained about 2.4 mg As/kg; water contained 0.42 mg As/L	12
Domestic sheep, *Ovis aries*		
Arsanilic acid	One-year-old castrates fed diets with 273 mg As/kg for 28 days had 0.54 mg As/L in blood, 29 mg/kg DW in liver, 24 in kidney, and 1.2 in muscle (vs. < 0.01 in all control tissues). After 6 days on an As-free diet, liver residues were <5 mg/kg DW. Maximum tissue levels in sheep fed diets containing 27 mg As/kg for 28 days was 3.2 mg/kg DW kidney; for an 144 mg/kg diet, the maximum tissue level was 27 mg/kg DW liver	7
Sodium arsenite	Single oral dose of 5 to 12 mg/kg BW (0.2 to 0.5 g) was acutely toxic	3
Soluble arsenic	Lambs fed supplemental arsenic for 3 months at 2 mg As/kg DW diet contained maximum concentrations of 2 μg/kg FW brain (vs. 1 in controls), 14 in muscle (2), 24 in liver (4), and 57 in kidney (10)	18
Total arsenic	Sheep were fed on diets containing lakeweed, *Lagarosiphon major* (288 mg As/kg DW), at 58 mg total As/kg diet, for 3 weeks without ill effect. Tissue residues increased during feeding, but rapidly declined when lakeweed was removed from diet	7

Table 6 *(Continued)*

Organism and Arsenical	Effect	Reference[a]
Rat, *Rattus* sp.		
Arsanilic acid	No teratogenesis observed in 7 generations at dietary level of 17.5 mg/kg; positive effect on litter size and survival	6
Arsenate	Fed diets containing 50 mg/kg for 10 weeks with no effect on serum uric acid levels	19
Arsenic trioxide	Single oral LD_{50} (96 hr) value of 15.1 mg/kg BW	12
Arsenic trioxide	Single dose of 17 mg/kg BW administered intratracheally was maximally tolerated nonlethal dose; 2 weeks later, blood As was elevated (36 mg/L) and lung histopathology evident	20
Arsenic trioxide	After 21 days on diet containing 50 mg/kg, tissue arsenic levels were elevated in blood (125 mg/L vs. 15 in controls), heart (43 mg/kg FW vs. 3.3), spleen (60 vs. < 0.7) and kidney (25 vs. 1.5)	7
Arsenite	Oral administration of 1.2 mg/kg BW daily for 6 weeks reduced uric acid levels in plasma by 67%	19
Cacodylic acid	Fetal and maternal deaths noted when pregnant rats were dosed by gavage at 50 to 60 mg/kg BW daily during gestation days 6 to 13. Fetal abnormalities observed when dams given oral dosages of 40 to 60 mg/kg BW on days 7 to 16 of gestation	5
3-Nitro-4-hydroxy-phenylarsonic acid	Single oral LD_{50} value of 44 mg/kg BW	12
Sodium arsenate	LD_{75} (48 hr) value of 14 to 18 mg/kg BW (intraperitoneal route)	12
Sodium arsenate	Single intraperitoneal injection of 5 to 12 mg/kg on days 7 to 12 of gestation produced eye defects, exencephaly, and faulty development of kidney and gonads	6
Sodium arsenite	LD_{75} (48 hr) value of 4.5 mg/kg BW (intraperitoneal injection)	12
Rodents, various species		
Cacodylic acid	LD_{50} (various routes) values range from 470 to 830 mg/kg BW	5
Sodium cacodylate	LD_{50} (various routes) values range from 600 to 2600 mg/kg BW	5
Domestic pig, *Sus* sp.		
Sodium arsenite	Drinking water containing 500 mg/L was lethal at 100 to 200 mg/kg BW	12

Table 6 *(Continued)*

Organism and Arsenical	Effect	Reference[a]
3-Nitro-4-hydroxy-phenylarsonic acid	Arsenosis documented after 2 months on diets containing 100 mg/kg, or after 3 to 10 days on diets containing 250 mg/kg	12
Rabbit, *Sylvilagus* sp.		
Cacodylic acid	Adverse effects at dermal dosages equivalent to 4 to 6 g/kg BW	5
Calcium arsenate	Single oral dose of 23 mg/kg BW fatal in 3 days	12
Copper acetoarsenite	Single oral dose of 10.5 mg/kg BW fatal in 50 hr	12
Inorganic arsenate	Single oral LD_{50} value of 8 mg/kg BW	3
Lead arsenate	Single oral dose of 40.4 mg/kg BW fatal in 52 hr	12

[a] 1, Vreman et al. (1986); 2, Thatcher et al. (1985); 3, NRCC (1978); 4, Robertson et al. (1984); 5, Hood (1985); 6, Pershagen and Vahter (1979); 7, Woolson (1975); 8, Pershagen and Bjorklund (1985); 9, Belton et al. (1985); 10, Yamauchi et al. (1986); 11, Charbonneau et al. (1978); 12, NAS (1977); 13, Nagymajtenyi et al. (1985); 14, Hood et al. (1987); 15, Stine et al. (1984); 16, Deknudt et al. (1986); 17, Hudson et al. (1984); 18, Veen and Vreman (1985); 19, Jauge and Del-Razo (1985); 20, Webb et al. (1986).

more toxic than organic arsenic compounds, and trivalent species are more so than pentavalent; (3) inorganic arsenicals can cross the placenta in most mammals; (4) early developmental stages are the most sensitive, and human beings appear to be one of the most susceptible species; (5) animal tissues usually contain low levels (< 0.3 mg As/kg fresh weight) of arsenic; after the administration of arsenicals these levels are elevated, especially in the liver, kidney, spleen, and lung; several weeks later, arsenic is translocated to ectodermal tissues (hair, nails) because of the high concentration of sulfur-containing proteins in these tissues; (6) inorganic arsenicals are oxidized in vivo, biomethylated, and usually excreted rapidly in the urine, but organoarsenicals are usually not subject to similar transformations; (7) acute or subacute arsenic exposure can lead to elevated tissue residues, appetite loss, reduced growth, hearing loss, dermatitis, blindness, degenerative changes in the liver and kidneys, cancer, chromosomal damage, birth defects, and death; (8) death or malformations have been documented at single oral doses of 2.5 to 33 mg As/kg body weight, at chronic doses of 1 to 10 mg As/kg body weight, and at dietary levels > 5 and < 50 mg As/kg diet.

Episodes of wildlife poisoning by arsenic are infrequent. White-tailed deer (*Odocoileus virginianus*) consumed, by licking, fatal amounts of sodium arsenite used to debark trees. The practice of debarking trees with arsenicals for commercial use has been almost completely replaced by mechanical debarking (NAS, 1977). Snowshoe hares (*Lepus* sp.) appear to be especially sensitive to methylated arsenicals; hares died after consuming plants heavily contaminated with monosodium methanearsonate as a result of careless silviculture practices (Hood, 1985).

Unlike wildlife, reports of arsenosis in domestic animals are common in bovines and felines, less common in ovines and equines, and rare in porcines and poultry (NAS, 1977). In practice, the most dangerous arsenic preparations are dips, herbicides, and defoliants in which the arsenical is in a highly soluble trivalent form, usually as trioxide or arsenite (Selby et al., 1977). Accidental poisoning of cattle with arsenicals, for example, is well documented. In one instance, more than 100 cattle died after accidental overdosing with arsenic trioxide applied topically to control lice. On necropsy, there were subcutaneous edematous swellings and petechial hemorrhages in the area of application, and histopathology of the intestine, mucosa, kidneys, and epidermis (Robertson et al., 1984). In Bangladesh, poisoned cattle showed depression, trembling, bloody diarrhea, restlessness, unsteady gait, stumbling, convulsions, groaning, shallow labored breathing, teeth grinding, and salivation (Samad and Chowdhury, 1984b). Cattle usually died 12 to 36 hr after the onset of signs; necropsy showed extensive submucosal hemorrhages of the gastrointestinal tract(Samad and Chowdhury, 1984a), and tissue residues > 10 mg/kg fresh weight in the liver and kidneys (Thatcher et al., 1985). It sometimes appears that animals, especially cattle, develop a preference for weeds sprayed with an arsenic weed killer, probably because arsenic compounds are salty and thus attractive to animals (Selby et al., 1977).

When extrapolating animal data from one species to another, the species tested must be considered. For example, the metabolism of arsenic in the rat (*Rattus* sp.) is unique, and very different from that in human beings and other animals. Rats store arsenic in blood hemoglobin, excreting it very slowly—unlike most mammals, which rapidly excrete ingested inorganic arsenic in the urine as methylated derivatives (NAS, 1977). Blood arsenic, whether given as As^{+3} or As^{+5}, rapidly clears from humans, mice, rabbits, dogs, and primates; half-life is 6 hr for the fast phase and about 60 hr for the slow phase (EPA, 1980). In rats, however, blood arsenic is mostly retained in erythrocytes, and clears slowly; half-life is 60 to 90 days (EPA, 1980). In rats, the excretion of arsenic into bile is 40 times faster than in rabbits and up to 800 times faster than in dogs (Pershagen and Vahter, 1979). Most researchers now agree that the rat is unsatisfactory for use in arsenic research (NAS, 1977; NRCC, 1978; Pershagen and Vahter, 1979; EPA, 1980; Webb et al., 1986).

Dimethylarsinic acid is the major metabolite of orally administered arsenic trioxide and is excreted rapidly in the urine (Yamauchi and Yamamura, 1985). The methylation process is true detoxification, since methanearsonates and cacodylates are about 200 times less toxic than sodium arsenite (NAS, 1977). The marmoset monkey (*Callithrix jacchus*), unlike all other animal species studied to date, was not able (for unknown reasons) to metabolize administered As^{+5} to demethylarsinic acid; most was reduced to As^{+3}. Only 20% of the total dose was excreted in urine as unchanged As^{+5}, and another 20% as As^{+3}. The rest was bound to tissues, giving distribution patterns similar to those of arsenite (Vahter and Marafante, 1985). Accordingly, the marmoset, like the rat, may be unsuitable for research with arsenicals.

Arsenicals were ineffective in controlling certain bacterial and viral infections. Mice experimentally infected with bacteria (*Klebsiella pneumoniae*) or viruses (pseudorabies, encephalitis, encephalmyocarditis) showed a significant increase in mortality when treated with large doses of arsenicals compared with nonarsenic-treated groups (NAS, 1977; Aranyi et al., 1985).

It has been suggested, but not yet verified, that many small mammals avoid arsenic-treated feeds and consume other foods if given the choice (NAS, 1977), and that cacodylic acid, which has negligible effects on wildlife, reduces species diversity due to selective destruction of vegetation (Hood, 1985). Both topics merit more research.

7. RECOMMENDATIONS

Numerous criteria for arsenic have been proposed to protect natural resources and human health (Table 7): But many authorities recognize that these criteria are not sufficient for adequate or (in some cases) reasonable protection, and that many additional data are required if meaningful standards are to be promulgated (NAS, 1977; NRCC, 1978; Pershagen and Vahter, 1979; EPA, 1980, 1985). Specifically, data are needed on the following subjects: (1) cancer incidence and other abnormalities in the natural resources of areas with elevated arsenic levels, and their relation to the potential carcinogenicity of arsenic compounds; (2) interaction effects of arsenic with other carcinogens, cocarcinogens, promoting agents, inhibitors, and common environmental contaminants; (3) controlled studies with aquatic and terrestrial indicator organisms on the physiological and biochemical effects of long-term, low-dose exposures to inorganic and organic arsenicals, including effects on reproduction and genetic makeup; (4) methodologies for establishing maximum permissible tissue concentrations of arsenic; (5) effects of arsenic in combination with infectious agents; (6) mechanisms of arsenical growth-promoting agents; (7) role of arsenic in nutrition; (8) extent of animal adaptation to arsenicals and the mechanisms of action; and (9) physicochemical processes influencing arsenic cycling. In addition, the following techniques and procedures should be developed and implemented: (1) more sophisticated measurements of the chemical forms of arsenic in plant and animal tissues, (2) correlation of biologically observable effects with particular chemical forms of arsenic, and (3) management of arsenical wastes that accommodates recycling, reuse, and long-term storage.

Some proposed arsenic criteria merit additional comment, such as those on aquatic life protection, levels in seafoods and drinking water, and use in food-producing animals as growth stimulants or for disease prevention and treatment.

For saltwater-life protection, the current water-quality criterion of 36 μg As^{+3}/L (EPA, 1985; Table 7) seems to offer a reasonable degree of safety; only a few species of algae show adverse effects at < 36 μg/L (e.g., reduced growth at 19

Table 7 Proposed Arsenic criteria for Protection of Selected Natural Resources and Human Health

Resource, Criterion, and Other Variables	Criterion or Effective Arsenic Concentration (reference)
AQUATIC LIFE	
Freshwater biota: medium concentrations	Four-day mean water concentration not to exceed 190 μg total recoverable inorganic As^{+3}/L more than once every 3 years; one-hour mean not to exceed 360 μg inorganic As^{+3}/L more than once every 3 years Insufficient data for criteria formulation for inorganic As^{+5}, or for any organoarsenical (EPA, 1985)
Freshwater biota: tissue residues	Diminished growth and survival reported in immature bluegills when total arsenic residues in muscle > 1.3 mg/kg fresh weight (FW) or > 5 mg/kg in adults (NRCC, 1978)
Saltwater biota: medium concentrations	Four-day average water concentration not to exceed 36 μg As^{+3}/L more than once every 3 years; one-hour mean not to exceed 69 μg As^{+3}/L more than once every 3 years; insufficient data for criteria formulation for inorganic As^{+5}, or for any organoarsenical (EPA, 1985)
Saltwater biota: tissue residues	Depending on chemical form of arsenic, certain marine teleosts may be unaffected at muscle total arsenic residues of 40 mg/kg FW (NRCC, 1978)
BIRDS	
Tissue residues	Residues, in mg total As/kg FW, in liver or kidney in the 2–10 range are considered elevated; residues > 10 are indicative of arsenic poisoning (Goede, 1985)
Mallard, *Anas platyrhynchos*	
Sodium arsenate in diet	Reduced growth in ducklings fed > 30 mg As/kg diet (Camardese et al., 1990)
Turkey, *Meleagris gallopavo*	
Arsanilic acid in diet	Maximum dietary concentration for turkeys < 28 days old is 300 to 400 mg/kg feed (NAS, 1977)
Phenylarsonic feed additives for disease control and improvement of weight gain in domestic poultry; safe dietary levels	Maximum levels in diets, in mg/kg feed, are 50 to 100 for arsanilic acid, 25 to 188 for 3-nitro-4-hydroxyphenylarsonic acid (for chickens and turkeys, not

Table 7 *(Continued)*

Resource, Criterion, and Other Variables	Criterion or Effective Arsenic Concentration (reference)
	recommended for ducks and geese), and 180 to 370 for others (NAS, 1977)
DOMESTIC LIVESTOCK	
Prescribed limits for arsenic in feedstuffs	
Straight feedstuffs, except those listed below	< 2 mg total As/kg FW (Vreman et al., 1986)
Meals from grass, dried lucerne, or dried clover	< 4 mg total As/kg FW (Vreman et al., 1986)
Phosphate mealstuffs	< 10 mg total As/kg FW (Vreman et al., 1986)
Fish meals	< 10 mg total As/kg FW (Vreman et al., 1986)
Tissue residues	
Poisoned	
Liver, kidney	5 to > 10 mg total As/kg FW (Thatcher et al., 1985; Vreman et al., 1986)
Normal, muscle	< 0.3 mg total As/kg FW (Veen and Vreman, 1985)
VEGETATION	
No observable effects	< 1 mg total water soluble soil As/L; < 25 mg total As/kg soil; < 3.9 μg As/m^3 air (NRCC, 1978)
HUMAN HEALTH	
Diet	
Permissible levels	
Total diet	< 0.5 As/kg dry weight diet (Sorensen et al., 1985)
Fruits, vegetables	Tolerance for arsenic residues as As_2O_3, resulting from pesticidal use of copper, magnesium, and sodium arsenates, is 3.5 mg/kg (Jelinek and Corneliussen, 1977)
Muscle of poultry and swine, eggs, swine edible byproducts	< 0.5 mg total As/kg FW (Jelinek and Corneliussen, 1977)
Edible byproducts of chickens and turkey, liver and kidney of swine	< 2 mg total As/kg FW (Jelinek and Corneliussen, 1977)
Seafood products	In Hong Kong, limited to 6 mg total As/kg FW for edible tissues of finfish, and 10 mg/kg for mollusks and crustaceans (Phillips et al., 1982); in Yugoslavia, these values are 2 for fish and 4 for mollusks and crustaceans (Ozretic et al., 1990)

Table 7 *(Continued)*

Resource, Criterion, and Other Variables	Criterion or Effective Arsenic Concentration (reference)
Adverse effects	
Consumption of aquatic organisms living in As-contaminated waters	
Cancer risk of	
10^{-5}	0.175 μg As/L (EPA, 1980)
10^{-6} [a]	0.0175 μg As/L (EPA, 1980)
10^{-7}	0.00175 μg As/L (EPA, 1980)
Drinking water	
Allowable concentrations	
Total arsenic	< 10 μg/L (NAS, 1977)
Total arsenic	50 μg/L (Pershagen and Vahter, 1979; EPA, 1980; Norin et al., 1985)
Adverse effects	
Cancer risk of	
10^{-5}	0.022 μg As/L (EPA, 1980)
10^{-6} [a]	0.0022 μg As/L (EPA, 1980)
10^{-7}	0.00022 μg As/L (EPA, 1980)
Symptoms of arsenic toxicity observed	9% incidence at 50 μg As/L, 16% at 50–100 μg As/L, and 44% at > 100 μg As/L (NRCC, 1978)
Harmful after prolonged consumption	> 50 to 960 μg As/L (NRCC, 1978)
Cancer	In Chile, cancer rate estimated at 0.01% at 82 μg As/L, 0.17% at 600 μg As/L (NRCC, 1978)
Skin Cancer	0.26% frequency at 290 μg/L and 2.14% at 600 μg/L (EPA, 1980)
Total intake	
No observable effect	
North America	0.007 to 0.06 mg As daily (Pershagen and Vahter, 1979)
Japan	0.07 to 0.17 mg As daily (Pershagen and Vahter, 1979)
USA	
1960s	0.05 to 0.1 mg As daily (Pershagen and Vahter, 1979)
1974	0.015 mg As daily (Pershagen and Vahter, 1979)
Canada	0.03 mg As daily (NRCC, 1978)
Netherlands	
Acceptable	2 μg total inorganic As/kg body weight (BW), or about 0.14 mg daily for 70-kg adult; 0.094 mg daily through fishery products (Vos and Hovens, 1986)

Table 7 *(Continued)*

Resource, Criterion, and Other Variables	Criterion or Effective Arsenic Concentration (reference)
Adverse effects (prolonged exposure)	
Subclinical symptoms	0.15 to 0.6 mg As daily (NRCC, 1978)
Intoxication	3 to 4 mg As daily (NRCC, 1978)
Blackfoot disease	Total dose of 20 g over several years increases prevalence of disease by 3% (Pershagen and Vahter, 1979)
Mild chronic poisoning	0.15 mg As daily or about 2 μg/kg BW daily (NRCC, 1978)
Chronic arsenicism	Lifetime cumulative absorption of 1 g As, or intake of 0.7 to 2.6 g/year for several years (in medications) can produce symptoms after latent period of 4 to 24 years (NRCC, 1978)
Tissue residues	
No observed effect levels	
Urine	< 0.05 mg As/L (NRCC, 1978)
Liver, kidney	< 0.5 mg As/kg FW (NRCC, 1978)
Blood	< 0.7 mg As/L (NRCC, 1978)
Hair	< 2 mg As/kg FW (NRCC, 1978)
Fingernail	< 5 mg As/kg FW (NRCC, 1978)
Arsenic-poisoned	
Liver, kidney	2 to 100 mg As/kg FW; confirmatory tests > 10 mg As/kg FW; residues in survivors several days later were 2 to 4 mg/kg FW (NAS, 1977)
Whole body	In child, symptoms of chronic arsenicism evident at 1 mg As/kg BW, equivalent to intake of about 10 mg/month for 3 months; for adults, values were 80 mg/kg BW, equivalent to about 2 g/year for 3 years (NRCC, 1978)
Air	
Allowable concentrations	
Arsine	< 200 μg/m³ for USA industrial workers; proposed mean arsine limit of < 4 μg/m³ in 8-hr period and < 10 μg/m³ maximum in 15 minutes (NAS, 1977)
Arsine	< 4 μg/m³ (NRCC, 1978)
Total As	< 3 μg/m³ in former USSR and Czechoslovakia; < 500 μg/m³ for USA industrial workers (NAS, 1977)
Total As (threshold limit value-time weighted average: 8 hr/day, 40 hr week)	Proposed limit of < 50 μg/m³, maximum of 2 μg/m³ in 15 minutes, < 10 μg airborne inorganic As/m³ (EPA, 1980)

Table 7 *(Continued)*

Resource, Criterion, and Other Variables	Criterion or Effective Arsenic Concentration (reference)
Arsenic trioxide	$<0.3\ \mu g/m^3$ in former USSR, $<0.1\ \mu g/m^3$ in USA (Nagymajtenyi et al., 1985)
Adverse effects	
Increased mortality	Associated with daily time-weighted average As exposure of $>3\ \mu g/m^3$ for one year (NRCC, 1978)
Respiratory cancer (increased risk)	Associated with chronic exposure $>3\ \mu g$ As/m^3, or occupational exposure (lifetime) of $>54.6\ \mu g\ As/m^3$ (NRCC, 1978)
Respiratory cancer (increased risk)	Exposure to $50\ \mu g\ As/m^3$ for more than 25 years associated with 3X increase (Pershagen and Vahter, 1979)
Skin diseases	Associated with ambient air concentrations of 60 to 13,000 $\mu g\ As/m^3$ (NRCC, 1978)
Dermatitis	Associated with ambient air concentrations of 300 to 81,500 $\mu g\ As/m^3$ (NRCC, 1978)

[a] One excess cancer per million population (10^{-6}) is estimated during lifetime exposure to 0.0022 μg arsenic per liter of drinking water, or to lifetime consumption of aquatic organisms residing in waters containing 0.0175 μg As/L (EPA, 1980).

to 22 $\mu g/L$). In 1980, this criterion was 508 $\mu g/L$ (EPA, 1980), about 14 times higher than the current criterion. The downward modification seems to be indicative of the increasingly stringent arsenic criteria formulated by regulatory agencies. The current criterion for freshwater-life protection of 190 $\mu g\ As^{+3}/L$ (EPA, 1985; Table 7), however, which is down from 440 $\mu g\ As^{+3}/L$ in 1980 (EPA, 1980), is unsatisfactory. Many species of freshwater biota are adversely affected at $<190\ \mu g/L$ of As^{+3}, As^{+5}, and various organoarsenicals (Table 4). These adverse effects include death and malformations of toad embryos at 40 $\mu g/L$, growth inhibition of algae at 48 to 75 $\mu g/L$, mortality of amphipods and gastropods at 85 to 88 $\mu g/L$, and behavioral impairment of goldfish (*Carassius auratus*) at 100 $\mu g/L$. A downward adjustment in the current freshwater aquatic-life protection criterion seems warranted.

In Hong Kong, permissible concentrations of arsenic in seafood destined for human consumption range from 6 to 10 mg/kg fresh weight (Table 7); however, these values are routinely exceeded in 22% of finfish, 20% of bivalve mollusks, 67% of gastropods, 29% of crabs, 21% of shrimp and prawns, and 100% of lobsters (Phillips et al., 1982). The highest arsenic concentrations recorded in Hong Kong seafood products were in gastropods (*Hemifusus* spp.), in which the concentrations of 152 to 176 mg/kg fresh weight were among the highest recorded in any species to date (Phillips et al., 1982). A similar situation exists in

Yugoslavia, where almost all seafoods exceed the upper limit prescribed by food quality regulations (Ozretic et al., 1990). Most of the arsenic in seafood products is usually arsenobetaine or some other comparatively harmless form. In effect, arsenic criteria for seafoods are neither enforced nor enforceable. Some toxicologists from the U.S. Food and Drug Administration believe that the average daily intake of arsenic in different foods does not pose a hazard to the consumer (Jelinek and Corneliussen, 1977).

For maximum protection of human health from the potential carcinogenic effects of exposure to arsenic through drinking water or contaminated aquatic organisms, the ambient water concentration should be zero, based on the nonthreshold assumption for arsenic. But zero level may not be attainable. Accordingly, the levels established are those that are estimated to increase cancer risk over a lifetime to only one additional case per 100,000 population. These values are estimated at 0.022 μg As/L for drinking water and 0.175 μg As/L for water containing edible aquatic resources (EPA, 1980; Table 7).

Various phenylarsonic acids—especially arsanilic acid, sodium arsanilate, and 3-nitro-4-hydroxyphenylarsonic acid—have been used as feed additives for disease control and for improvement of weight gain in swine and poultry for almost 40 years (NAS, 1977). The arsenic is present as As^{+5} and is rapidly excreted; present regulations require withdrawal of arsenical feed additives 5 days before slaughter for satisfactory depuration (NAS, 1977). Under these conditions, total arsenic residues in edible tissues do not exceed the maximum permissible limit of 2 mg/kg fresh weight (Jelinek and Corneliussen, 1977). Organoarsenicals probably will continue to be used as feed additives unless new evidence indicates otherwise.

8. CONCLUDING REMARKS

Arsenic is a teratogen and carcinogen that can traverse placental barriers and produce fetal death and malformations in many species of mammals. Although it is carcinogenic in humans, evidence for arsenic-induced carcinogenicity in other mammals is scarce. Paradoxically, evidence is accumulating that arsenic is nutritionally essential or beneficial. Arsenic deficiency effects, such as poor growth, reduced survival, and inhibited reproduction, have been recorded in mammals fed diets containing < 0.05 mg As/kg, but not in those fed diets with 0.35 mg As/kg. At comparatively low doses, arsenic stimulates growth and development in various species of plants and animals.

Most arsenic produced domestically is used in the manufacture of agricultural products such as insecticides, herbicides, fungicides, algicides, wood preservatives, and growth stimulants for plants and animals. Living resources are exposed to arsenic via atmospheric emissions from smelters, coal-fired power plants, and arsenical herbicide sprays; via water contaminated by mine tailings, smelter wastes, and natural mineralization; and via diet, especially consumption of marine biota. Arsenic concentrations are usually low (< 1.0 mg/kg fresh weight) in most living organisms, but they are elevated in marine biota (in which arsenic occurs as arsenobetaine and poses little risk to organisms or their consumer) and in plants and animals from areas that are naturally arseniferous or that are near industrial manufacturers and agricultural users of arsenicals. Arsenic is bioconcentrated by organisms but is not biomagnified in the food chain.

Arsenic exists in four oxidation states, as inorganic or organic forms. Its bioavailability and toxic properties are significantly modified by numerous biological and abiotic factors, including the physical and chemical forms of arsenic tested, the route of administration, the dose, and the species of animal. In general, inorganic arsenic compounds are more toxic than organic compounds, and trivalent species are more toxic than pentavalent species. Arsenic may be absorbed by ingestion, inhalation, or through permeation of the skin or mucous membranes; cells take up arsenic through an active transport system normally used in phosphate transport. The mechanisms of arsenic toxicity differ greatly among chemical species, although all appear to cause similar signs of poisoning. Biomethylation is the preferred detoxification mechanism for absorbed inorganic arsenicals; methylated arsenicals usually clear from tissues within a few days.

Episodes of arsenic poisoning are either acute or subacute; chronic cases of arsenosis are seldom encountered in any species except human beings. Single oral doses of arsenicals fatal to 50% of the sensitive species tested ranged from 17 to 48 mg/kg body weight in birds and from 2.5 to 33 mg/kg body weight in mammals. Susceptible species of mammals were adversely affected at chronic doses of 1 to 10 mg As/kg body weight, or 50 mg As/kg diet; mallard ducklings were negatively affected at dietary concentrations as low as 30 mg/kg ration. Sensitive aquatic species were damaged at water concentrations of 19 to 48 μg As/L (the U.S. Environmental Protection Agency drinking-water criterion for human health protection is 50 μg/L), 120 mg As/kg diet, or (in the case of freshwater fish) tissue residues > 1.3 mg/kg fresh weight. Adverse effects on crops and vegetation were recorded at 3 to 28 mg of water-soluble As/L (equivalent to about 25 to 85 mg total As/kg soil) and at atmospheric concentrations > 3.9 μg As/m^3.

Numerous and disparate arsenic criteria have been proposed for the protection of sensitive natural resources; however, the consensus is that many of these criteria are inadequate and that additional information is needed in at least five categories: (1) developing standardized procedures to permit correlation of biologically observable effects with particular chemical forms of arsenic; (2) conducting studies under controlled conditions with appropriate aquatic and terrestrial indicator organisms to determine the effects of chronic exposure to low doses of inorganic and organic arsenicals on reproduction, genetic makeup, adaptation, disease resistance, growth, and other variables; (3) measuring interaction effects of arsenic with other common environmental contaminants, including carcinogens, cocarcinogens, and promoting agents; (4) monitoring the incidence of cancer and other abnormalities in the natural resources of areas with relatively high arsenic levels, and correlating these with the possible carcinogenicity of arsenic compounds; and (5) developing appropriate models of arsenic cycling and budgets in natural ecosystems.

ACKNOWLEDGMENTS

I thank Marcia Holmes for secretarial help, and James R. Zuboy for editorial services.

REFERENCES

Aggett, J., and O'Brien, G. A. (1985). Detailed model for the mobility of arsenic in lacustrine sediments based on measurements in Lake Ohakuri. *Environ. Sci. Technol.* **19**, 231–238.

Aggett, J., and Roberts, L. S. (1986). Insight into the mechanism of accumulation of arsenate and phosphate in hydro lake sediments by measuring the rate of dissolution with ethylenediaminetetraacetic acid. *Environ. Sci. Technol.* **20**, 183–186.

Andreae, M. O. (1986). Organoarsenic compounds in the environment. In P. J. Craig (Ed.), *Organometallic Compounds in the Environment. Principles and Reactions.* Wiley, New York, pp. 198–228.

Aranyi, C., Bradof, J. N., O'Shea, W. J., Graham, J. A., and Miller, F. J. (1985). Effects of arsenic trioxide inhalation exposure on pulmonary antibacterial defenses in mice. *J. Toxicol. Environ. Health* **15**, 163–172.

Austin, L. S., and Millward, G. E. (1984). Modelling temporal variations in the global tropospheric arsenic burden. *Atmos. Environ.* **18**, 1909–1919.

Belton, J. C., Benson, N. C., Hanna, M. L., and Taylor, R. T. (1985). Growth inhibitory and cytotoxic effects of three arsenic compounds on cultured Chinese hamster ovary cells. *J. Environ. Sci. Health* **20A**, 37–72.

Benson, A. A. (1984). Phytoplankton solved the arsenate-phosphate problem. In O. Holm-Hansen, L. Bolis, and R. Gilles (Eds.), *Lecture Notes on Coastal and Estuarine Ecology. 8. Marine Phytoplankton and Productivity*. Springer-Verlag, Berlin, pp. 55–59.

Blus, L. J., Neely, B. S., Jr., Lamont, T. G., and Mulhern, B. (1977). Residues of organochlorines and heavy metals in tissues and eggs of brown pelicans 1969–73. *Pestic. Monit. J.* **11**, 40–53.

Bottino, N. R., Newman, R. D., Cox, E. R., Stockton, R., Hoban, M., Zingaro, R. A., and Irgolic, K. J. (1978). The effects of arsenate and arsenite on the growth and morphology of the marine unicellular algae *Tetraselmis chui* (Chlorophyta) and *Hymenomonas carterae* (Chrysophyta). *J. Exp. Mar. Biol. Ecol.* **33**, 153–168.

Bryant, V., Newbery, D. M., McLusky, D. S., and Campbell, R. (1985). Effect of temperature and salinity on the toxicity of arsenic to three estuarine invertebrates (*Corophium volutator, Macoma balthica, Tubifex costatus*). *Mar. Ecol.: Prog. Ser.* **24**, 129–137.

Camardese, M. B., Hoffman, D. J., LeCaptain, L. J., and Pendleton, G. W. (1990). Effects of arsenate on growth and physiology in mallard ducklings. *Environ. Toxicol. Chem.* **9**, 785–795.

Charbonneau, S. M., Spencer, K., Bryce, F., and Sandi, E. (1978). Arsenic excretion by monkeys dosed with arsenic-containing fish or with inorganic arsenic. *Bull. Environ. Contam. Toxicol.* **20**, 470–477.

Cockell, K. A., and Hilton, J. W. (1985). Chronic toxicity of dietary inorganic and organic arsenicals to rainbow trout (*Salmo gairdneri* R.). *Fed. Proc., Fed. Am. Soc. Exp. Biol.* **44**(4), 938.

Deknudt, G., Leonard, A., Arany, J., Du Buisson, G. J., and Delavignette, E. (1986). *In vivo* studies in male mice on the mutagenic effects of inorganic arsenic. *Mutagenesis* **1**, 33–34.

Denham, D. A., Oxenham, S. L., Midwinter, I., and Friedheim, E. A. H. (1986). The antifilarial activity of a novel group of organic arsenicals upon *Brugia pahangi*. *J. Helminthol.* **60**, 169–172.

Dudas, M. J. (1984). Enriched levels of arsenic in post-active acid sulfate soils in Alberta. *Soil Sci. Soc. Am. J.* **48**, 1451–1452.

Edmonds, J. S., and Francesconi, K. A. (1987). Trimethylarsine oxide in estuary catfish (*Cnidoglanis macrocephalus*) and school whiting (*Sillago bassensis*) after oral administration of sodium arsenate; and as a natural component of estuary catfish. *Sci. Total Environ.* **64**, 317–323.

Eisler, R. (1981). *Trace Metal Concentrations in Marine Organisms.* Pergamon, New York.

Eisler, R. (1988). Arsenic hazards to fish, wildlife, and invertebrates: A synoptic review. *U.S. Fish Wildl. Serv., Biol. Rep.* **85**(1.12), 1–92.

Environmental Protection Agency (EPA) (1980). Ambient water quality criteria for arsenic. *U.S. Environ. Prot. Agency Rep.* **440/5-80-021**, 1–205.

Environmental Protection Agency (EPA) (1985). Ambient water quality criteria for arsenic—1984. *U.S. Environ. Prot. Agency Rep.* **440/5-84-033**, 1–66.

Farmer, J. G., and Lovell, M. A. (1986). Natural enrichment of arsenic in Loch Lomond sediments. *Geochim. Cosmochim. Acta* **50**, 2059–2067.

Ferm, V. H., and Hanlon, D. P. (1985). Constant rate exposure of pregnant hamsters to arsenate during early gestation. *Environ. Res.* **37**, 425–432.

Ferm, V. H., and Hanlon, D. P. (1986). Arsenate-induced neural tube defects not influenced by constant rate administration of folic acid. *Pediatr. Res.* **20**, 761–762.

Fischer, A. B., Buchet, J. P., and Lauwerys, R. R. (1985). Arsenic uptake, cytotoxicity and detoxification studied in mammalian cells in culture. *Arch. Toxicol.* **57**, 168–172.

Fowler, S. W., and Unlu, M. Y. (1978). Factors affecting bioaccumulation and elimination of arsenic in the shrimp *Lysmata seticaudata. Chemosphere* **9**, 711–720.

Francesconi, K. A., Micks, P., Stockton, R. A., and Irgolic, K. J. (1985). Quantitative determination of arsenobetaine, the major water-soluble arsenical in three species of crab, using high pressure liquid chromatography and an inductively coupled argon plasma emission spectrometer as the arsenic-specific detector. *Chemosphere* **14**, 1443–1453.

Freeman, M. C. (1985). The reduction of arsenate to arsenite by an *Anabaena*-bacteria assemblage isolated from the Waikato River. *N.Z. J. Mar. Freshwater Res.* **19**, 277–282.

Freeman, M. C., Aggett, J., and O'Brien, G. (1986). Microbial transformations of arsenic in Lake Ohakuri, New Zealand. *Water Res.* **20**, 283–294.

Froelich, P. N., Kaul, L. W., Byrd, J. T., Andreae, M. O., and Roe, K. K. (1985). Arsenic, barium, germanium, tin, dimethylsulfide and nutrient biogeochemistry in Charlotte Harbor, Florida, a phosphorus-enriched estuary. *Estuarine Coastal Shelf Sci.* **20**, 239–264.

Fullmer, C. S., and Wasserman, R. H. (1985). Intestinal absorption of arsenate in the chick. *Environ. Res.* **36**, 206–217.

Goede, A. A. (1985). Mercury, selenium, arsenic and zinc in waders from the Dutch Wadden Sea. *Environ. Pollut.* **37A**, 287–309.

Hall, R. A., Zook, E. G., and Meaburn, G. M. (1978). National Marine Fisheries Service survey of trace elements in the fishery resources. *NOAA Tech. Rep.* **NMFS SSRF-721**, 1–313.

Hall, R. J. (1980). Effects of environmental contaminants on reptiles: A review. *U.S. Fish Wildl. Serv., Spec. Sci. Rep.: Wildl.* **228**, 1–12.

Hallacher, L. E., Kho, E. B., Bernard, N. D., Orcutt, A. M., Dudley, Jr. W. C., and Hammond, T. M. (1985). Distribution of arsenic in the sediments and biota of Hilo Bay, Hawaii. *Pac. Sci.* **39**, 266–273.

Hanaoka, K., and Tagawa, S. (1985a). Isolation and identification of arsenobetaine as a major water soluble arsenic compound from muscle of blue pointer *Isurus oxyrhincus* and whitetip shark *Carcharhinus longimanus. Bull. Jpn. Soc. Sci. Fish.* **51**, 681–685.

Hanaoka, K., and Tagawa, S. (1985b). Identification of arsenobetaine in muscle of roundnose flounder *Eopsetta grigorjewi. Bull. Jpn. Soc. Sci. Fish.* **51**, 1203.

Hanlon, D. P., and Ferm, V. H. (1986a). Teratogen concentration changes as the basis of the heat stress enhancement of arsenite teratogenesis in hamsters. *Teratology* **34**, 189–193.

Hanlon, D. P., and Ferm, V. H. (1986b). Concentration and chemical status of arsenic in the blood of pregnant hamsters during critical embryogenesis. 1. Subchronic exposure to arsenate using constant rate administration. *Environ. Res.* **40**, 372–379.

Hanlon, D. P., and Ferm, V. H. (1986c). Concentration and chemical status of arsenic in the blood of pregnant hamsters during critical embryogenesis. 2. Acute exposure. *Environ. Res.* **40**, 380–390.

Haswell, S. J., O'Neill, P., and Bancroft, K. C. (1985). Arsenic speciation in soil-pore waters from mineralized and unmineralized areas of south-west England. *Talanta* **32**, 69–72.

Hood, R. D. (1985). *Cacodylic Acid: Agricultural Uses, Biologic Effects, and Environmental Fate,* VA Monogr. Superintendent of Documents, U.S. Govt. Printing Office, Washington, DC.

Hood, R. D., Vedel-Macrander, G. C., Zaworotko, M. J., Tatum, F. M., and Meeks, R. G. (1987). Distribution, metabolism, and fetal uptake of pentavalent arsenic in pregnant mice following oral or intraperitoneal administration. *Teratology* **35**, 19–25.

Howard, A. G., Arbab-Zavar, M. H., and Apte, S. (1984). The behaviour of dissolved arsenic in the estuary of the River Beaulieu. *Estuarine, Coastal Shelf Sci.* **19**, 493–504.

Hudson, R. H., Tucker, R. K., and Haegèle, M. A. (1984). Handbook of toxicity of pesticides to wildlife. *U.S. Fish Wildl. Serv., Resour. Publ.* **153**, 1–90.

Irgolic, K. J., Woolson, E. A., Stockton, R. A., Newman, R. D., Bottino, N. R., Zingaro, R. A., Kearney, P. C., Pyles, R. A., Maeda, S., McShane, W. J., and Cox, E. R. (1977). Characterization of arsenic compounds formed by *Daphnia magna* and *Tetraselmis chuii* from inorganic arsenate. *Environ. Health Perspect.* **19**, 61–66.

Jauge, P., and Del-Razo, L. M. (1985). Uric acid levels in plasma and urine in rats chronically exposed to inorganic As(III) and As(V). *Toxicol. Lett.* **26**, 31–35.

Jelinek, C., and Corneliussen, P. E. (1977). Levels of arsenic in the United States food supply. *Environ. Health Perspect.* **19**, 83–87.

Jenkins, D. W. (1980). Biological monitoring of toxic trace metals. Vol. 2. Toxic trace metals in plants and animals of the world. Part I. *U.S. Environ. Prot. Agency Rep.* **600/3-80-090**, 30–138.

Johns, C., and Luoma, S. N. (1990). Arsenic in benthic bivalves of San Francisco Bay and the Sacramento/San Joaquin River delta. *Sci. Total Environ.* **97/98**, 673–684.

Johnson, D. L. (1972). Bacterial reduction of arsenate in seawater. *Nature (London)* **240**, 44–45.

Johnson, D. L., and Burke, R. M. (1978). Biological mediation of chemical speciation. II. Arsenate reduction during marine phytoplankton blooms. *Chemosphere* **8**, 645–648.

Johnson, W., and Finley, M. T. (1980). Handbook of acute toxicity of chemicals to fish and aquatic invertebrates. *U.S. Fish Wildl. Serv. Resour. Publ.* **137**, 1–98.

Jongen, W. M. F., Cardinaals, J. M., and Bos, P. M. J. (1985). Genotoxocity testing of arsenobetaine, the predominant form of arsenic in marine fishery products. *Food Chem. Toxicol.* **23**, 669–673.

Kaise, T., Watanabe, S., and Itoh, K. (1985). The acute toxicity of arsenobetaine. *Chemosphere* **14**, 1327–1332.

Knowles, F. C., and Benson, A. A. (1984a). The mode of action of arsenical herbicides and drugs. *Z. Gesamte Hyg. Ihre Grenzgeb.* **30**, 407–408.

Knowles, F. C., and Benson, A. A. (1984b). The enzyme inhibitory form of inorganic arsenic. *Z. Gesamte Hyg. Ihre Grenzgeb.* **30**, 625–626.

Lee, T. C., Oshimura, M., and Barrett, J. C. (1985). Comparison of arsenic-induced cell transformation, cytotoxicity, mutation and cytogenetic effects in Syrian hamster embryo cells in culture. *Carcinogenesis* (*London*) **6**, 1421–1426.

Lee, T. C., Lee, K. C. C., Chang, C., and Jwo, W. L. (1986a). Cell-cycle dependence of the cytotoxicity and clastogenicity of sodium arsenate in Chinese hamster ovary cells. *Bull. Inst. Zool., Acad. Sin.* **25**, 91–97.

Lee, T. C., Wang-Wuu, S., Huang, R. Y., Lee, K. C. C., and Jan, K. Y. (1986b). Differential effects of pre- and posttreatment of sodium arsenite on the genotoxicity of methyl methanesulfonate in Chinese hamster ovary cells. *Cancer Res.* **43**, 1854–1857.

Lima, A. R., Curtis, C., Hammermeister, D. E., Markee, T. P., Northcott, C. E., and Brooke, L. T. (1984). Acute and chronic toxicities of arsenic(III) to fathead minnows, flagfish, daphnids, and an amphipod. *Arch. Environ. Contam. Toxicol.* **13**, 595–601.

Lunde, G. (1973). The synthesis of fat and water soluble arseno organic compounds in marine and limnetic algae. *Acta Chem. Scand.* **27**, 1586–1594.

Lunde, G. (1977). Occurrence and transformation of arsenic in the marine environment. *Environ. Health Perspect.* **19**, 47–52.

Maeda, S., Nakashima, S., Takeshita, T., and Higashi, S. (1985). Bioaccumulation of arsenic by freshwater algae and the application to the removal of inorganic arsenic from an aqueous phase. Part II. By *Chlorella vulgaris* isolated from arsenic-polluted environment. *Sep. Sci. Technol.* **20**, 153–161.

Maher, W. A. (1985a). Arsenic in coastal waters of South Australia. *Water Res.* **19**, 933–934.

Maher, W. A. (1985b). The presence of arsenobetaine in marine animals. *Comp. Biochem. Physiol.* **80C**, 199–201.

Mankovska, B. (1986). Accumulation of As, Sb, S, and Pb in soil and pine forest. *Ekologia (CSSR)* **5**, 71–79.

Marafante, E., Vahter, M., and Envall, J. (1985). The role of the methylation in the detoxication of arsenate in the rabbit. *Chem. -Biol. Interact.* **56**, 225–238.

Marques, I. A., and Anderson, L. E. (1986). Effects of arsenite, sulfite, and sulfate on photosynthetic carbon metabolism in isolated pea (*Pisum sativum* L., cv Little Marvel) chloroplasts. *Plant Physiol.* **82**, 488–493.

Matsuto, S., Kasuga, H., Okumoto, H., and Takahashi, A. (1984). Accumulation of arsenic in blue-green alga, *Phormidium* sp. *Comp. Biochem. Physiol.* **78C**, 377–382.

Matsuto, S., Stockton, R. A., and Irgolic, K. J. (1986). Arsenobetaine in the red crab, *Chionoecetes opilio*. *Sci. Total Environ.* **48**, 133–140.

Merry, R. H., Tiller, K. G., and Alston, A. M. (1986). The effects of contamination of soil with copper, lead and arsenic on the growth and composition of plants. I. Effects of season, genotype, soil temperature and fertilizers. *Plant Soil* **91**, 115–128.

Michnowicz, C. J., and Weaks, T. E. (1984). Effects of pH on toxicity of As, Cr, Cu, Ni, and Zn to *Selenastrum capricornutum* Printz. *Hydrobiologia* **118**, 299–305.

Nagymajtenyi, L., Selypes, A., and Berencsi, G. (1985). Chromosomal aberrations and fetotoxic effects of atmospheric arsenic exposure in mice. *J. Appl. Toxicol.* **5**, 61–63.

Naqvi, S. M., and Flagge, C. T. (1990). Chronic effects of arsenic on American red crayfish, *Procambarus clarkii*, exposed to monosodium methanearsonate (MSMA) herbicide. *Bull. Environ. Contam. Toxicol.* **45**, 101–106.

Naqvi, S. M., Flagge, C. T., and Hawkins, R. L. (1990). Arsenic uptake and depuration by red crayfish, *Procambarus clarkii*, exposed to various concentrations of monosodium methanearsonate (MSMA) herbicide. *Bull. Environ. Contam. Toxicol.* **45**, 94–100.

National Academy of Sciences (NAS) (1977). *Arsenic*. NAS, Washington, DC.

National Research Council of Canada (NRCC) (1978). Effects of arsenic in the Canadian environment. *Natl. Res. Counc. Can. Publ.* **NRCC 15391**, 1–349.

Nevill, E. M. (1985). The effect of arsenical dips on *Parafilaria bovicola* in artificially infected cattle in South Africa. *Onderstepoort J. Vet. Res.* **52**, 221–225.

Norin, H., Vahter, M., Christakopoulos, A., and Sandström, M. (1985). Concentration of inorganic and total arsenic in fish from industrially polluted water. *Chemosphere* **14**, 325–334.

Nyström, R. R. (1984). Cytological changes occurring in the liver of coturnix quail with an acute arsenic exposure. *Drug Chem. Toxicol.* **7**, 587–594.

Ozretic, B., Krajnovic-Ozretic, M., Santin, J., Medjugorac, B., and Kras, M. (1990). As, Cd, Pd, and Hg in benthic animals from the Kvarner-Rijeka Bay region, Yugoslovia. *Mar. Pollut. Bull.* **21**, 595–597.

Passino, D. R. M., and Novak, A. J. (1984). Toxicity of arsenate and DDT to the cladoceran *Bosmina longirostris*. *Bull. Environ. Contam. Toxicol.* **33**, 325–329.

Pershagen, G., and Bjorklund, N. E. (1985). On the pulmonary tumorogenicity of arsenic trisulfide and calcium arsenate in hamsters. *Cancer Lett.* **27**, 99–104.

Pershagen, G., and Vahter, M. (1979). *Arsenic–A Toxicological and Epidemiological Appraisal*, Naturvardsverket Rapp. SNV PM 1128. Liber Tryck, Stockholm.

Phillips, D. J. H. (1990). Arsenic in aquatic organisms: A review, emphasizing chemical speciation. *Aquat. Toxicol.* **16**, 151–186.

Phillips, D. J. H., and Depledge, M. H. (1986). Chemical forms of arsenic in marine organisms, with emphasis on *Hemifusus* species. *Water Sci. Technol.* **18**, 213–222.

Phillips, D. J. H., Thompson, G. B., Gabuji, K. M., and Ho, C. T. (1982). Trace metals of toxicological significance to man in Hong Kong seafood. *Environ. Pollut.* **3B**, 27–45.

Ramelow, G. J., Webre, C. L., Mueller, C. S., Beck, J. N., Young, J. C., and Langley, M. P. (1989). Variations of heavy metals and arsenic in fish and other organisms from the Calcasieu River and Lake, Louisiana. *Arch. Environ. Contam. Toxicol.* **18**, 804–818.

Reinke, J., Uthe, J. F., Freeman, H. C., and Johnston, J. R. (1975). The determination of arsenite and arsenate ions in fish and shellfish by selective extraction and polarography. *Environ. Lett.* **8**, 371–380.

Robertson, I. D., Harms, W. E., and Ketterer, P. J. (1984). Accidental arsenical toxicity of cattle. *Aust. Vet. J.* **61**, 366–367.

Robertson, J. L., and McLean, J. A. (1985). Correspondence of the LC50 for arsenic trioxide in a diet-incorporation experiment with the quantity of arsenic ingested as measured by X-ray, energy-dispersive spectrometry. *J. Econ. Entomol.* **78**, 1035–1036.

Rosemarin, A., Notini, M., and Holmgren, K. (1985). The fate of arsenic in the Baltic Sea *Fucus vesiculosus* ecosystem. *Ambio* **14**, 342–345.

Sadiq, M. (1986). Solubility relationships of arsenic in calcareous soils and its uptake by corn. *Plant Soil* **91**, 241–248.

Samad, M. A., and Chowdhury, A. (1984a). Clinical cases of arsenic poisoning in arsenic trioxide in a diet-incorporation experiment with the quantity of arsenic ingested as measured by X-ray, energy-dispersive spectrometry. *J. Econ. Entomol.* **78**, 1035–1036.

Samad, M. A., and Chowdhury, A. (1984b). Clinical cases of arsenic poisoning in cattle. *Indian J. Vet. Med.* **4**, 107–108.

Sanders, J. G. (1980). Arsenic cycling in marine systems. *Mar. Environ. Res.* **3**, 257–266.

Sanders, J. G. (1985). Arsenic geochemistry in Chesapeake Bay: Dependence upon anthropogenic inputs and phytoplankton species composition. *Mar. Chem.* **17**, 329–340.

Sanders, J. G. (1986). Direct and indirect effects of arsenic on the survival and fecundity of estuarine zooplankton. *Can. J. Fish. Aquat. Sci.* **43**, 694–699.

Sanders, J. G., and Cibik, S. J. (1985). Adaptive behavior of euryhaline phytoplankton communities to arsenic stress. *Mar. Ecol.: Prog. Ser.* **22**, 199–205.

Sanders, J. G., and Osman, R. W. (1985). Arsenic incorporation in a salt marsh ecosystem. *Estuarine, Coastal Shelf Sci.* **20**, 387–392.

Schmitt, C. J., and Brumbaugh, W. G. (1990). National Contaminant Biomonitoring Program: Concentrations of arsenic, cadmium, copper, lead, mercury, selenium, and zinc in U.S. freshwater fish, 1976–1984. *Arch. Environ. Contam. Toxicol.* **19**, 731–747.

Selby, L. A., Case, A. A., Osweiler, G. D., and Hages, H. M., Jr. (1977). Epidemiology and toxicology of arsenic poisoning in domestic animals. *Environ. Health Perspect.* **19**,183–189.

Sheppard, M. I., Thibault, D. H., and Sheppard, S. C. (1985). Concentrations and concentration ratios of U, As and Co in Scots Pine grown in a waste-site soil and an experimentally contaminated soil. *Water, Air, Soil Pollut.* **26**, 85–94.

Shiomi, K., Shinagawa, A, Hirota, K., Yamanaka, H., and Kikuchi, T. (1984a). Identification of arsenobetaine as a major arsenic compound in the ivory shell *Buccinum striatissimum. Agric. Biol. Chem.* **48**, 2863–2864.

Shiomi, K., Shinagawa, A., Igarashi, T., Yamanaka, H., and Kikuchi, T. (1984b). Evidence for the presence of arsenobetaine as a major arsenic compound in the shrimp *Sergestes lucens*. *Experientia* **40**, 1247–1248.

Sloot, H. A., Hoede, D., Wijkstra, J., Duinker, J. C., and Nolting, R. F. (1985). Anionic species of V, As, Se, Mo, Sb, Te and W in the Scheldt and Rhine estuaries and the southern bight (North Sea). *Estuarine, Coastal Shelf Sci.* **21**, 633–651.

Smiley, R. W., Fowler, M. C., and O'Knefski, R. C. (1985). Arsenate herbicide stress and incidence of summer patch on Kentucky bluegrass turfs. *Plant Dis.* **69**, 44–48.

Smith, R. A., Alexander, R. B., and Wolman, M. G. (1987). Water-quality trends in the Nation's rivers. *Science* **235**, 1607–1615.

Sorensen, E. M. B., Mitchell, R. R., Pradzynski, A., Bayer, T. L., and Wenz, L. L. (1985). Stereological analyses of hepatocyte changes parallel arsenic accumulation in the livers of green sunfish. *J. Environ. Pathol. Toxicol. Oncol.* **6**, 195–210.

Spehar, R. L., Fiandt, J. T., Anderson, R. L., and DeFoe, D. L. (1980). Comparative toxicity of arsenic compounds and their accumulation in invertebrates and fish. *Arch. Environ. Contam. Toxicol.* **9**, 53–63.

Stine, E. R., Hsu, C. A., Hòovers, T. D., Aposhian, H. V., and Carter, D. E. (1984). *N*-(2,3-dimercaptopropyl) phthalamidic acid: Protection, *in vivo* and *in vitro*, against arsenic intoxication. *Toxicol. Appl. Pharmacol.* **75**, 329–336.

Takamatsu, T., Kawashima, M., and Koyama, M. (1985). The role of Mn^{2+} -rich hydrous manganese oxide in the accumulation of arsenic in lake sediments. *Water Res.* **19**, 1029–1032.

Taylor, D., Maddock, B. G., and Mance, G. (1985). The acute toxicity of nine "grey list" metals (arsenic, boron, chromium, copper, lead, nickel, tin, vanadium and zinc) to two marine fish species: Dab (*Limanda limanda*) and grey mullet (*Chelon labrosus*). *Aquat. Toxicol.* **7**, 135–144.

Thanabalasingam, P., and Pickering, W. F. (1986). Arsenic sorption by humic acids. *Environ. Pollut.* **12B**, 233–246.

Thatcher, C. D., Meldrum, J. B., Wikse, S. E., and Whittier, W. D. (1985). Arsenic toxicosis and suspected chromium toxicosis in a herd of cattle. *J. Am. Vet. Med. Assoc.* **187**, 179–182.

Thursby, G. B., and Steele, R. L. (1984). Toxicity of arsenite and arsenate to the marine macroalgae *Champia parvula* (Rhodophyta). *Environ. Toxicol. Chem.* **3**, 391–397.

Tremblay, G. H., and Gobeil, C. (1990). Dissolved arsenic in the St. Lawrence estuary and the Saguenay Fjord, Canada. *Mar. Pollut. Bull.* **21**, 465–468.

Unlu, M. Y., and Fowler, S. W. (1979). Factors affecting the flux of arsenic through the mussel *Mytilus galloprovincialis*. *Mar. Biol.* **51**, 209–219.

Vahter, M., and Marafante, E. (1985). Reduction and binding of arsenate in marmoset monkeys. *Arch. Toxicol.* **57**, 119–124.

Veen, N. G., and Vreman, K. (1985). Transfer of cadmium, lead, mercury and arsenic from feed into various organs and tissues of fattening lambs. *Neth. J. Agric. Sci.* **34**, 145–153.

Vos, G., and Hovens, J. P. C. (1986). Chromium, nickel, copper, zinc, arsenic, selenium, cadmium, mercury and lead in Dutch fishery products 1977–1984. *Sci. Total Environ.* **52**, 25–40.

Vreman, K., Veen, N. G., Molen, E. J., and de Ruig, W. B. (1986). Transfer of cadmium, lead, mercury and arsenic from feed into milk and various tissues of dairy cows: Chemical and pathological data. *Neth. J. Agric. Sci.* **34**, 129–144.

Wang, D. S., Weaver, R. W., and Melton, J. R. (1984). Microbial decomposition of plant tissue contaminated with arsenic and mercury. *Environ. Pollut.* **34A**, 275–282.

Webb, D. R., Wilson, S. E., and Carter, D. E. (1986). Comparative pulmonary toxicity of gallium arsenide, gallium(III) oxide, or arsenic(III) oxide intratracheally instilled into rats. *Toxicol. Appl. Pharmacol.* **82**, 405–416.

Wells, J. M., and Richardson, D. H. S. (1985). Anion accumulation by the moss *Hylocomium splendens*: Uptake and competition studies involving arsenate, selenate, selenite, phosphate, sulphate and sulphite. *New Phytol.* **101**, 571–583.

White, D. H, King, K. A., and Prouty, R. M. (1980). Significance of organochlorine and heavy metal residues in wintering shorebirds at Corpus Christi, Texas, 1976–77. *Pestic. Monit J.* **14**, 58–63.

Wiemeyer, S. N., Lamont, T. G., and Locke, L. N. (1980). Residues of environmental pollutants and necropsy data for eastern United States ospreys, 1964–1973. *Estuaries* **3**, 155–167.

Wiener, J. G., Jackson, G. A., May, T. W., and Cole, B. P. (1984). Longitudinal distribution of trace elements (As, Cd, Cr, Hg, Pb, and Se) in fishes sediments in the upper Mississippi River. In J. G. Wiener, R. V. Anderson, and D. R. McConville (Eds.), *Contaminants in Upper Mississippi River*. Butterworth, Stoneham, MA, pp. 139–170.

Windom, H., Stickney, R., Smith, R, White, D., and Taylor, F. (1973). Arsenic, cadmium, copper, mercury, and zinc in some species of North Atlantic finfish. *J. Fish. Res. Board Can.* **30**, 275–279.

Woolson, E. A. (1975). Arsenical pesticides. *ACS Ser.* **7**, 1–176.

Wrench, J., Fowler, S. W., and Unlu, M. Y. (1979). Arsenic metabolism in a marine food chain. *Mar. Pollut. Bull.* **10**, 18–20.

Yamaoka, Y., and Takimura, O. (1986). Marine algae resistant to inorganic arsenic. *Agric. Biol. Chem.* **50**, 185–186.

Yamauchi, H., and Yamamura, Y. (1985). Metabolism and excretion of orally administered arsenic trioxide in the hamster. *Toxicology* **34**, 113–121.

Yamauchi, H., Takahashi, K., and Yamamura, Y. (1986). Metabolism and excretion of orally and intraperitoneally administered gallium arsenide in the hamster. *Toxicology* **40**, 237–246.

Zaroogian, G. E., and Hoffman, G. L. (1982). Arsenic uptake and loss in the American oyster, *Crassostrea virginica. Environ. Monit. Assess.* **1**, 345–358.

12

ARSENIC IN MARINE ORGANISMS: CHEMICAL FORMS AND TOXICOLOGICAL ASPECTS

Kazuo Shiomi

Tokyo University of Fisheries, Konan, Minato-ku, Tokyo 108, Japan

Arsenic in the Environment, Part II: Human Health and Ecosystem Effects,
Edited by Jerome O. Nriagu.
ISBN 0-471-30436-0

1. INTRODUCTION

Arsenic is an interesting element in that it has long affected human life in two contradictory ways: it is an essential as well as a toxic element. It was widely used as a tonic in the eighteenth and nineteenth centuries. The use of several arsenic compounds as insecticides and herbicides has continued from ancient times to the present. Gallium arsenide is now expected to become a new semiconductor material because it is superior to silicon in several points. In spite of arsenic's excellent properties, its high toxicity has given it a bad reputation. During the Middle Ages, arsenic compounds such as arsenic trioxide were often used as homicidal and suicidal agents. In modern times, arsenic has become associated with occupational hazards to workers at smelters and environmental pollution from smelters and insecticides. Incidents of severe arsenic poisoning via food have also occurred. The Japanese have experienced three cases of large-scale poisoning due to accidental arsenic contamination during food processing. In the two cases that occurred in Tsu-city, Mie Prefecture, in September 1948, and outside Ube-city, Yamaguchi Prefecture, from December 1955 to January 1956, soy sauce was the contaminated food. The other one, which involved ingestion of powdered milk, occurred in the western part of Japan, especially Okayama Prefecture, from June to July 1955. In this incident, all of the victims were babies; more than 12,000 victims and as many as 130 deaths were recorded.

Arsenic is ubiquitous in the natural environment and its concentrations in marine organisms are especially high, independent of pollution. In the first published scientific paper on arsenic, by Chapman (1926), arsenic concentrations were reported to be 3 to 10 ppm in oysters and 174 ppm in shrimp. According to the extensive work performed by Lunde (1967, 1968a, 1970a, b, 1972, 1973a, b, 1977), who used a variety of animals and plants, terrestrial organisms generally contain arsenic at the ppb level, while marine organisms such as crustaceans and seaweeds contain arsenic at the ppm level. The high level of arsenic in marine organisms has since been pointed out by many researchers. For example, an extremely high arsenic content of 340 ppm was observed in the midgut gland of a certain gastropod (Shiomi et al., 1987a). It is assumed that the powdered milk poisoning described involved arsenic in the inorganic form at around 20 to 30 ppm. People such as the Japanese who daily consume a lot of marine organisms as food may therefore be exposed to the risk of acute or chronic arsenic poisoning. However, there have been no reports of arsenic poisoning by ingestion of marine organisms, suggesting that the arsenic in marine organisms exists in much less toxic forms than the inorganic arsenic in the powdered milk. Is the arsenic in marine organisms really not hazardous to our health? To answer this question, it is essential to identify the chemical forms of arsenic contained in marine organisms and to determine their toxicity and metabolism. Much information has been accumulated in the last two decades. The present paper reviews the chemical forms and toxicological aspects of arsenic contained in marine organisms, especially from the viewpoint of food hygiene.

2. CHEMICAL FORMS

2.1. Organic or Inorganic?

Inorganic arsenic is said to be more toxic than organic arsenic. Therefore, whether arsenic in marine organisms exists in organic or inorganic forms is significant. Lunde (1969, 1973a, b) first determined inorganic and organic arsenic in some marine animals by means of ion-exchange chromatography and distillation in hydrochloric acid and found that most of the arsenic is in the organic form. This finding was later confirmed in a variety of marine organisms by Flanjak (1982), using the method of Lunde, and by Reinke et al. (1975), Maher (1983, 1985b), and Shinagawa et al. (1983), who used selective extraction methods. The results that Shinagawa et al. (1983) obtained with 20 species of fish, invertebrates, and brown algae are shown in Table 1. Total arsenic concentrations are high, while the concentrations of inorganic arsenic are insignificant in all organisms but the brown alga *Hizikia fusiforme*, in which inorganic arsenic accounts for as much as 60% of the total arsenic. Some species, such as *Limanda herzensteini* and *Penaeus orientalis*, contain no detectable inorganic arsenic.

Edmonds and Francesconi (1977), Yamauchi and Yamamura (1980), and Kaise et al. (1988a, b) determined inorganic arsenic and methylated arsenic in marine organisms by the hydride generation technique, in which arsines produced by alkaline treatment, followed by reduction with sodium borohydride, are analyzed by gas chromatography (GC) or GC–mass spectrometry (MS). These investigators also noted the low concentration of inorganic arsenic in marine organisms compared to the total arsenic. The significance of their studies lies in the finding that most of the arsenic in marine organisms exists in the methylated form; the major arsenicals in animals and seaweeds are trimethylated and dimethylated compounds, respectively. This finding agrees with the current knowledge that the major arsenic compound in marine animals is arsenobetaine, a trimethylated compound, whereas the compounds in seaweeds are arsenosugars, which are dimethylated compounds.

2.2. Water-Soluble or Lipid-Soluble?

Lunde (1968a, 1969, 1973a, 1977) has shown that arsenic compounds in marine organisms are largely water-soluble. Kobayashi et al. (1980) extracted the muscles of several species of fish with Tris-HCl buffer, 5% perchloric acid, or ether and determined the arsenic in each extract. Especially high amounts of arsenic were found in the perchloric acid extract containing water-soluble, low-molecular-weight compounds. It is interesting to note with respect to this result that heavy metals such as mercury and cadmium are known to be strongly bound to proteins in living organisms. Differing from the heavy metals, arsenic may be characterized by no strong binding to proteins. We further examined whether the arsenic in marine fish and invertebrates is water-soluble or lipid-

Table 1 Total, Inorganic [(III) and (V)], Organic, Water-Soluble, and Lipid-Soluble Arsenic Content of Marine Organisms

		Arsenic content (μg/g of dry tissue)					
Species	Tissue	Total	Inorganic (III)	Inorganic (V)	Organic	Water-soluble	Lipid-soluble
Fish							
Limanda herzensteini	Muscle	36.0	0.00	0.00	34.2	34.4	0.22
Seriola quinqueradiata	Muscle	5.0	0.05	0.12	4.2	4.2	0.24
Trachurus japonicus	Muscle	25.6	0.00	0.06	24.0	24.3	0.18
Pneumatophorus japonicus	Muscle	5.4	0.00	0.00	5.1	4.6	0.54
Cololabis saira	Muscle	5.5	0.05	0.17	4.8	5.1	0.31
Sardinops melanosticta	Muscle	17.3	0.00	0.28	15.0	15.1	0.23
Sea squirt							
Halocynthia roretzi	Muscle	25.0	0.00	0.05	24.3	17.3	7.6
Sea cucumber							
Stichopus japonicus	Muscle	12.4	0.00	0.10	11.3	7.2	1.0
Sea urchin							
Anthocidaris crassipina	Gonad	7.3[a]	0.16[a]	0.22[a]	7.0[a]	5.1[a]	1.8[a]

Crustacean							
Penaeus orientalis	Muscle	41.3	0.00	0.00	39.2	39.8	1.0
Sergestes lucens	Whole	7.6	0.07	0.00	7.2	6.0	1.0
Shellfish							
Turbo cornutus	Muscle	15.0	0.00	0.02	14.1	9.0	4.9
Tapes japonica	Soft tissue	17.5	0.04	0.01	15.9	11.7	5.0
Cephalopod							
Paroctopus dofleini	Muscle	49.0	0.00	0.00	48.8	47.3	0.20
Todarodes pacificus	Muscle	17.2	0.00	0.00	16.1	15.9	0.22
Illex argentinus	Muscle	9.5	0.00	0.00	9.0	9.0	0.26
Polychaete							
Perinereis sp.	Whole	5.1	0.00	0.00	5.1	3.3	1.5
Brown alga							
Hizikia fusiforme	Whole	61.3	36.7[b]		15.2		
Laminaria japonica	Whole	25.4	0.8[b]		20.2		
Undaria pinnatifida	Whole	8.3	0.6[b]		6.5		

[a] Wet-weight basis.

[b] Inorganic arsenic (III) and (V).

Source: Shinagawa et al. (1983).

soluble. As shown in Table 1, the level of water-soluble arsenic was much higher in all species of marine animals than that of lipid-soluble arsenic; the water-soluble arsenic accounted for more than 80% of the total arsenic in many species (Shinagawa et al., 1983). Similar results were recently reported by Kaise et al. (1988b), who examined a variety of marine organisms including seaweeds. Thus, it can be concluded that most of the arsenic in marine organisms exists in the water-soluble, low-molecular-weight, organic (methylated) form.

2.3. Chemical Structure

Arsenobetaine was the first arsenical to be definitively identified in marine organisms. In the momentous study of Edmonds et al. (1977), a water-soluble major arsenic compound was isolated from the tail muscle of the western rock lobster (*Panulirus longipes cygnus*) by ion-exchange chromatography and preparative thin layer chromatography (TLC); it was then identified as arsenobetaine by comparison with a synthesized preparation as well as by spectral methods. Since this study, much effort has been concentrated by researchers on the identification of the chemical forms of arsenic contained in marine organisms. The chemical structures of the identified organoarsenicals are illustrated in Figure 1. It is worth mentioning that all arsenicals are methylated compounds. In addition, all arsenicals other than the arsenolipids are water-soluble. These facts agree with the previous finding that the arsenic in marine organisms is mainly present in the water-soluble, methylated form.

2.3.1. Analytical Methods

The identification of arsenic compounds in marine organisms has progressed in parallel with the development of analytical methods, especially high-performance liquid chromatography (HPLC) and inductively coupled argon plasma emission spectrometry (ICP). The chemical structure of an unknown compound in biological samples is commonly determined by spectrometric analyses such as nuclear magnetic resonance (NMR) and MS following extraction and laborious chromatographic isolation. This was the case in earlier studies on the chemical structure of arsenic compounds in marine organisms. As knowledge about the chemical structure of arsenic compounds in marine organisms began to accumulate, a rapid analytical method using HPLC was devised. Separation of arsenic compounds by HPLC is based on ion exchange, ion pairing, or gel filtration. The eluate from HPLC is specifically analyzed for arsenic by a graphite-furnace atomic absorption spectrometer (GFAA) or an ICP. The ICP is applicable to a much wider range of arsenic concentrations than the GFAA. Furthermore, on-line analysis is easily achieved by introducing the eluate from HPLC into the nebulizer of ICP. Thus, the HPLC–ICP system seems to be particularly convenient in the analysis of biological samples for arsenic. The chromatograms of some standard arsenic compounds analyzed by our HPLC–ICP method (Shiomi et al., 1987a) are shown in Figure 2. Recently, the HPLC–ICP–MS system consisting of HPLC and ICP–MS was developed (Shibata and Morita, 1989) and

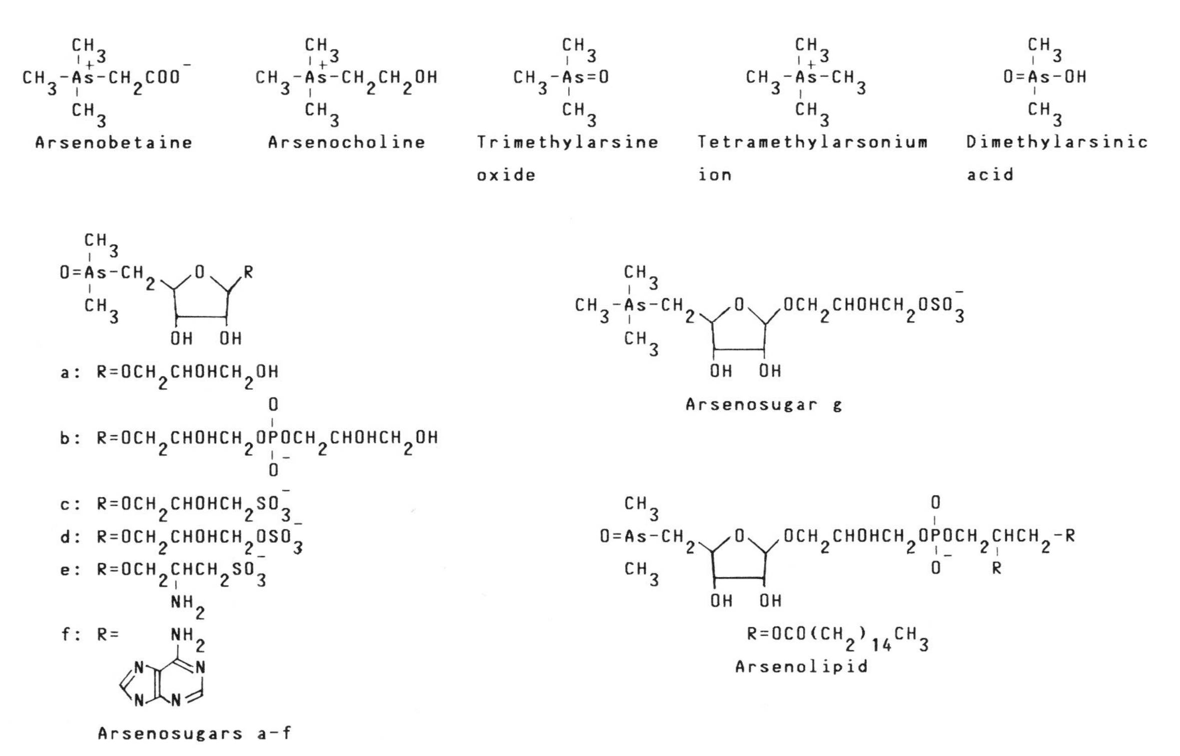

Figure 1. Structure of organoarsenicals identified in marine organisms.

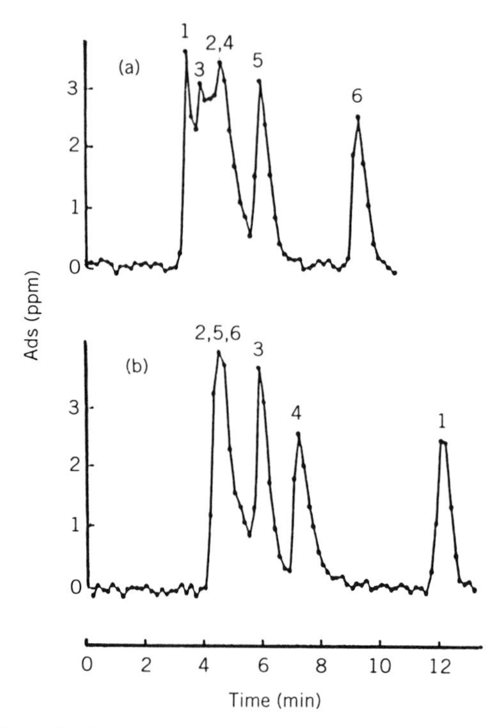

Figure 2. Analysis of standard arsenic compounds by HPLC–ICP: (a) column, Nucleosil 10SA (4.6 × 250 mm); solvent, 0.1 M pyridine-formate buffer (pH 3.1); flow rate, 1 mL/min. (b) column, Nucleosil 10SB (4.6 × 250 mm); solvent, 0.02 M phosphate buffer (pH 7.0); flow rate, 1 mL/min. Arsenic compounds: 1, arsenate; 2, arsenite; 3, methylarsonic acid; 4, dimethylarsinic acid; 5, arsenobetaine; 6, arsenocholine. Both trimethylarsine oxide and tetramethylarsonium ion can be analyzed under the conditions of (a). The former compound is eluted between arsenobetaine and arsenocholine and the latter after arsenocholine. (Shiomi et al., 1987c.)

effectively applied to the analysis of arsenic compounds in seaweeds (Morita and Shibata, 1990). Another system (an LC–MS system) consisting of HPLC and MS has also been reported by Cullen and Dodd (1989). These improved systems will find wide use because they make it possible to analyze arsenic compounds with greater sensitivity than the HPLC–ICP system previously used.

2.3.2. *Arsenobetaine*

Arsenobetaine is structurally related to glycine betaine, which is contained at high levels in marine invertebrates, especially in crustaceans and mollusks. In arsenobetaine, the nitrogen of glycine betaine is replaced by arsenic. Since the discovery of arsenobetaine in the western rock lobster (Edmonds et al., 1977), many researchers have established the presence of arsenobetaine in fish (Kurosawa et al., 1980; Cannon et al., 1981; Edmonds and Francesconi, 1981a, 1987a; Luten et al., 1983; Shiomi et al., 1983b; Hanaoka and Tagawa, 1985a, b;

Lawrence et al., 1986; Hanaoka et al., 1986, 1987a, b, 1988), sea squirt (Hanaoka et al., 1988), echinoderms (Shiomi et al., 1983b; Hanaoka et al., 1988), crustaceans (Edmonds and Francesconi, 1981b; Norin and Christakopoulos, 1982; Luten et al., 1983; Norin et al., 1983; Shiomi et al., 1984a; Francesconi et al., 1985; Maher, 1985a; Matsuto et al., 1986; Francesconi and Edmonds, 1987), mollusks (Shiomi et al., 1983a, 1984b, 1987a, b, 1988a; Maher, 1985a; Lawrence et al., 1986; Hanaoka et al., 1988), sea anemone (Shiomi et al., 1988b), and sponge (Shiomi et al., 1988b). Some animals have been shown to contain other types of arsenicals together with arsenobetaine. Even in those animals, however, the major arsenical is usually arsenobetaine. These results lead us to conclude that arsenobetaine is ubiquitously distributed, mainly as the major arsenic constituent, in marine animals at different trophic levels, and that it is the end product of arsenic cycling in the marine ecosystem. It is particularly interesting to note that arsenobetaine has not yet been detected in any marine macroalgae. The presence of arsenobetaine in the brown alga *Hizikia fusiforme* was previously suggested by Morita et al. (1981). In their study, however, evidence for the presence of arsenobetaine was based only on HPLC. Judging from the more detailed work of Edmonds et al. (1987), arsenobetaine appears to be absent in *H. fusiforme*.

2.3.3. Arsenocholine

Arsenocholine has the reduced structure of arsenobetaine. In marine animals, there is a metabolic pathway in which glycine betaine is reduced to choline or the reverse. If this pathway is applicable to the corresponding arsenic compounds, the fact that arsenobetaine is widely found in marine animals may suggest a wide distribution of arsenocholine. At present, however, the detection of arsenocholine is limited to several species. Norin and coworkers (Norin and Christakopoulos, 1982; Norin et al., 1983) found an unknown arsenic compound more basic than arsenobetaine in the shrimp *Pandalus borealis* and identified it as arsenocholine by means of TLC, electrophoresis, MS, and other methods. Lawrence et al. (1986) also reported the presence of arsenocholine, together with arsenobetaine, in two species of (unidentified) shrimps. Although many shrimps seem to contain arsenocholine, Shiomi et al. (1984a) and Luten et al. (1983) failed to detect it in the shrimps *Sergestes lucens* and *Nephrops norvegicus*, respectively. Besides shrimps, the midgut gland of the marine gastropod *Charonia sauliae* was determined to contain arsenocholine as a minor component by the HPLC–ICP method (Shiomi et al., 1987a).

2.3.4. Trimethylarsine Oxide

Trimethylarsine oxide has been detected in the following animals: several species of marine fish (Norin et al., 1985), an estuary catfish (Edmonds and Francesconi, 1987a), marine gastropods (Hanaoka et al., 1988), and bivalves (Cullen and Dodd, 1989). Norin et al. (1985) noted that the amount of trimethylarsine oxide in fresh samples of fish was several percent of the total arsenic at most. In frozen samples, however, as much as 40% of the total arsenic was believed to be trimethylarsine oxide in one sample. The results of Norin et al. suggest the increase of trimethylarsine oxide occurred during storage of the fish. Trimethylamine oxide, in which the arsenic of trimethylarsine oxide is replaced by nitrogen, is a compound peculiar to marine animals and its concentration is especially high in sharks, cods, and cephalopods. This compound is known to be biosynthesized from choline and trimethylamine. On the other hand, trimethylarsine seems to be produced from arsenobetaine and arsenocholine. If this is so, it is very likely that trimethylarsine oxide in fish comes from the major arsenical, arsenobetaine, via trimethylarsine.

2.3.5. *Tetramethylarsonium Ion*

In our attempts to identify the chemical forms of arsenic contained in different tissues of marine animals, we first selected the clam *Meretrix lusoria* because its various tissues were easily obtained. As expected, the major arsenical in five tissues, including the adductor muscle and midgut gland, was demonstrated to be arsenobetaine by the HPLC–ICP method (Shiomi et al., 1987b). In addition to the interesting finding that arsenic content was significantly high in the gill compared to the other tissues examined, the major arsenical in the gill was shown not to be arsenobetaine but a much more cationic compound. This unknown arsenic compound was isolated by successive column chromatography and identified as the tetramethylarsonium ion by TLC, NMR, and MS. In similar studies performed with four species of bivalves (short-necked clam, oyster, arkshell, and scallop), neither the especially high arsenic content in the gill nor the presence of tetramethylarsonium in the gill was found (Shiomi et al., 1988a). It may safely be said that the initial fortuitous selection of *M. lusoria* as a sample resulted in the discovery of a new arsenic compound, the tetramethylarsonium ion. The tetramethylarsonium ion has now been identified in the gastropod *Tectus pyramis* (Francesconi et al., 1988), in several species of bivalves (Cullen and Dodd, 1989), in the sea anemone *Parasicyonis actinostoloides* (Shiomi et al., 1988b) and in the sea hare *Aplysia kurodai* (Shiomi et al., 1988b).

2.3.6. *Arsenosugars*

The major arsenicals in seaweeds are arsenosugars, sugar (ribofuranose)-containing compounds with a more complicated architecture than the other types of arsenic compounds so far identified in marine fauna. Edmonds and Francesconi (1981c, 1983) laboriously isolated three water-soluble arsenic compounds from the brown alga *Ecklonia radiata* and identified them as arsenosugars a–c, all of which are dimethylated arsenicals. Since then, various types of arsenosugars have successively been found in one species of green alga, *Codium fragile* (Jin et al., 1988a); seven species of brown algae, *Eisenia bicyclis* (Morita and Shibata, 1990), *Hizikia fusiforme* (Edmonds et al., 1987), *Laminaria japonica* (Shibata et al., 1987), *Sargassum lacerifolium* (Francesconi et al., 1991a), *Sargassum thunbergii* (Shibata and Morita, 1988), *Sphaerotrichia divaricata* (Jin et al., 1988b), and *Undaria pinnatifida* (Morita and Shibata, 1990); and one species of red alga, *Porphyra tenera* (Shibata et al., 1990). Arsenosugar g in *S. thunbergii* is unique in being the only trimethylated arsenical identified in seaweeds. *Codium fragile* and *S. lacerifolium* also contain dimethylarsinic acid, and *H. fusiforme* and

Table 2 Arsenosugar Composition of Seaweeds

Seaweed	Arsenosugar[a]						Reference
	a	b	c	d	e	g	
Green alga							
Codium fragile	○	○					Jin et al. (1988a)
Brown alga							
Ecklonia radiata	○	○	○				Edmonds and Francesconi (1983)
Eisenia bicyclis	○	○	○				Morita and Shibata (1990)
Hizikia fusiforme	○		○	○	○		Edmonds et al. (1987)
Laminaria japonica	○	○	○				Shibata et al. (1987)
Sargassum lacerifolium[b]	○	○	○	○	○		Francesconi et al. (1991a)
Sargassum thunbergii	○	○		○		○	Shibata and Morita (1988)
Sphaerotrichia divaritica	○	○	○		○		Jin et al. (1988b)
Undaria pinnatifida	○	○	○				Morita and Shibata (1990)
Red alga							
Porphyra tenera	○	○					Shibata et al. (1990)

[a] See Figure 1.
[b] Three arsenosugars distinguishable from arsenosugars a–e and g are minor constituents of *Sargassum lacerifolium.*

S. laceriforme contain arsenate. The arsenosugar content of seaweeds is summarized in Table 2. Arsenosugars appear to be of taxonomical importance for seaweeds. Four brown algae, *E. bicyclis*, *E. radiata*, *L. japonica*, and *U. pinnatifida*, belong to the order Laminariales and apparently contain the same arsenosugars (a–c). Three brown algae, *H. fusiforme*, *S. lacerifolium*, and *S. thunbergii*, belonging to the order Fucales commonly possess arsenosugar d as the major component, although their arsenosugar constituents are not identical. Moreover, all the brown algae contain at least one of the sulfur-containing arsenosugars (c–e, g), while none of these arsenosugars is present in green and red algae.

Of marine animals, only the giant clam *Tridacna maxima* has been shown to contain arsenosugars. Edmonds et al. (1982b) were the first to isolate two water-soluble arsenicals from the kidney of the giant clam and spectrometrically identified them as arsenosugars a and d. The accumulation of arsenosugars in the giant clam is assumed to result from the biosynthesis of arsenate in seawater by symbiotic unicellular algae. Very recently, Francesconi et al. (1991b) detected an unusual type of arsenosugar (f, arsenic-containing nucleoside) as a minor constituent in the giant clam. Arsenosugar f is a key intermediate in the proposed biosynthetic pathway for the dimethylarsinylriboside types of arsenosugars (a–e), the dominant arsenosugars in seaweeds, which would be produced by glycosidation of arsenosugar f (Francesconi et al., 1991b).

It should be pointed out that arsenosugars may be involved in the formation of arsenobetaine, the major arsenical in marine animals, in the natural environment. Edmonds et al. (1982a) confirmed the production of dimethylarsinoyl ethanol [$OAs(CH_3)_2CH_2CH_2OH$] after anaerobic incubation of the brown alga *E. radiata*. Dimethylarsinoyl ethanol would give rise to arsenobetaine following oxidation and reductive methylation or the reverse. Various plausible pathways for converting arsenosugars in seaweeds to arsenobetaine in marine animals have been postulated by Edmonds and coworkers (Edmonds et al., 1982a; Edmonds and Francesconi, 1988).

2.3.7. *Arsenolipids*

Studies on lipid-soluble arsenic compounds (arsenolipids) are very few compared to those on water-soluble compounds. Based on chromatographic behavior, Lunde (1968b, 1973a) suggested that arsenolipids are relatively polar compounds resembling the phospholipids. An arsenic analogue of phosphatidylcholine was assumed to be a possible arsenolipid by Irgolic et al. (1977). After arsenosugar b was discovered as a water-soluble compound in the brown alga *Ecklonia radiata* (Edmonds and Francesconi, 1981c, 1983), Knowles and Benson (1983) and Edmonds and Francesconi (1987b) suggested that its diacylated derivatives were arsenolipids. This was experimentally demonstrated by Morita and Shibata (1988), who isolated an arsenolipid from the brown alga *Undaria pinnatifida* and elucidated its structure as a dipalmitoyl ester of arsenosugar b. Considering that arsenosugar b is widely distributed in seaweeds, the arsenolipid identified in *U. pinnatifida* is likely to occur in many seaweeds. On the other hand, arsenosugar b has not yet been detected in marine animals. Arsenosugars a, d, and f found in the giant clam are assumed to be produced not by the animal but by symbiotic unicellular algae. Consequently, lipid-soluble arsenicals in marine animals may be different from the identified arsenolipid.

2.3.8. *Dimethylarsinic Acid*

During the examination of arsenic compounds in the green alga *Codium fragile*, Jin et al. (1988a) identified dimethylarsinic acid as a minor arsenical and arsenosugars as the major constituents. Small amounts of dimethylarsinic acid were also detected in the brown alga *Sargassum lacerifolium* by Francesconi et al. (1991a). Previously, it was suggested that dimethylarsinic acid was present in another brown alga, *Hizikia fusiforme*, by Tagawa (1980). In this study, samples were exposed to both strongly acidic and alkaline solutions. Since arsenosugars decompose into dimethylarsinic acid at extreme pH values, the dimethylarsinic acid detected in *H. fusiforme* may be an artifact of the decomposition of arsenosugars, the major arsenicals of the alga. No information about the presence of dimethylarsinic acid in marine animals is available.

2.3.9. *Arsenate*

Levels of inorganic arsenic in marine animals are generally negligible. The exception is the brown alga *Hizikia fusiforme*, which contains appreciable amounts (usually about 50% of the total arsenic) of inorganic arsenic. The inorganic arsenic compound in *H. fusiforme* has been isolated and demonstrated to be arsenate by Edmonds et al. (1987). The presence of arsenate as a minor constituent in the brown alga *Sargassum lacerifolium* has also been established by Francesconi et al. (1991a). Arsenate appears to be a typical, though not abundant, inorganic arsenical in marine organisms.

3. ACUTE TOXICITY

The LD_{50} (oral administration to rats) of arsenate, methylarsonic acid, and dimethylarsinic acid are 0.041, 0.79, and 2.6 g/kg, respectively (Ishinishi et al., 1985). Acute toxicity increases in the following order: arsenite > arsenate > methylarsonic acid > dimethylarsinic acid, suggesting that the acute toxicity of arsenic compounds diminishes with the progress of methylation. If this is universally valid for arsenic compounds, the methylated arsenic compounds detected in marine organisms would show low acute toxicity.

Arsenobetaine (Kaise et al., 1985), arsenocholine (Kaise et al., 1992), trimethylarsine oxide (Kaise et al., 1989), tetramethylarsonium (Shiomi et al., 1988c), and dimethylarsinic acid (Kaise et al., 1989) have so far been fed to mice to estimate their toxicity, as summarized in Table 3. At present, there are no data regarding the acute toxicity of the arsenosugars a–g and the arsenolipid, because sufficient amounts of these compounds are not available for toxicity experiments. For comparison, the LD_{50} of arsenic trioxide (Kaise et al., 1988a) and methylarsonic acid (Kaise et al., 1989) is also included in Table 3. As expected, the acute toxicity of the methylated arsenic compounds in marine organisms is considerably weak compared to that of arsenic trioxide. For arsenobetaine, the most ubiquitous arsenical in marine animals, an accurate LD_{50} cannot be obtained, because of its extremely low toxicity. The other trimethylated compounds, arsenocholine and trimethylarsine oxide, are 200 to 300 times less toxic than arsenic trioxide. Hence, the trimethylated arsenicals are judged to be virtually nontoxic. In this regard, Irvin and Irgolic (1988) have revealed by experiments using rat embryos that both arsenobetaine and arsenocholine have no subacute or acute embryotoxicity.

Table 3 LD_{50} (Oral Administration to Mice) of Arsenic Compounds

Arsenic Compound	LD_{50} (g/kg)	Reference
Arsenobetaine	> 10.0	Kaise et al. (1985)
Arsenocholine	6.5	Kaise et al. (1992)
Trimethylarsine oxide	10.6	Kaise et al. (1989)
Tetramethylarsonium iodide	0.89	Shiomi et al. (1988c)
Dimethylarsinic acid	1.2	Kaise et al. (1989)
Methylarsonic acid	1.8	Kaise et al. (1989)
Arsenic trioxide	0.0345	Kaise et al. (1985)

Contrary to the results obtained with the trimethylated arsenicals, the tetramethylarsonium ion—the most highly methylated arsenical—exhibits a considerable lethality; its acute toxicity against mice is even higher than that of methylarsonic acid and dimethylarsinic acid. The general opinion on arsenic compounds that advanced methylation means detoxification is not true for the tetramethylarsonium ion. Of the organoarsenicals in marine organisms, the tetramethylarsonium ion appears to be the only compound that needs special attention from the viewpoint of acute toxicity.

In connection with the tetramethylarsonium ion, it should be pointed out that its nitrogenous analogue, the tetramethylammonium ion (tetramine), is contained at high levels in the salivary glands of some whelks such as *Neptunea arthritica*. Occasional outbreaks of poisoning from ingestion of whelks have been a public health problem in Japan. In whelk poisoning, the principal symptoms are intense headache, dizziness, nausea, and vomiting. Asano and Ito (1959, 1960) have identified the causative compound of whelk poisoning as tetramine. Our comparative study (Shiomi et al., 1988c) demonstrated that lethal doses of the tetramethylarsonium ion and tetramine induce essentially the same symptoms in mice, including tonic convulsion followed by death, suggesting that these two related compounds exert their toxicity by the same mode of action. Judging from the estimated LD_{50}, however, the acute toxicity of the tetramethylarsonium ion is 20 times weaker than that of tetramine. Therefore, much larger doses of the tetramethylarsonium ion than of tetramine seem to be needed for acute poisoning in humans. In the case of tetramine, the dose required to cause poisoning in humans is estimated to be 350 to 450 mg. Since the concentrations of the tetramethylarsonium ion in marine organisms are assumed to be 10 to 20 mg/kg at most, acute poisoning by the tetramethylarsonium ion is unlikely to result from ingestion of marine organisms.

4. METABOLISM

As for the metabolism of arsenic compounds in mammals, extensive studies have thus far been carried out using inorganic arsenicals, methylarsonic acid, and dimethylarsinic acid. Accumulated information may be summarized as follows: (1) inorganic arsenicals are consecutively converted to monomethylated and dimethylated compounds by mammals; (2) part of both methylarsonic acid and dimethylarsinic acid undergoes further methylation, giving rise to dimethylated and trimethylated compounds, respectively; however, these two compounds never undergo demethylation; (3) all of the arsenic compounds administered show a tendency to be eliminated with or without biotransformation in a relatively short time; (4) methylated arsenicals are apparently excreted more rapidly than inorganic compounds; and (5) the excretion of arsenic occurs mainly through urine. Judging from these facts, methylated arsenicals, in particular, seem to be rapidly excreted in urine following administration, probably because of their low affinity for organs and tissues.

Are the methylated arsenicals contained in marine organisms also easily eliminated in a short time after administration? Vahter et al. (1983) reported that the greater part (98.1%) of the arsenic administered was recovered in urine within 72 hr when arsenobetaine was orally administered to mice, and that arsenobetaine was the only arsenic compound detected in urine. These results indicate that orally administered arsenobetaine is almost completely absorbed from the gastrointestinal tract and then rapidly excreted in urine without biotransformation. Intravenous injection of arsenobetaine into mice, rats, or rabbits also resulted in rapid excretion in urine without biotransformation; the urinary excretion within 72 hr was more than 98% of the injected dose in both mice and rats and about 75% in rabbits. The results of Vahter et al. were later reconfirmed by Kaise et al. (1985) using mice and by Yamauchi et al. (1986) using hamsters.

The tendency for a large part of the arsenic administered to appear in the urine in a relatively short time after administration is also commonly observed in animal experiments performed with arsenocholine, trimethylarsine oxide, and the tetramethylarsonium ion. However, the major arsenical detected in urine following oral or intravenous administration of arsenocholine to mice, rats, or rabbits was not arsenocholine but arsenobetaine; arsenocholine was found in the urine only during the first 24 hr (Marafante et al., 1984). This indicates that appreciable amounts of arsenocholine are oxidized to arsenobetaine in vivo and then excreted in urine. Probably because of the time necessary for oxidation, the rate of arsenic excretion was somewhat lower after administration of arsenocholine than after administration of arsenobetaine. When trimethylarsine oxide was administered orally or intraperitoneally to hamsters, the only arsenic compound excreted in urine was spectrometrically identified as trimethylarsine oxide (Yamauchi et al., 1989). In this case, however, the expired air of the animals had a garlicky odor characteristic of trimethylarsine. Analysis by GC–MS revealed the presence of trimethylarsine in the expired air, suggesting that trimethylarsine oxide was partly reduced in vivo. The metabolism of the tetramethylarsonium ion having a considerable acute toxicity was examined using mice and tetramethylarsonium iodide as a model compound (Shiomi et al., 1988c). This compound was found to be neither converted to other forms nor accumulated in the body. Tetramine, the nitrogenous analogue of the tetramethylarsonium ion, is also known to be rapidly eliminated from the body, which accounts for the transient symptoms of food poisoning.

As for arsenosugars, the major arsenicals in marine macroalgae, no metabolic experiments with pure substances have yet been carried out. We examined the metabolism of arsenosugars in mice using a partially purified preparation from the red alga *Porphyra yezoensis* (Shiomi et al., 1990). Another red alga, *Porphyra tenera*, closely related to *P. yezoensis* has been reported to contain two kinds of arsenosugars (a and b) by Shibata et al. (1990). It may be relevant that the same arsenosugars as those found in *P. tenera* were probably contained in our partially purified preparation from *P. yezoensis*. After mice were orally or intravenously administered an arsenosugar preparation, the arsenic was rapidly eliminated from their bodies, which is similar to what occurred with other arsenicals.

However, the excretion route largely depended on the administration method, which is not the case with other arsenic compounds. The arsenic was excreted mainly through feces in the case of oral administration and through urine in the case of intravenous administration. Based on chromatographic analyses, the arsenic compounds recovered in feces following oral administration were identified as arsenosugars, suggesting that most orally administered arsenosugars are not absorbed from the gastrointestinal tract. On the other hand, three arsenicals—methylarsonic acid, dimethylarsinic acid, and arsenobtaine—were detected in urine following both oral and intravenous administration. As mentioned, the preparation used in our study probably contained dimethylated arsenosugars. If so, methylarsonic acid detected in urine must have been formed by demethylation of dimethylated arsenicals. This contradicts the general rule that methylated arsenic compounds undergo no demethylation. It would be very interesting to isolate the arsenic compounds excreted in urine after administration of arsenosugars and determine their chemical structures by spectrometric analysis.

Several reports are available concerning the human metabolism of arsenobetaine. Cannon et al. (1981) detected arsenobetaine in the urine of humans who had ingested muscles of the western rock lobster containing arsenobetaine as the major arsenical. Luten et al. (1982) showed that 69 to 85% of ingested arsenic was excreted in urine within five days following ingestion of the muscle of the plaice *Pleuronectes platessa*, whose major arsenical is arsenobetaine. In this case, the major urinary metabolite was assumed to be arsenobetaine. The rapid excretion in urine of arsenic was also observed by Tam et al. (1982) using the muscle of flounder and by Yamauchi and Yamamura (1984) using the muscle of shrimp; between 76 and 90% of the ingested arsenic was excreted in urine within eight days in the former study and within four days in the latter, respectively. In both experiments, the arsenical recovered in urine was a trimethylated form (probably arsenobetaine). Judging from these results, it is likely that orally ingested arsenobetaine is rapidly excreted in human urine without biotransformation, implying that the human body, like experimental animals, does not retain arsenobetaine.

Fukui et al. (1981) examined the human metabolism of arsenic compounds contained in two species of brown algae, *Laminaria japonica* and *Hizikia fusiforme*. Following ingestion of these algae, the excretion of arsenic in urine was relatively rapid. However, prolonged excretion of arsenic was observed in the case of *H. fusiforme*. The major urinary metabolites were dimethylated arsenicals in both cases. Inorganic arsenicals detected in urine were negligible in the case of *L. japonica* but accounted for more than 20% of the arsenic recovered in the case of *H. fusiforme*. These observations may be the reflection of the fact that arsenosugars are major arsenicals in *L. japonica* (Shibata et al., 1987), whereas *H. fusiforme* contains appreciable amounts of inorganic arsenic as arsenate (Edmonds et al., 1987), which is known to be rather slowly eliminated from the body compared to methylated arsenicals. Considering the results with *L. japonica* containing arsenosugars as the major arsenical, orally ingested arsenosugars seem to be rapidly excreted in the urine of humans. In our work using mice, orally administered arsenosugars were shown to be excreted not in urine but in feces without biotransformation (Shiomi et al., 1990). The discrepancy between humans and mice regarding the metabolism of arsenosugars remains unexplained.

5. SUMMARY AND CONCLUSIONS

Marine organisms contain arsenic at much higher levels than terrestrial organisms. To our knowledge, however, no arsenic poisoning by ingestion of marine organisms has so far been recorded in the literature. This can be largely explained by the accumulated information regarding the chemical form, toxicity, and metabolism of arsenic contained in marine organisms. The major arsenical in marine animals is arsenobetaine, a water-soluble and trimethylated compound; the major arsenicals in seaweeds are several types of arsenosugars, which are water-soluble and dimethylated compounds. Other methylated arsenicals, such as arsenocholine and trimethylarsine oxide, occur usually as minor constituents in a few species. Trimethylated arsenicals exhibit insignificant toxicity and are easily eliminated from the body in a relatively short time, even when ingested. Although the tetramethylarsonium ion has a considerably acute toxicity, it is also rapidly eliminated from the body. The following problems concerning the safety of arsenic in marine organisms remain to be solved:

1. The chemical forms of unknown arsenic compounds that may be present in marine organisms should be identified.
2. The acute toxicity of arsenosugars and arsenolipids is unclear. The acute toxicity of the other arsenic compounds should be examined using experimental animals other than mice in order to assess their effects on humans, who are generally more sensitive than rodents to arsenicals.
3. Metabolic studies using pure arsenosugars and arsenolipids are needed.
4. The brown alga *H. fusiforme* is an important edible species in Japan. However, it contains an appreciable amount of inorganic arsenic (arsenate) with higher toxicity than that of organic arsenicals. The determination of its safety in food awaits further detailed toxicological studies.
5. Marine organisms are often utilized as food after undergoing processing such as heating and fermentation. During processing, the arsenic in marine organisms may undergo conversion of its chemical form. Chemical and toxicological studies on arsenic compounds in processed food are especially needed.

REFERENCES

Asano, M., and Ito, M. (1959). Occurrence of tetramine and choline compounds in the salivary gland of a marine gastropod, *Neptunea arthritica* Bernandi. *Tohoku J. Agric. Res.* **10**, 209–227.

Asano, M., and Ito, M. (1960). Salivary poison of a marine gastropod, *Neptunea arthritica* Bernandi, and seasonal variation of its toxicity. *Ann. N. Y. Acad. Sci.* **90**, 674–688.

Cannon, J. R., Edmonds, J. S., Francesconi, K. A., Raston, C. L., Sanders, J. B., Skelton, B. W., and White, A. H. (1981). Isolation, crystal structure and synthesis of arsenobetaine, a constituent of the western rock lobster, the dusky shark and some samples of human urine. *Aust. J. Chem.* **34**, 787–798.

Chapman, A. C. (1926). On the presence of compounds of arsenic in marine crustaceans and shellfish. *Analyst* **51**, 548–563.

Cullen, W. R., and Dodd, M. (1989). Arsenic speciation in clams of British Columbia. *Appl. Organomet. Chem.* **3**, 79–88.

Edmonds, J. S., and Francesconi, K. A. (1977). Methylated arsenic from marine fauna. *Nature (London)* **265**, 436.

Edmonds, J. S., and Francesconi, K. A. (1981a). The origin and chemical form of arsenic in the school whiting. *Mar. Pollut. Bull.* **12**, 92–96.

Edmonds, J. S., and Francesconi, K. A. (1981b). Isolation and identification of arsenobetaine from the American lobster *Homarus americanus. Chemosphere* **10**, 1041–1044.

Edmonds, J. S., and Francesconi, K. A. (1981c). Arseno-sugars from brown kelp (*Ecklonia radiata*) as intermediates in cycling of arsenic in a marine ecosystem. *Nature* (*London*) **289**, 602–604.

Edmonds, J. S., and Francesconi, K. A. (1983). Arsenic-containing ribofuranosides: Isolation from brown kelp *Ecklonia radiata* and nuclear magnetic resonance spectra. *J. Chem. Soc., Perkin Trans. 1*, pp. 2375–2382.

Edmonds, J. S., and Francesconi, K. A. (1987a). Trimethylarsine oxide in estuary catfish (*Cnidoglanis macrocephalus*) and school whiting (*Sillago bassensis*) after oral administration of sodium arsenate, and as a natural component of estuary catfish. *Sci. Total Environ.* **64**, 317–323.

Edmonds, J. S., and Francesconi, K. A. (1987b). Transformation of arsenic in the marine environment. *Experientia* **43**, 553–557.

Edmonds, J. S., and Francesconi, K. A. (1988). The origin of arsenobetaine in marine animals. *Appl. Organomet. Chem.* **2**, 297–302.

Edmonds, J. S., Francesconi, K. A., Cannon, J. R., Raston, C. L., Skelton, B. W., and White, A. H. (1977). Isolation, crystal structure and synthesis of arsenobetaine, the arsenical constituent of the western rock lobster *Panulirus longipes cygnus* George. *Tetrahedron Lett.*, pp. 1543–1546.

Edmonds, J. S., Francesconi, K. A., and Hansen, J. A. (1982a). Dimethyloxarsylethanol from anaerobic decomposition of brown kelp (*Ecklonia radiata*): A likely precursor of arsenobetaine in marine fauna. *Experientia* **38**, 643–644.

Edmonds, J. S., Francesconi, K. A., Healy, P. C., and White, A. H. (1982b). Isolation and crystal structure of an arsenic-containing sugar sulfate from the kidney of the giant clam, *Tridacna maxima*. X-ray crystal structure of (2*S*) -3-[5-deoxy-5(dimethylarsinoyl)-β-D- ribofuranosyloxyl]-2-hydroxypropyl hydrogen sulfate. *J. Chem. Soc., Perkin Trans. 1*, pp. 2989–2993.

Edmonds, J. S., Morita, M., and Shibata, Y. (1987). Isolation and identification of arsenic-containing ribofuranosides and inorganic arsenic from Japanese edible seaweed *Hizikia fusiforme. J. Chem. Soc., Perkin Trans. 1*, pp. 577–580.

Flanjak, J. (1982). Inorganic and organic arsenic in some commercial east Australian crustacea. *J. Sci. Food Agric.* **33**, 579–583.

Francesconi, K. A., and Edmonds, J. S. (1987). The identification of arsenobetaine as the sole water-soluble arsenic constituent of the tail muscle of the western king prawn *Penaeus latisulcatus. Comp. Biochem. Physiol.* **87C**, 345–347.

Francesconi, K. A., Micks, P., Stockton, R. A., and Irgolic, K. J. (1985). Quantitative determination of arsenobetaine, the major water-soluble arsenical in three species of crab, using high performance liquid chromatography and inductively coupled argon plasma emission sepctrometer as the arsenic-specific detector. *Chemosphere* **14**, 1443–1453.

Francesconi, K. A., Edmonds, J. S., and Hatcher, B. G. (1988). Examination of the arsenic constituents of the herbivorous marine gastropod *Tectus pyramis*: Isolation of tetramethylarsonium ion. *Comp. Biochem. Physiol.* **90C**, 313–316.

Francesconi, K. A., Edmonds, J. S., Stick, R. V., Skelton, B. W., and White, A. H. (1991a). Arsenic-containing ribosides from the brown alga *Sargassum lacerifolium*: X-ray molecular structure of 2-amino-3-[5′-deoxy-5′-(dimethylarsinoyl)-ribosyloxy] propane-1-sulphonic acid. *J. Chem. Soc., Perkin Trans. 1*, pp. 2707–2716.

Francesconi, K. A., Stick, R. V., and Edmonds, J. S. (1991b). An arsenic-containing nucleoside from the kidney of the giant clam, *Tridacna maxima*. *J. Chem. Soc., Chem. Commun.*, pp. 928–929.

Fukui, S., Hirayama, T., Nohara, M., and Sakagami, Y. (1981). Studies on the chemical forms of arsenic in some sea foods and in urine after ingestion of these foods. *J. Food Hyg. Soc. Jpn.* **22**, 513–519.

Hanaoka, K., and Tagawa, S. (1985a). Isolation and identification of arsenobetaine as a major water-soluble arsenic compound from muscle of blue pointer *Isurus oxyrhinchus* and whitetip shark *Carcarhunus longimanus*. *Nippon Suisan Gakkaishi* **51**, 681–685.

Hanaoka, K., and Tagawa, S. (1985b). Identification of arsenobetaine in muscle of roundnose flounder *Eopsetta grigorjewi*. *Nippon Suisan Gakkaishi* **51**, 1203.

Hanaoka, K., Matsuda, H., Kaise, T., and Tagawa, S. (1986). Identification of arsenobetaine as a major arsenic compound in the muscle of a pelagic shark, *Carcharodon carcharias*. *J. Shimonoseki Univ. Fish.* **35**, 37–40.

Hanaoka, K., Fujita, T., Matsuura, M., Tagawa, S., and Kaise, T. (1987a). Identification of arsenobetaine as a major arsenic compound in muscle of two demersal sharks, shortnose dogfish *Squalus brevirostris* and starspotted shark *Mustelus manazo*. *Comp. Biochem. Physiol.* **86B**, 681–682.

Hanaoka, K., Kobayashi, H., Tagawa, S., and Kaise, T. (1987b). Identification of arsenobetaine as a major water-soluble arsenic compound in the liver of two demersal sharks, shortnose dogfish *Squalus brevirostris* and starspotted shark *Mustelus manazo*. *Comp. Biochem. Physiol.* **88C**, 189–191.

Hanaoka, K., Yamamoto, H., Kawashima, K., Tagawa, S., and Kaise, T. (1988). Ubiquity of arsenobetaine in marine animals and degradation of arsenobetaine by sedimentary microorganisms. *Appl. Organomet. Chem.* **2**, 371–376.

Irgolic, K. J., Woolson, E. A., Stockton, R. A., Newman, R. D., Bottino, N. R., Zingaro, R. A., Kearney, P. C., Pyles, R. A., Maeda, S., McShane, W. J., and Cox, E. R. (1977). Characterization of arsenic compounds formed by *Daphnia magna* and *Tetrasemilis chuii* from inorganic arsenate. *Environ. Health Perspect.* **19**, 61–66.

Irvin, T. R., and Irgolic, K. J. (1988). Arsenobetaine and arsenocholine: Two marine arsenic compounds without embryotoxicity. *Appl. Organomet. Chem.* **2**, 509–514.

Ishinishi, N., Hisanaga, A., Inamasu, T., Yamamoto, A., and Hirata, M. (1985). Biological effects of arsenic. Toxicity and carcinogenecity. In N. Ishinishi, S. Okabe, and T. Kikuchi (Eds.), *Arsenic—Chemistry, Metabolism and Toxicity*. Koseisha Koseikaku, Tokyo, pp. 5-26.

Jin, K., Hayashi, T., Shibata, Y., and Morita, M. (1988a). Arsenic-containing ribofuranosides and dimethylarsinic acid in green seaweed, *Codium fragile*. *Appl. Organomet. Chem.* **2**, 365–369.

Jin, K., Shibata, Y., and Morita, M. (1988b). Isolation and identification of arsenic-containing ribofuranosides from the edible brown seaweed, *Sphaerotrichia divaricata* (Ishimozuku). *Agric. Biol. Chem.* **52**, 1965–1971.

Kaise, T., Watanabe, S., and Itoh, K. (1985). The acute toxicity of arsenobetaine. *Chemosphere* **14**, 1327–1332.

Kaise, T., Yamauchi, H., Hirayama, T., and Fukui, S. (1988a). Determination of inorganic arsenic and organic arsenic compounds in marine organisms by hydride generation/cold trap/gas chromatography-mass spectrometry. *Appl. Organomet. Chem.* **2**, 339–347.

Kaise, T., Hanaoka, K., Tagawa, S., Hirayama, T., and Fukui, S. (1988b). Distribution of inorganic arsenic and methylated arsenic in marine organisms. *Appl. Organomet. Chem.* **2**, 539–546.

Kaise, T., Yamauchi, H., Horiguchi, Y., Tani, T., Watanabe, S., Hirayama, T., and Fukui, S. (1989). A comparative study on acute toxicity of methylarsonic acid, dimethylarsinic acid and trimethylarsine oxide in mice. *Appl. Organomet. Chem.* **3**, 273–277.

Kaise, T., Horiguchi, Y., Fukui, S., Shiomi, K., Chino, M., and Kikuchi, T. (1992). Acute toxicity and metabolism of arsenocholine in mice. *Appl. Organomet. Chem.* **6**, 369–373.

Knowles, F. C., and Benson, A. A. (1983). The biochemistry of arsenic. *Trends Biochem. Sci.* **8**, 178–180.

Kobayashi, R., Shiomi, K., Yamanaka, H., and Kikuchi, T. (1980). Distribution of arsenic in fishes. *Nippon Suisan Gakkaishi* **46**, 1265–1268.

Kurosawa, S., Yasuda, K., Taguchi, M., Yamazaki, S., Toda, S., Morita, M., Uehiro, T., and Fuwa, K. (1980). Identification of arsenobetaine, a water soluble organo-arsenic compound in muscle and liver of a shark, *Prionace glaucus. Agric. Biol. Chem.* **44**, 1993–1994.

Lawrence, J. F., Michalik, P., Tam, G., and Conacher, H. B. S. (1986). Identification of arsenobetaine and arsenocholine in Canadian fish and shellfish by high-performance liquid chromatography with atomic absorption detection and confirmation by fast atom bombardment mass spectrometry. *J. Agric. Food Chem.* **34**, 315–319.

Lunde, G. (1967). Activation analysis of bromine, iodide, and arsenic in oils from fishes, whales, phyto- and zooplankton of marine and limnetic biotopes. *Int. Rev. Gesamten Hydrobiol.* **52**, 265–279.

Lunde, G. (1968a). Activation analysis of trace elements in fishmeal. *J. Sci. Food Agric.* **19**, 432–434.

Lunde, G. (1968b). Analysis of arsenic in marine oils by neutron activation. Evidence of arseno organic compounds. *J. Am. Oil Chem. Soc.* **45**, 331–332.

Lunde, G. (1969). Water soluble arseno-organic compounds in marine fishes. *Nature* (*London*) **224**, 186–187.

Lunde, G. (1970a). Analysis of arsenic and selenium in marine raw materials. *J. Sci. Food Agric.* **21**, 242–247.

Lunde, G. (1970b). Analysis of trace elements in seaweed. *J. Sci. Food Agric.* **21**, 416–418.

Lunde, G. (1972). Analysis of arsenic and bromine in marine and terrestrial oils. *J. Am. Oil Chem. Soc.* **49**, 44–47.

Lunde, G. (1973a). Separation and analysis of organic-bound and inorganic arsenic in marine organisms. *J. Sci. Food Agric.* **24**, 1021–1027.

Lunde, G. (1973b). The synthesis of fat and water soluble arseno-organic compounds in marine and limnetic algae. *Acta Chem. Scand.* **27**, 1586–1594.

Lunde, G. (1977). Occurrence and transformation of arsenic in the marine environment. *Environ. Health Perspect.* **19**, 47–52.

Luten, J. B., Riekwel-Booy, G., and Rauchbaar, A. (1982). Occurrence of arsenic in plaice (*Pleuronectes platessa*), nature of organo-arsenic compound present and its excretion by man. *Environ. Health Perspect.* **45**, 165–170.

Luten, J. B., Biekwel-Booy, G., V. d. Greef, J., and ten Noever de Brauw, M. C. (1983). Identification of arsenobetaine in sole, lemon sole, flounder, dab, crab and shrimps by field desorption and fast atom bombardment mass spectrometry. *Chemosphere* **12**, 131–141.

Maher, W. A. (1983). Inorganic arsenic in marine organisms. *Mar. Pollut. Bull.* **14**, 308–310.

Maher, W. A. (1985a). The presence of arsenobetaine in marine animals. *Comp. Biochem. Physiol.* **80C**, 199–201.

Maher, W. A. (1985b). Distribution of arsenic in marine animals: Relationship to diet. *Comp. Biochem. Physiol.* **82C**, 433–434.

Marafante, E., Vahter, M., and Dencker, L. (1984). Metabolism of arsenocholine in mice, rats and rabbits. *Sci. Total Environ.* **34**, 223–240.

Matsuto, S., Stockton, K. A., and Irgolic, K. J. (1986). Arsenobetaine in the red crab, *Chionoecetes opilio*. *Sci. Total Environ.* **48**, 133–140.

Morita, M., and Shibata, Y. (1988). Isolation and identification of arseno-lipid from a brown alga, *Undaria pinnatifida* (Wakame). *Chemosphere* **17**, 1147–1152.

Morita, M., and Shibata, Y. (1990). Chemical form of arsenic in marine macroalgae. *Appl. Organomet. Chem.* **4**, 181–190.

Morita, M., Uehiro, T., and Fuwa, K. (1981). Determination of arsenic compounds in biological samples by liquid chromatography with inductively coupled argon plasma atomic emission spectrometric detection. *Anal. Chem.* **53**, 1806–1808.

Norin, H., and Christakopoulos, A. (1982). Evidence for the presence of arsenobetaine and another organoarsenical in shrimps. *Chemosphere* **11**, 287–298.

Norin, H., Ryhage, R., Christakopoulos, A., and Sandström, M. (1983). New evidence for the presence of arsenocholine in shrimps (*Pandalus borealis*) by use of pyrolysis gas chromatography—atomic absorption spectrometry/mass spectrometry. *Chemosphere* **12**, 299–315.

Norin, H., Christakopoulos, A., Sandström, M., and Ryhage, R. (1985). Mass fragmentographic estimation of trimethylarsine oxide in aquatic organisms. *Chemosphere* **14**, 313–323.

Reinke, J., Uthe, J. F., Freeman, H. C., and Johnson, J. R. (1975). The determination of arsenite and arsenate ions in fish and shellfish by selective extraction and polarography. *Environ. Lett.* **8**, 371–380.

Shibata, Y., and Morita, M. (1988). A novel, trimethylated arseno-sugar isolated from the brown alga *Sargassum thunbergii*. *Agric. Biol. Chem.* **52**, 1087–1089.

Shibata, Y., and Morita, M. (1989). Speciation of arsenic by reversed-phase high performance liquid chromatography–inductively coupled plasma mass spectrometry. *Anal. Sci.* **5**, 107–109.

Shibata, Y., Morita, M., and Edmonds, J. S. (1987). Purification and identification of arsenic-containing ribofuranosides from the edible brown seaweed, *Laminaria japonica* (Makonbu). *Agric. Biol. Chem.* **51**, 391–398.

Shibata, Y., Jin, K., and Morita, M. (1990). Arsenic compounds in the edible red alga, *Porphyra tenera*, and in *nori* and *yakinori*, food items produced from red algae. *Appl. Organomet. Chem.* **4**, 255–260.

Shinagawa, A., Shiomi, K., Yamanaka, H., and Kikuchi, T. (1983). Selective determination of inorganic arsenic(III), (V) and organic arsenic in marine organisms. *Nippon Suisan Gakkaishi* **49**, 75–78.

Shiomi, K., Shinagawa, A., Yamanaka, H., and Kikuchi, T. (1983a). Purification and identification of arsenobetaine from the muscle of an octopus *Paroctopus dofleini*. *Nippon Suisan Gakkaishi* **49**, 79–83.

Shiomi, K., Shinagawa, A., Azuma, M., Yamanaka, H., and Kikuchi, T. (1983b). Purification and comparison of water-soluble arsenic compounds in a flatfish *Limanda herzensteini*, sea squirt *Halocynthia roretzi*, and sea cucumber *Stichopus japonicus*. *Comp. Biochem. Physiol.* **74C**, 393–396.

Shiomi, K., Shinagawa, A., Igarashi, T., Yamanaka, H., and Kikuchi, T. (1984a). Evidence for the presence of arsenobetaine as a major arsenic compound in the shrimp *Sergestes lucens*. *Experientia* **40**, 1247–1248.

Shiomi, K., Shinagawa, A., Hirota, K., Yamanaka, H., and Kikuchi, T. (1984b). Identification of arsenobetaine as a major arsenic compound in the ivory shell *Buccinum striatissimum*. *Agric. Biol. Chem.* **48**, 2863–2864.

Shiomi, K., Orii, M., Yamanaka, H., and Kikuchi, T. (1987a). The determination method of arsenic compounds by high performance liquid chromatography with inductively coupled argon plasma emission spectrometry and its application to shellfishes. *Nippon Suisan Gakkaishi* **53**, 103–108.

Shiomi, K., Kakehashi, Y., Yamanaka, H., and Kikuchi, T. (1987b). Identification of arsenobetaine and a tetramethylarsonium salt in the clam *Meretrix lusoria*. *Appl. Organomet. Chem.* **1**, 177–183.

Shiomi, K., Sakamoto, Y., Yamanaka, H., and Kikuchi, T. (1988a). Arsenic concentrations in various tissues of bivalves and arsenic species in gills. *Nippon Suisan Gakkaishi* **54**, 539.

Shiomi, K., Aoyama, M., Yamanaka, H., and Kikuchi, T. (1988b). Chemical forms of arsenic in sponges, sea anemones and sea hare. *Comp. Biochem. Physiol.* **90C**, 361–365.

Shiomi, K., Horiguchi, Y., and Kaise, T. (1988c). Acute toxicity and rapid excretion in urine of tetramethylarsonium salts found in some marine animals. *Appl. Organomet. Chem.* **2**, 385–389.

Shiomi, K., Chino, M., and Kikuchi, T. (1990). Metabolism in mice of arsenic compounds contained in the red alga *Porphyra yezoensis*. *Appl. Organomet. Chem.* **4**, 281–286.

Tagawa, S. (1980). Confirmation of arsenate, arsenite, methylarsonate and dimethylarsinate in an aqueous extract from a brown seaweed, *Hizikia fusiforme*. *Nippon Suisan Gakkaishi* **46**, 1257–1259.

Tam, G. K. H., Charbonneau, S. M., Bryce, F., and Sandi, E. (1982). Excretion of a single oral dose of fish-arsenic in man. *Bull. Environ. Contam. Toxicol.* **28**, 669–673.

Vahter M., Marafante, E., and Dencker, L. (1983). Metabolism of arsenobetaine in mice, rats and rabbits. *Sci. Total Environ.* **30**, 197–211.

Yamauchi, H., and Yamamura, Y. (1980). Arsenite (As III), arsenate (As V) and methylarsenic in raw food. *Nippon Koshu Eisei Zasshi* **27**, 647–653.

Yamauchi, H., and Yamamura, Y. (1984). Metabolism and excretion of orally ingested trimethylarsenic in man. *Bull. Environ. Contam. Toxicol.* **32**, 682–687.

Yamauchi, H., Kaise, T., and Yamamura, Y. (1986). Metabolism and excretion of orally administered arsenobetaine in the hamster. *Bull. Environ. Contam. Toxicol.* **36**, 350–355.

Yamauchi, H., Takahashi, K., Yamamura, T., and Kaise, T. (1989). Metabolism and excretion of orally and intraperitoneally administered trimethylarsine oxide in the hamster. *Toxicol. Environ. Chem.* **22**, 69–76.

Index